Global Land Cover Validation

This book aims to summarize and report the major research achievements and validation results under the global land cover (GLC) initiative led by the Group of Earth Observation (GEO). The first part of the book introduces the major tasks and challenges facing the validation of finer-resolution GLC maps and presents the concepts and overall framework of the GEO-led initiative. Chapters 2–5 provide systematic introductions to the major methodology of finer-resolution GLC map validation, including sampling design, reference data collection, sample labeling, and accuracy assessment. Chapter 6 introduces the online validation tools that have been developed, including their design, considerations, and functionalities. Chapter 7 presents the international validation practices and the results of validating GlobeLand30 at country, regional, and global scales. Future directions are also discussed in the Conclusion.

Features

- Presents complete coverage of land cover validation, from concepts, methodology, and collaborative tools to applications.
- Details algorithms, techniques, and methods for land cover validation, including sampling, judgment, and accuracy assessment.
- Reviews some of the software tools that can be used for GLC validation and discusses the issues related to the design, usability, efficiency, and limitations of these tools.
- Highlights case studies of validation at global, regional, and national scales, which can serve as great references for researchers.
- Provides an extensive bibliography covering the whole scope of land cover validation.

Global Land Cover Validation: Methodology, Tools, and Practices serves as a reference for those engaged in land cover validation, especially at a global scale, including students, professionals, researchers, and general practitioners. It is also an excellent resource for professionals involved in sustainable development monitoring, environmental change studies, natural resource management, disaster assessment and mitigation, and many other applications.

International Society for Photogrammetry and Remote Sensing (ISPRS)
Book Series

ISSN: 1572–3348

Book Series Editor
Zhilin Li
Faculty of Geosciences and Engineering
Southwest Jiaotong University
Chengdu, China

Global Land Cover Validation
Methodology, Tools, and Practices

Edited by
Jun Chen, Xiaohua Tong, Huan Xie, Lijun Chen and
Songnian Li

CRC Press is an imprint of the
Taylor & Francis Group, an **informa** business

Designed cover image: © The Editors

First edition published 2025
by CRC Press
2385 NW Executive Center Drive, Suite 320, Boca Raton FL 33431

and by CRC Press
4 Park Square, Milton Park, Abingdon, Oxon, OX14 4RN

CRC Press is an imprint of Taylor & Francis Group, LLC

© 2025 ISPRS

Reasonable efforts have been made to publish reliable data and information, but the author and publisher cannot assume responsibility for the validity of all materials or the consequences of their use. The authors and publishers have attempted to trace the copyright holders of all material reproduced in this publication and apologize to copyright holders if permission to publish in this form has not been obtained. If any copyright material has not been acknowledged please write and let us know so we may rectify in any future reprint.

Except as permitted under U.S. Copyright Law, no part of this book may be reprinted, reproduced, transmitted, or utilized in any form by any electronic, mechanical, or other means, now known or hereafter invented, including photocopying, microfilming, and recording, or in any information storage or retrieval system, without written permission from the publishers.

For permission to photocopy or use material electronically from this work, access www.copyright.com or contact the Copyright Clearance Center, Inc. (CCC), 222 Rosewood Drive, Danvers, MA 01923, 978-750-8400. For works that are not available on CCC please contact mpkbookspermissions@tandf.co.uk

Trademark notice: Product or corporate names may be trademarks or registered trademarks and are used only for identification and explanation without intent to infringe.

Library of Congress Cataloging-in-Publication Data
Names: Chen, Jun, 1956- editor. | Tong, Xiaohua, editor. |
Xie, Huan, editor. | Chen, Lijun, editor. | Li, Songnian, 1963- editor.
Title: Global land cover validation : methodology, tools, and practices /
Edited by Jun Chen, Xiaohua Tong, Huan Xie, Lijun Chen and Songnian Li.
Description: Boca Raton, FL : CRC Press, 2025. | Series: ISPRS book series
| Includes bibliographical references and index.
Identifiers: LCCN 2024031511 (print) | LCCN 2024031512 (ebook) |
ISBN 9781032903989 (hardback) | ISBN 9781032903996 (paperback) |
ISBN 9781003557791 (ebook)
Subjects: LCSH: Land cover--Remote sensing--Research. | Globalization. |
Geographic information systems--Research. | Photogrammetry--Research.
Classification: LCC GB400.42.R4 G56 2025 (print) | LCC GB400.42.R4
(ebook) | DDC 550.72/3--dc23/eng/20240902
LC record available at https://lccn.loc.gov/2024031511
LC ebook record available at https://lccn.loc.gov/2024031512

ISBN: 978-1-032-90398-9 (hbk)
ISBN: 978-1-032-90399-6 (pbk)
ISBN: 978-1-003-55779-1 (ebk)

DOI: 10.1201/9781003557791

Typeset in Times
by SPi Technologies India Pvt Ltd (Straive)

Contents

Preface...vi
About the Editors vii
Contributors ix
Acknowledgments...xi

Chapter 1   Introduction 1

Jun Chen and Xiaohua Tong

Chapter 2   Sampling Design 23

*Huan Xie, Xiaohua Tong, Fei Chen, Chao Wei, Zhenhua Wang,
Shicheng Liao, and Jun Chen*

Chapter 3   Reference Data 65

*Xiaohua Tong, Chao Wei, Lijun Chen, Dan Liang, Zhengxing Wang,
Maria Antonia Brovelli, and Hanfa Xing*

Chapter 4   Sample Judgment 104

*Lijun Chen, Zhengxing Wang, Maria Antonia Brovelli, Gorica Bratic,
Hanfa Xing, Wen Meng, and Chao Wei*

Chapter 5   Accuracy Assessment 136

*Xiaohua Tong, Chao Wei, Jingxiong Zhang, Maria Antonia Brovelli,
Zhenhua Wang, Yanmin Jin, and Gorica Bratic*

Chapter 6   Software Tools for Validation 163

*Shishuo Xu, Maria Antonia Brovelli, Gang Han, Yang Zhao, Songnian Li,
Gorica Bratic, and Candan Eylul Kilsedar*

Chapter 7   Collaborative GlobeLand30 Validation 199

*Fei Chen, Huan Xie, Jingxiong Zhang, Zhengxing Wang, Lijun Chen,
Vanya Stamenova, Stefan Stamenov, Thomas Katagis, Ioannis Gitas,
Maria Antonia Brovelli, Yifang Ban, and Monia Elisa Molinari*

Chapter 8   Conclusions 241

Xiaohua Tong and Jun Chen

Index...245

Preface

Accurate characterization of land cover at global, regional, and local scales is essential for sustainable development monitoring, environmental change studies, natural resource management, disaster assessment and mitigation, and many other applications. The thematic accuracy of these land cover data products can be documented through sample-based validation or comparison with existing reference data. The accuracy assessment results will not only help the users to understand the uncertainty and the scope of data applications but also enable the producers to examine and analyze the types, sources, and spatial distribution of errors.

Over the past ten years, a variety of GLC data products with finer spatial resolutions (30 m) have been developed. In order to address the challenges facing global-scale accuracy assessment of 30 m resolution GLC data products, a collaborative validation initiative was launched by the Global Land Cover Community Activity of the Group of Earth Observation (GEO) in 2014. With the joint effort of experts and users from GEO, United Nations Committee of Experts on Global Geospatial Information Management (UN-GGIM) member countries, and participating organizations, a technical specification for 30 m GLC validation was developed based on the findings and experiences. An online validation tool, GLCVal, was developed by integrating land cover validation procedures with service computing technologies. About 20 countries (regions) have completed the accuracy assessment of GlobeLand30 for their territories with the guidance of the technical specification and the support of the GLCVal tool.

This book introduces the motivation, methodology development, and collaborative validation practices of this GEO-led collaborative GLC validation and summarizes its major outcomes. The methodology development includes the sampling design strategy and different sampling methods, the selection of different types of reference data, sample judgment, and accuracy assessment methods, as well as the development of online validation tools. The collaborative validation practices consist of validating Globeland30 datasets at global, regional, and national scales, which serve as the showcase of how collaborative validation works at these different levels.

The book can be used as a reference book by those engaged in land cover data validation, especially at a global scale, including students, professionals, researchers, and general practitioners.

About the Editors

Dr. Jun Chen is a leading scientist in geomatics at the Ministry of Natural Resources of China. He became an associate professor at Wuhan Technical University in 1987 and then worked as an executive director of the National Laboratory on Information Engineering of Surveying and Mapping. In 1995, he joined the National Geomatics Center of China (NGCC) and served as the Vice President, President, and Chief Scientist of NGCC. He has led several national geomatics projects, including the establishment and annual updating of national 1:50,000 topographic databases, the development of the world's first 30 m GLC map (GlobeLand30), and the first comprehensive assessment of the local progress toward Sustainable Development Goals with geospatial and statistical information. He has published more than 350 publications and 4 books. During his 40-year professional career, Jun Chen has served as the President of the International Society for Photogrammetry and Remote Sensing (ISPRS), Co-lead of the GEO Global Land Cover Task, and President of the Chinese Association of GISs. He received many international and national scientific awards, including the Geospatial World Innovation Award, Geospatial World Excellence Awards, Asia Geospatial Lifetime Achievement Award, and National Scientific and Technological Prizes. He is an elected academician of the Chinese Academy of Engineering, an honorary member of ISPRS, and a fellow of the International Scientific Council.

Dr. Xiaohua Tong is a Professor at the College of Surveying and Geo-Information and the Vice President of the Tongji University, China. He received the Ph.D. degree from Tongji University, Shanghai, China, in 1999. From 2001 to 2003, he was a Post-Doctoral Researcher with the State Key Laboratory of Information Engineering in Surveying, Mapping, and Remote Sensing, Wuhan University, Wuhan, China. He was a Research Fellow with Hong Kong Polytechnic University, Hong Kong, in 2006 and a Visiting Scholar with the University of California, Santa Barbara, California, USA, from 2008 to 2009. He received the National Science and Technology Progress Award in 2017 and the National Natural Science Award in 2007. He served as an associate editor or editorial board member for several international journals, such as Marine Geodesy and IEEE JSTARS. He was the Co-Chair of the ISPRS WGII/4 on Spatial Statistics and Uncertainty Modelling for the period 2012–2016. He is an elected member of the Chinese Academy of Engineering. His research interests include the theory of trust in spatial data, earth and planetary photogrammetry, and remote sensing.

Dr. Huan Xie is Professor and Dean of the College of Surveying and Geo-Information at the Tongji University, China. She received her B.S. degree in Surveying Engineering and her M.S. and Ph.D. degrees in cartography and geoinformation from Tongji University, Shanghai, China, in 2003, 2006, and 2009, respectively. From 2006 to 2007, she was with the Department of Land Surveying and Geo-Informatic and the Hong Kong Polytechnic University, Hong Kong, as a research assistant. From 2007 to 2008, she was with the Institute of Photogrammetry and GeoInformation, Leibniz Universität Hannover, Germany, as a visiting scholar. Her research interests include spatial data sampling, data quality of satellite laser altimetry, etc. She has published more than 100 research papers in SCI journals. She has been an IEEE Senior Member since 2019 and served as an ExCom Member in the IEEE GRSS Shanghai Chapter since 2015. She was the Secretary of the ISPRS WGII/4 Spatial Statistics and Uncertainty Modelling for the period 2012–2016.

Dr. Lijun Chen graduated from Lanzhou University in 1991 with a bachelor's degree. In the same year, he was admitted to the Xinjiang Institute of Geography of the Chinese Academy of Sciences to study for a master's degree. In 1994, he received his master's degree and stayed at this institute to work. In 1998, he was admitted to the State Key Laboratory of Resources and Information Systems, Institute of Geographic Science and Resources, Chinese Academy of Sciences to study for a doctor's degree. In 2001, he received a doctor's degree in cartography and geographic information systems and joined the NGCC. At present, he is a senior engineer, and his main research focuses on remote sensing information extraction and analysis. He has undertaken tasks such as global high-resolution land cover remote sensing mapping, global high-precision digital elevation model construction, and global geographic information mapping capacity building. From 2012 to 2016, he joined the ISPRS Global Land Cover Mapping and GEO SB-02 Global Land Cover and Change working groups and launched a GlobeLand30 validation event organized by GEO with nearly 40 countries, aiming to promote the international applications of GlobeLand30. He currently holds positions such as the Deputy Director of the Geographic Big Data Working Committee of the Chinese Geographical Society and the Deputy Editor in Chief of the Global Change Science Research Data Publishing System.

Dr. Songnian Li is a Professor of Geomatics Engineering and Associate Chair for Graduate Studies in the Department of Civil Engineering at Toronto Metropolitan University. He holds a B.Eng. degree in Surveying Engineering from the Wuhan Technical University of Surveying and Mapping (now Wuhan University) and a Ph.D. degree from the Department of Geodesy and Geomatics Engineering of the University of New Brunswick in New Brunswick, Canada. Before joining Toronto Metropolitan University, Dr. Li taught at Wuhan Technical University of Surveying and Mapping and Ningxia Institute of Technology (now Ningxia University) in China for 14 years and spent one year at Lakehead University as a visiting scholar. Prof. Li has served as the Deputy Chair of the Academic Network of the UN-GGIM (2021–2024), the Council Member of the *ISPRS* (2016–2022), and the Executive Committee Director of the Canadian Remote Sensing Society (2022–2024), among other professional services. He was a twice-invited Researcher by the *Japan Society for the Promotion of Science* and received the 2016 ISPRS U.V. Helava Best Paper Award. He is an ISPRS Fellow and the Executive Editor in Chief of *Big Earth Data* journal and has served on the editorial board of many international journals such as *ISPRS Journal of Photogrammetry and Remote Sensing*, *ISPRS International Journal of Geo-Information*, and *Smart Cities*. His current research focuses on geospatial big data, spatiotemporal analysis, human mobility, geo-collaboration, smart cities (digital twins), and web services for land cover/use mapping.

Contributors

Yifang Ban
KTH Royal Institute of Technology
Stockholm, Sweden

Gorica Bratic
Politecnico Milano
Milan, Italy

Maria Antonia Brovelli
Politecnico Milano
Milan, Italy

Fei Chen
East China University of Technology
Jiangxi, China

Jun Chen
National Geomatics Center of China
Beijing, China

Lijun Chen
National Geomatics Center of China
Beijing, China

Ioannis Gitas
Aristotle University
Thessaloniki, Greece

Gang Han
National Geomatics Center of China
Beijing, China

Yanmin Jin
Tongji University
Shanghai, China

Thomas Katagis
Aristotle University
Thessaloniki, Greece

Candan Eylul Kilsedar
Politecnico Milano
Milan, Italy

Songnian Li
Toronto Metropolitan University
Toronto, Canada

Shicheng Liao
Tongji University
Shanghai, China

Dan Liang
Zhejiang A&F University
Hangzhou, China

Shicheng Liao
Tongji University
Shanghai, China

Wen Meng
Tongji University
Shanghai, China

Monia Elisa Molinari
Politecnico Milano
Milan, Italy

Stefan Stamenov
Space Research and Technology Institute
Bulgarian Academy of Sciences
Sofia, Bulgaria

Vanya Stamenova
Bulgarian Academy of Sciences
Sofia, Bulgaria

Xiaohua Tong
Tongji University
Shanghai, China

Zhengxing Wang
Chinese Academy of Sciences
Beijing, China

Zhenhua Wang
Shanghai Ocean University
Shanghai, China

Chao Wei
Tongji University
College of Surveying and Geo-informatics
Shanghai, China

Huan Xie
Tongji University
College of Surveying and
  Geo-informatics
Shanghai, China

Hanfa Xing
South China Normal University
Guangdong, China

Shishuo Xu
Beijing University of Civil Engineering
  and Architecture
Beijing, China

Jingxiong Zhang
Wuhan University
Hubei, China

Yang Zhao
Beijing University of Civil Engineering
  and Architecture
Beijing, China

Acknowledgments

The authors would like to acknowledge many colleagues who have contributed to the GEO-led 30 m GLC validation initiative and to the accomplishment of preparing and authoring the book, as well as reviewing chapters and providing administrative assistance in the final compilation. Without their invaluable help and support, the book would not have been completed and published.

All chapters included in the book have gone through a few rounds of review and editing process. We are grateful to all our colleagues, including chapter authors and many external researchers in the field, who kindly agreed to help review and edit the chapters. Our special thanks go to the following experts and colleagues who are not chapter authors but have contributed to the book: Yali Gong, Chuang Liu Phoebe Oduor, Shu Peng, Wenzhong Shi, Fang Wang, Ang Zhao, Xiaoyu Zhao, and as well as Jingxiong Zhang, who is an author but also made a great effort to review and edit a few other chapters.

We wish to acknowledge the editorial and professional guidance given by the editors, Ms. Irma Shagla Britton and Ms. Chelsea Reeves, and other colleagues at CRC Press/Balkema, Taylor & Francis Group. Both assisted us by kindly answering questions during the book editing and publication processes.

The GEO-led 30 m GLC validation initiative and the preparation of this book have been funded by the Ministry of Science and Technology of China (Project # 2015DFA11360) and the National Science Foundation of China (Project #41930650, #41631178, and #42221002).

Finally, all the chapter authors deserve our special thanks for their time and effort spent on chapter proposals, full chapter preparations, and many rounds of chapter revisions. We thank them for their patience in this long and sometimes arduous editing process. It was our great pleasure to work with them on different aspects of GLC validation, to collectively produce this book and make it available on the desks of many readers.

1 Introduction

Jun Chen
National Geomatics Center of China, Beijing, China

Xiaohua Tong
Tongji University, Shanghai, China

1.1 GLOBAL LAND COVER MAPPING

1.1.1 LAND COVER AND CHANGE INFORMATION

Land cover refers to the biophysical material covering Earth's surface and immediate sub-surfaces and man-made structures (Lambin and Helmut 2008; Chen et al. 2017). Owing to the dramatic increasing population and economic growth, humans have significantly modified Earth's land cover over the past three centuries, especially in recent decades. While these changes have facilitated the provision of essential resources, such as food, fiber, shelter, and freshwater to meet immediate human needs, they have also degraded environmental conditions, ecosystem services, and human welfare (Foley et al. 2005; Pielke 2005; Sterling and Ducharne 2008). For instance, the growth of urban or cropland areas at the expense of forests has led to increased atmospheric carbon dioxide levels and the creation of urban heat islands (Verburg et al. 2011). Land cover changes have notable impacts on greenhouse gases (GHGs) emissions, ecosystem structures and functions, continental and global atmospheric circulation, nutrient and hydrological cycles, biogeochemical cycles, and biodiversity (Sterling and Ducharne 2008; Verburg et al. 2011). The reduction in negative impacts on land cover change while maintaining the production of essential resources has developed into a significant concern and obstacle for policymakers and the scientific community worldwide (Foley et al. 2005; Reid et al. 2010; Grekousis et al. 2015).

Information regarding land cover and change over time is crucial for understanding the state, trends, drivers, and impacts of various land activities on social and natural processes, along with creating pathways toward sustainable evolution (Verburg et al. 2006; Zell et al. 2012). It is useful for various societal and scientific purposes, ranging from natural resource management, environmental studies, and urban planning to sustainable evolution (Foley et al. 2005; Roger and Pielke 2005; Running 2008; Zell et al. 2012; Sterling et al. 2013). Generic statistical analysis, cause and consequence analysis, and coupling analysis with Earth system modeling are three typical applications of land cover and change information (Chen et al. 2017). The first one aims to derive spatially referenced and quantitative information from land cover datasets, such as acreage statistics, geographic distributions (i.e., the extent and patterns), and the magnitude and type of change (i.e., expansion, shrinkage, or intensification, and specific areas altered; Sterling and Ducharne 2008; Arsanjani et al. 2016a). These studies offer insights into the state, patterns, and changes of the main land cover classes. The second group of applications focuses on how land cover patterns form—that is, global divergence of artificial surfaces—and how land cover changes impact the environment, such as the evaluation of ecosystem services, estimates of carbon dioxide emissions, and impacts on the terrestrial water cycle, among other factors (Brown 2013; Nagendra et al. 2013; Raymond et al. 2013; Sterling et al. 2013). The third kind of application incorporates land cover information into Earth system modeling to simulate climate, biological, and geochemical processes, as well as to predict future environmental conditions and their consequences (Reid et al. 2010; Lu et al. 2016; Shi et al. 2016).

DOI: 10.1201/9781003557791-1

Reliable land cover and change (LCC) information is the key to the success of these applications. An accurate characterization or mapping of LCC on both global and local scales is therefore becoming a critically challenging task for international, regional, and national communities. Remote sensing has long been used as an efficient tool for large-area land cover mapping (Cihlar 2000; Hansen and Loveland 2012). Over the last 20 years, a number of LCC data products have been produced at different geographical scales, including national, regional, and global, with varied spatial resolutions, temporal coverage, thematic details, and mapping accuracies (Liu et al. 1999; Grekousis et al. 2015; Diogo and Koomen 2016). For example, at continental (regional) scale, several land cover data products were produced, including a 100–250 m resolution Corine dataset (Büttner et al. 2004), 250 m resolution NLCD dataset in North America (Latifovic et al. 2012), 500 m resolution in South America and the Caribbean (Blanco et al. 2013), and 30 m resolution dataset in South America in 2010 (Giri and Long 2014). At the national scale, the United States, Canada, Brazil, India, Australia, and other countries have all developed multi-resolution land cover data products covering their entire country (Olthof et al. 2009, Deng and Liu 2012; Jin et al. 2013; Lymburner et al. 2013; Macedo et al. 2013; NGCC 2014). China has completed the national monitoring project of its fundamental geographic condition and produced a time-series, national-wide detailed land cover data products with 10 first-level classes, 59 second-level classes, and 143 third-level classes from the year 2015 to 2021 (Zhang et al. 2015, 2016).

1.1.2 Finer-Resolution Land Cover Mapping at Global Scale

Global land cover (GLC) mapping is a complicated process that is likely to be impacted by many factors, such as image quality, data preprocessing, training sample quality, classification approaches, post-classification processing, and accuracy assessment (Lu and Weng 2007). In other words, it is much more difficult than at a local scale due to the need for high-quality imagery covering Earth's entire land surface (approximately 150 million km^2), the complex spectral and textual characterization of global landscapes, and many other reasons. This makes the production of reliable GLC datasets a highly challenging task, requiring substantial technical innovation as well as significant human and financial resources (Townshend et al. 1991; Chen et al. 2015). Since the end of the 1990s, numerous GLC datasets have been developed with resolutions ranging from 300 m to 1 km, utilizing coarse-resolution satellite imagery such as AVHRR, MODIS, and MERIS (Hansen et al. 2000; Loveland et al. 2000; Friedl et al. 2002; Bontemps et al. 2011). DISCover, UMD, MODIS, GLC2000, and Globcover 2005 and 2009, with their basic information listed in Table 1.1, are among the coarse-resolution GLC datasets developed since 1990.

Though these GLC data products have been found useful, their quality is far from satisfactory for many applications (Herold et al. 2008; Verburg et al. 2011). These coarse-resolution GLC data products did not provide sufficient spatial and thematic details of land activities, and has limited their usability in scientific analysis, forecasting, policy debate, and political decisions (Scoones et al. 2013). Additional shortfalls have also been highlighted, such as significantly low accuracies and

TABLE 1.1
The Coarse-Resolution GLC Datasets

Program	Period	Resolution	Organization
DISCover	1992–1993	1 km	USGS
UMD	1992–1993	1 km	University of Maryland
GLC2000	1999–2000	1 km	JRC
MOD12	2000–2001	1 km	Boston University
Globcover 2005 and 2009	circa 2005/circa 2009	300 m	ESA
MCD12Q1(collection 6)	2001–2016	500 m	Boston University

Introduction   **3**

poor agreement among different datasets (Iwao et al. 2006; Gong 2009). As a result, there is a growing demand for new GLC products with enhanced spatial resolution and accuracy, which has been progressively recognized by both the remote sensing community and user societies, such as the Group on Earth Observations (GEO) and the International Society for Photogrammetry and Remote Sensing (ISPRS) (Zell et al. 2012; Giri et al. 2013).

The development of data products at finer resolution (such as 30 m) has been regarded as a good choice for the new generation of GLC maps. One of its motivations is that most human activities on land systems with high impacts can be captured at 30 m resolution (Giri et al. 2013). The other one is the free availability and the long-term archive of Landsat and Landsat-like image data, as well as the progress made in the automated and semiautomated methods for extracting land cover information from digital imagery (Lu et al. 2004; Lu and Weng 2007). For example, a worldwide forest data product at 30 m resolution was first generated by the University of Maryland (Townshend et al. 2012). However, it is a single-class GLC, and 30m GLC products with all kinds of land cover types are still very desirable.

Tremendous efforts have been devoted to the development of GLC products with a resolution of 30 m in the past ten years, owing to the public assessment of fine-resolution satellite remote sensing data (particularly the Landsat archive and Sentinel-2 data) and the enhanced computational and storage resources. These products include GlobeLand30 (Chen et al. 2015), FROM_GLC (Gong et al. 2012), GLC_FCS30 (Zhang et al. 2020), ESA-S2-LC20 (CCI_Land-Cover 2017), FROM_GLC10 (Gong et al. 2019), and ESA WorldCover 2020/2021 (Zanaga et al. 2021). More and more global thematic fine-resolution land-cover products have also been developed, including NUACI (Liu et al. 2018), ESA GHSL (Florczyk et al. 2019), GAIA (Gong et al. 2020), and MSMT_IS30 (Zhang et al. 2020); global inland water body datasets, such as GLCFGIW (Feng et al. 2014), G3WBM (Yamazaki et al. 2015), JRC_GSW (Pekel et al. 2016), and GLAD_Water (Pickens et al. 2020); global cropland datasets, including GFSAD30 (Oliphant et al. 2017), GCAD30 (Thenkabail 2012), and FROM_GC (Yu et al. 2013); and global forest cover datasets, including TreeCover-2010 (Hansen et al. 2013), GFCC30TC (Sexton et al. 2016), and GFC30 (Zhang et al. 2020).

There are two different strategies for such global-scale land cover mapping and monitoring—i.e., the experimental and operational classifications (Hansen and Loveland 2012). While the former emphasizes the development and testing of new algorithms and models, the latter focuses on the development and delivery of reliable data products. Table 1.2 lists the major finer-resolution (10–50 m) GLC mapping products.

1.1.3 GLOBELAND30 PRODUCTS

There are two different strategies for such global-scale land cover mapping and monitoring—i.e., the experimental and operational classifications (Hansen and Loveland 2012). While the former emphasizes the development and testing of new algorithms and models, the latter focuses on the development and delivery of reliable data products.

GlobeLand30 is an open-access 30m resolution GLC data product developed by the National Geomatics Center of China following an operational classification strategy. A pixel-object-knowledge (POK)-based approach was proposed to tackle the two critical issues facing the operational production of reliable datasets—i.e., the establishment of suitable classifiers for spectral and textual characterization of complex landscapes and the assurance of data product quality (Chen et al. 2015). Its key components include the optimal coverage of 30 m imagery for two baseline years (Tang et al. 2014), web-service-oriented integration of reference data (Han et al. 2015), integration of pixel- and object-based classification methods, and knowledge-based validation to ensure the consistency and accuracy of 30 m GLC data products (Zhang et al. 2016). The main elements of the POK-based operational mapping approach are illustrated in Figure 1.1.

GlobeLand30 comprises ten land cover types, including cultivated land, forest, grassland, shrubland, wetland, water bodies, tundra, artificial surface, bare land, and permanent snow/ice, for the

TABLE 1.2
Finer-Resolution Land Cover Mapping Products at a Global Scale

Name	Res. (m)	Period	Number of Classes	Literature
GlobeLand 30	30	2000, 2010, 2020	10	Chen et al. (2015)
FROM-GLC	30	2010	10/25	Gong et al. (2012)
GLC_FCS30	30	2015, 2020	10/25	Zhang et al. (2020)
ESA-S2-LC20	20	2016	10/22	CCI_Land-Cover (2017)
FROM_GLC10	10	2018	10	Gong et al. (2019)
ESA WolrdCover2020	10	2020	11	Zanaga et al. (2021)
ESA WolrdCover2021	10	2021	11	Zanaga et al. (2022)
MSMT_IS30	30	2015, 2020	impervious surface	Zhang et al. (2020)
NUACI	30	1980–2015 (5 year)	impervious surface	Liu et al. (2018)
GHSL	30	1975, 1990, 2000, 2015	impervious surface	Florczyk et al. (2019)
GAIA	30	1985–2018 (annual)	impervious surface	Gong et al. (2020)
HBASE/GMIS	30	2010	impervious surface	Wang et al. (2017)
GUF	30	2011–2012	impervious surface	Esch et al. (2013); Esch et al. (2017)
Global Human Settlement Layer (GHS BUILT-UP GRID S1)	20	2016	impervious surface	Corbane et al. (2017)
Global Urban Footprint	12	2011	impervious surface	Esch et al. (2017)
GLCFGIW	30	2000	inland water body	Feng et al. (2014)
G3WBM	30	1990, 2000, 2005, 2010	inland water body	Yamazaki et al. (2015)
JRC_GSW	30	1984–2015	inland water body	Pekel et al. (2016)
GLAD_Water	30	1999–2018	inland water body	Pickens et al. (2020)
GFSAD30	30	2015	cropland	Oliphant et al. (2017)
GCAD30	30	2010	cropland	Thenkabail (2012)
FROM_GC	30	2010	cropland	Yu et al. (2013)
TreeCover-2010	30	2010	forest	Hansen et al. (2013)
GFCC30TC	30	2000, 2005, 2010, 2015	forest	Sexton et al. (2013)
GFC30	30	2018	forest	Zhang et al. (2020)
GLADForest	30	2000–2019 (annual)	forest	Hansen et al. (2013)
FNF	25	2007–2010, 2015–2016	2(forest/non-forest)	Shimada et al. (2014)

years 2000, 2010, and 2020 (see the 2020 map in Figure 1.2). The codes and definitions of the ten classes are listed in Table 1.3. These land cover types were extracted from over 30,000 Landsat and Chinese HJ-1 satellite images using the POK-based operational mapping approach, achieving an overall classification accuracy of over 80% (Chen et al. 2015). GlobeLand30 data is stored in raster format, utilizing the non-destructive GeoTIFF compression format and a 256-color, 8-bit indexed pattern. It adopts the WGS84 coordinate system, Universal Transverse Mercator (UTM) projection, and six-degree zoning. GlobeLand30 is organized into data tiles following two different latitude situations—that is, a size of 5° (latitude) × 6° (longitude) within 60°N and 60°S and a size of 5° (latitude) × 12° (longitude) within the area of 60° to 80° north and south of the equator.

On September 15, 2000, China donated GlobeLand30 (version 2000 and 2010) to the United Nations (UN) as a contribution toward global sustainable development and combat of climate change (Chen et al. 2014), followed by the donation of version 2020 on the 75th Anniversary of the UN.

Introduction

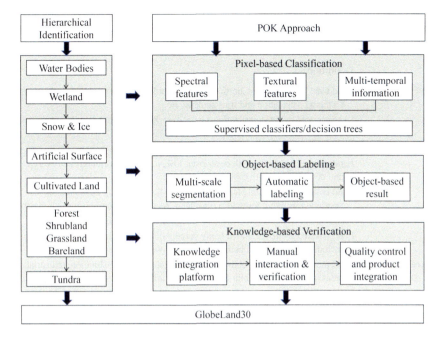

FIGURE 1.1 The GlobeLand30 data development POK-approach flowchart (Chen et al. 2017).

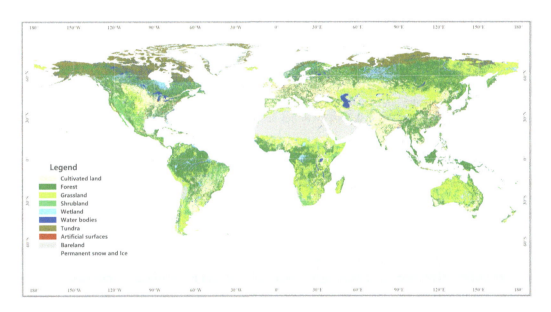

FIGURE 1.2 Map of GlobeLand30 (2020).

Over 50,000 users have been registered in the GlobeLand30 data platform (www.globeland30.org). The other users downloaded data from the Global Change Research Data Publishing and Repository. The top ten countries with the highest number of users are China, the United States, India, the United Kingdom, Germany, Canada, France, Brazil, the Netherlands, and Russia. The majority of users are from universities, research institutions, and government departments. More specifically, over half of the users are from university laboratories, including those at Harvard University, Princeton University, University of Heidelberg, and Peking University, among others. One in every four GlobeLand30

TABLE 1.3

Classification and Definition of Land Cover Types of GlobeLand30 (Chen et al. 2017)

Type	Definition
Cultivated land	Land used for agriculture, horticulture, and gardens, including paddy fields, irrigated and dry farmland, vegetation and fruit gardens, etc.
Forest	Land covered by trees, vegetation covers over 30%, including deciduous and coniferous forests, and sparse woodland with cover 10%–30%, etc.
Grassland	Land covered by natural grass with cover over 10%, etc.
Shrub land	Land covered by shrubs with cover over 30%, including deciduous and evergreen shrubs, and desert steppe with cover over 10%, etc.
Wetland	Land covered by wetland plants and water bodies, including inland marsh, lake marsh, river floodplain wetland, forest/shrub wetland, peat bogs, mangrove and salt marsh, etc.
Water bodies	Water bodies in land areas, including rivers, lakes, reservoirs, fish ponds, etc.
Tundra	Land covered by lichen, moss, hardy perennial herb, and shrubs in the polar regions, including shrub tundra, herbaceous tundra, wet tundra, and barren tundra, etc.
Artificial surface	Land modified by human activities, including all kinds of habitation, industrial and mining areas, transportation facilities, and interior urban green zones and water bodies, etc.
Bare land	Land with vegetation cover lower than 10%, including desert, sandy fields, Gobi, bare rocks, saline and alkaline land, etc.
Permanent snow and ice	Lands covered by permanent snow, glaciers, and ice caps.

users comes from scientific research institutions, such as the Joint Research Centre of the European Commission, the Institute for Global Environmental Strategies, and the Helmholtz-Centre Potsdam-German Research Centre for Geosciences, among others. Many UN agencies and non-governmental organizations also use GlobeLand30. For instance, the UN Field Operation Department has used GlobeLand30 to aid in the analysis and development of peacekeeping action plans in 18 countries. The UN Economic and Social Commission for Asia and the Pacific used GlobeLand30 to support drought management and land degradation. The UN Environment Programme's World Conservation Monitoring Centre (UNEP-WCMC) used GlobeLand30 for the land cover analysis for the protected areas worldwide.

GlobeLand30 data has been widely utilized in various Social Benefits Areas (SBAs), such as biodiversity and ecosystem sustainability, sustainable urban development, disaster resilience, food security and sustainable agriculture, water resources management, climate change adaptation, energy and mineral resources management, infrastructure and transportation management, public health surveillance, and more (Chen et al. 2017). Table 1.4 summarizes its applications in these SBAs based on the user registration information. Biodiversity and ecosystem sustainability are the largest application areas, accounting for over 26% of all applications. The second largest application area of GlobeLand30 is sustainable urban evolution, which accounts for nearly 16%. The applications in disaster resilience, food security and sustainability, and water resources management constitute more than 10%.

GlobeLand30 offers useful information with higher-resolution spatial data for various research fields, including deriving the status and changes in land cover, investigating their causes and effects, and exploring future development scenarios. For instance, Cao et al. (2014) utilized GlobeLand30's water layer data to analyze the distribution and changes in global open water from 2000 to 2010. They calculated two indicators: water body percentage and coefficient of spatial variation, to reflect the spatial distribution patterns and dynamic changes of global open water resources. The results indicated that the total area of land surface water in 2010 was approximately 3.68 million km^2, accounting for 2.73% of Earth's land surface. Similarly, Chen et al. (2014) calculated global artificial surface areas at both country and continental scales and analyzed changes between 2000 and 2010. The results showed that the total area of global built-up areas was 1.1875 million km^2 in 2010,

Introduction

TABLE 1.4
Major Application Fields of GlobeLand30 (Chen et al. 2017)

Research Fields	Proportion of Each Field (%)	University (%)	Institute (%)	Government (%)	NGO (%)	UN (%)	Other (%)
Climate Change	7.51	38.62	31.29	7.32	9.32	3.06	10.39
Biodiversity and Ecosystem	26.94	48.85	32.26	0.59	3.19	1.15	13.96
Disaster Resilience	13.69	73.78	10.30	9.72	1.68	2.26	2.26
Energy and Mineral Resources Management	5.33	29.46	29.46	20.64	4.32	0.00	16.14
Food Security and Sustainability Agriculture	10.09	48.86	16.25	3.87	12.39	6.24	12.39
Infrastructure and Transportation Management	3.84	48.96	26.56	2.08	12.24	0.00	10.16
Public Health Surveillance	4.06	40.39	38.67	3.94	5.67	5.67	5.67
Sustainable Urban Development	15.98	64.21	19.59	6.38	2.44	1.00	6.38
Water Resources Management	12.53	59.38	19.39	7.50	1.84	0.64	11.25
Proportion of Each Organization	100.00	53.88	23.81	5.72	4.62	1.96	10.02

covering 0.88% of Earth's land surface. From 2000 to 2010, the global built-up areas increased by 57,400 km^2, with a variation rate of 5.08%. China and the United States were the top two countries with the largest increases in built-up areas, accounting for approximately 50% of the global total. Additionally, 50.26% of the increased built-up areas came from arable lands.

One example of the cause-and-effect analysis is the spatio-temporal pattern analysis of artificial surface use efficiency based on GlobeLand30 (Li et al. 2016). Several indicators were extracted from the artificial surface layer of GlobeLand30, including artificial surface area per capita, population per unit area of artificial surface, GDP per unit area of artificial surface, population increase relative to artificial surface increase, and GDP increase relative to artificial surface increase. These indicators were used to reveal patterns of artificial surface use efficiency, relationship with population and GDP, and changes during 2000–2010 at the country level. The results showed distinct regional discrepancies in the use efficiency of artificial surfaces. Canada and the United States are categorized as having abundant resources but low land use efficiency, whereas South Korea, Japan, and Switzerland are characterized by limited resources but high land use efficiency.

1.2 VALIDATION OF GLC PRODUCTS

As with many other geospatial data, validation is an important process in land cover mapping. It can not only help data users to understand the uncertainty and the scope of data application but also enable data producers to examine and analyze the types, sources, and spatial distribution of errors (Strahler et al. 2008; Rwanga and Ndambuki 2017; Cao et al. 2016).

1.2.1 UNCERTAINTY OF SPATIAL DATA

The uncertainty of spatial data refers to the extent that the "true value" of the data cannot be confirmed. In a broad sense, uncertainty includes data error and the incompleteness and ambiguity of the concept and the data (Caspary and Scheuring 1993; Shi 1994 2005; Guptill and Morrison 1995; Heuvelink 1998; Zhang and Goodchild 2002; Ge and Wang 2003; Bai and Wang 2003; Cheng et al. 2004;

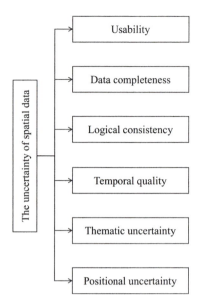

FIGURE 1.3 Components of uncertainty of spatial data.

Liu et al. 2010; Li et al. 2013; Tong et al. 2013). Usually, the error of spatial data is defined as the difference between the real value and its observation, which is classified into three categories (Huang et al. 1995; Liu et al. 1996). The first is systematic error, which has a regular magnitude and sign (positive and negative). The second is random error, which is random in magnitude and sign but with a statistical regularity. The third is gross error, which represents the outliers in a set of data.

Since the 1960s, researchers have found that errors occur in spatial data during map data acquisition, processing, and spatial analysis (Perkal 1956; Blakemore 1983; Burrough 1986; Goodchild and Gopal 1989; National Center of Geographical Information Analysis (NCGIA) 1989; MODIS land cover team 2003). Since the 1990s, researchers have gradually realized that the uncertainty of spatial data mainly involves the generation mechanism of the uncertainty, the mathematical expression model, and the propagation mechanism of spatial analysis and data mining. More attention should be paid to uncertainty in spatial data, analysis, and processing (Goodchild 1992; Shi 2005). The International Organization for Standardization (ISO) has described the uncertainty of spatial data as positional uncertainty, thematic uncertainty, temporal quality, logical consistency, data completeness, and usability, according to the elements of geographic information data quality (Figure 1.3) (ISO 19157: 2013(E) 2023). For instance, temporal quality is defined as the quality of the temporal attributes and temporal relationship of features.

In recent years, with the development of geographic information science and Earth observation system applications, it has been found that the trust, reliability, and usability of spatial data are the important foundation of spatial data quality control, analysis, and applications (Shi et al. 2017; Tong et al. 2017). The concept of trust in spatial data refers to the expectation of some match between the observed and the real object. Spatial data with a matching degree within the expectation range are called trusted spatial data. As a result, the definitions of the basic concepts, scientific connotation, and theoretical framework of trust in spatial data according to the actual requirements of applications have become the latest development in spatial data quality and represent a shift from the understanding of the problem to solving the problem (Tong et al. 2017; Tong et al. 2023).

1.2.2 Lessons Learned from Validating Existing GLC Data Products

Land cover validation aims to estimate the uncertainty of land cover data products and document their accuracy by using sample-based validation or by comparing them with existing reference

Introduction

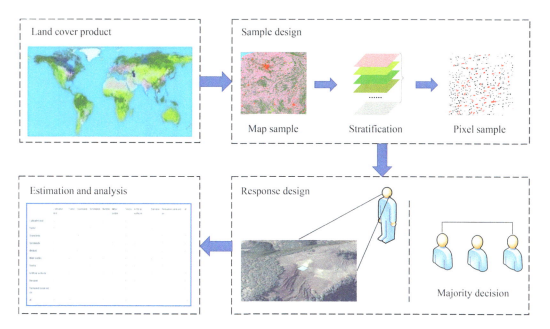

FIGURE 1.4 The process of validating a land cover product.

datasets (Strahler et al. 2006; Zimmerman et al. 2013; Olofsson et al. 2014; Chen et al. 2016; Chen et al. 2019; Yang et al. 2017). Regarding the sample-based validation, a set of appropriate samples is selected, from which reference data are collected and used for estimating the overall and class-specific accuracies of the given land cover data product in the target region(s), as shown in Figure 1.4.

Several sampling design approaches have been developed to determine the sample size and spatial distribution of samples based on fundamental criteria such as probability, cost-effectiveness, and spatial balance. The stratified random sampling and two-stage cluster sampling are two commonly used methods in land cover validation practices (Stehman and Czaplewski 1998; Tong et al. 2011; Stehman et al. 2012; Sulla-Menashe et al. 2019).

The GLC products at coarse resolution mentioned in Section 1.1.2 have been validated in the past two decades. The details of each validation are shown in Table 1.5.

(1) DISCover: The DISCover product with a resolution of 1 km is the research result of the International Geosphere and Biosphere Project (IGBP) established by the US Geological Survey (USGS) in 1992. The global classification is divided according to the IGBP classification system and unsupervised classification methods with 17 classes. DISCover products use stratified random sampling with the classes. The sample size of each stratum is a fixed value of 25. The determination of the sample size is based on 0.85 as the expected accuracy and obtained through a standard binomial confidence interval look-up table with 95% confidence. Although the sample size causes the sampling error interval to be ±14.3%, taking 25 samples can reduce costs based on practical principles. In order to avoid the situation that there is no reference data corresponding to the samples taken during the actual operation, 50 samples were actually taken for each land type for backup. DISCover validation used TM images as reference data. Since remote sensing technology was in its infancy at the time, two types of samples lacked corresponding reference data: ice and water bodies, and the other land types had uneven sample sizes due to reference data. DISCover product validation drew a total of 379 samples for 15 of the 17 land types.

(2) UMD: The validation of UMD products does not take the form of absolute verification of quantitative estimates. Due to cost reasons, a relative verification mode for consistency

TABLE 1.5

A Comparison of Validation Schemes Employed on Some Existing GLC Products

	DISCover	UMd	MODIS	GLC2000	Globcover 2005	CCI	GLC-SHARE	MCD12Q1 (Collection 6)
Validation principle	Practical purpose	Cost control	Cost control	Priority to ecological types	Joint validation by international experts			Priority to distinct classification schemes
Sampling strategy/ scheme	A two-tiered sampling strategy	Comparison approach	Random sampling	A two-stage stratified clustered sampling	Validation data sets	A two-rank stratified random sampling	Stratified random sampling	High-quality training examples
Sampling method	Random sampling	none		Stratified sampling Clustered sampling		Random sampling	Random sampling	
Sample size	379	None	1,370 (39,472)	253/1,265	3,167	13,000	1,000	3,095
Reference data	TM images/ SPOT images	DISCover	Training data	TM images/aerial photographs	Google Earth/ SPOT	Google Earth/ NDVI		STEP database
Assessment method	Error matrix	Area agreement	Confusion matrix	Confusion matrix	Confusion matrix	Confusion matrix	Confusion matrix	Tenfold cross-validated
Accuracy	Globally 66.9%	unknown	Globally 78.3%	Globally 68.6%	Globally 73.14%	Globally 74%	Globally 80%	Globally 67%–87%

Introduction

comparison with DISCover products was used. According to the verification results, a global area consistency comparison was performed for the same land category (14 classes) of UMD and DISCover, and the area mapping of the two products was consistent. The accuracy of 74% of the single categories, such as Forest, Grassland, Barren, and Cropland Consistency, is 80.32%.

(3) GLC2000: A two-stage sampling strategy was used for the validation of this product. The first stage of stratified sampling used the center of gravity of the Landsat-2 world reference system to calculate the global Tyson polygon as the first-level sampling unit. Then, the priority of each sampling unit was calculated as long as the area of the category forest within the sampling unit is greater than 30%, or cropland is greater than 10%, or wetland is greater than 10% to confirm that the sampling unit is priority. Otherwise, it is considered to be nonpriority. The second-stage sampling method is cluster sampling, setting the sampling window, and the fixed window sample location. In each sampling window, there were several Tyson polygon samples on different layers. Each layer is given a different sampling rate to extract Tyson polygon samples, and five sample points are set at fixed positions in the polygons that are selected. A total of 253 sampling units and 1,265 samples were selected, and the overall accuracy was reported as 68.6% globally.

(4) MOD12: For cost reasons, MODIS products used the training samples for classification as the samples for validation—i.e., all classification samples were considered as a sampling population and divided into ten groups evenly. Nine groups of samples were used for classification, and one group of samples was used for inspection—that is, the stratified random sampling method. Therefore, the validation samples of MODIS products consisted of 1370 regions of interest (ROI), which contained 39,472 pixels in total. The sample size of each land category was randomly determined by stratification. The overall accuracy was calculated based on area weights by the confusion matrix was 78.3%.

(5) Globcover 2005 and 2009: Globcover used zone sample verification, which is divided into five regions around the world: Africa, Australia and the Pacific, Eurasia, North America, and South America. The judgment of the familiar area by the division expert helped improve the accuracy of the judgment result. Globcover2005's global validation sample set contained 4,258 samples, of which 3,167 samples were suitable for product validation. After expert judgment, of the 3,167 samples, 2,115 samples were judged to be correct and homogeneous samples—that is, there were no other land types in the sample area, and 1,052 samples were judged to be wrong or heterogeneous samples.

(6) MCD12Q1: To evaluate the quality of the MCD12Q1, they used the STEP (the System for Terrestrial Ecosystem Parameterization database) training database to calculate tenfold cross-validated accuracies for the IGBP and LCCS maps. In total, 3,095 training samples were used. Each sample is a polygon (~4 sq km) drawn on Google Earth that is considered a stable example of a specific land cover type. They partitioned the complete set of training data into ten distinct and mutually exclusive subsets, with 90% allocated for training and 10% for testing. The results from this cross-validation process were subsequently used to generate confusion matrices and calculate area-adjusted classification accuracies. The overall accuracy of the MCD12Q1 LCCS3 layer was 87%, followed by LCCS2 (81%), LCCS1 (74%), and the IGBP layer at 67%.

The previous discussions show that different GLC products may adopt different methods for their validation, particularly in terms of validation principle, sample design, sample collections, and accuracy assessment. The details can be described as follows.

(1) Validation principle

For the DISCover product, validation was first defined as "the process that verifies the accuracy of the land cover information provided in as independent a means as practical"

(Belward et al. 1999). As a result, the practical principle is used for validation of the DISCover product. Due to the limitation on validation costs, the UMD product was assessed by a comparison approach which is to analyze the agreement with the product of DISCover, rather than by an absolute quantitative assessment approach (Hansen and Reed 2000). The MODIS product used the same validation principle as UMD, which is sampling from training data. The validation priority of the GLC2000 product was for vegetation that belongs to ecological types. As for the Globcover 2005 and 2009 products, an international expert network was established for their validation.

(2) Sampling design

The sampling strategy adopted in DISCover was random sampling in each stratified class, and the sample size of each class was calculated by the binomial confidence intervals (Pearson and Hartley 1966) based upon an expected accuracy of 85% at a 95% confidence level (Belward et al. 1999). At the same time, a minimum of 25 samples in each class was guaranteed. Therefore, a total of 379 samples were selected from 15 of the 17 classes (the classes "snow and ice" and "water" were not included to be verified). Each sample point corresponded to a 1 km pixel, and a majority interpretation rule was used to determine the correctness of the samples. The MODIS product used training samples for validation, with a random stratified sample and tenfold cross-validation. As a result, a total of 1,370 sites (39,472 pixels) were selected as validation samples. For the validation of GLC2000, a two-stage stratified clustered sampling strategy was adopted for the 23 land cover classes. The stratification was based on the proportion of priority classes and on the landscape complexity because GLC2000 was an eco-based global land cover product. Four strata were defined from the two criteria and sampled in each stratum with different sampling rates. Each sample size was a 3×3 pixel block, and the sample size was 1,265.

For the validation of Globcover 2005, samples were selected using stratified random sampling from the validation dataset. Each sample unit corresponded to 5×5 pixels. CCI land cover used a two-rank stratified random sampling to select samples. In the first stage, primary sampling units (PSUs) are selected by a simple random sampling protocol, leading to 2,600 PSUs (10×10 km). In the second stage, five secondary sampling units (SSUs) were selected at the corners and center of each PSU. Stratified sampling strategy by class was used for GLC-SHARE validation. Approximately 1,000 samples were estimated to be required for the quality assessment. MCD12Q1 product utilized the STEP database, which offers high-quality training samples representative of each land cover type in the MODIS Land Cover product. The C6 MLCT classifications were conducted using the Random Forests algorithm.

(3) Reference data

Samples are usually judged by using remote sensing imagery, such as TM, SPOT, and so on. The reference data adopted in DISCover were TM and SPOT images. For UMD validation, the DISCover product was used as reference data. In addition to TM images, aerial photographs, and NDVI profiles were used as reference data for GLC2000. In Globcover validation, an area-dominant method was used to distinguish if a sample was homogeneous or heterogeneous. That is, if a land cover type covered more than 75% of the observational unit, it was then considered to be the dominant land cover class. In addition, Google Earth was first used as reference data. Google Earth and NDVI profiles were the reference data in CCI validation. In MCD12Q1 validation, each site in the database was evaluated using high-resolution imagery from Google Earth™ and MODIS NBAR time-series data.

(4) Accuracy assessment

The validation results of the accuracy assessment for different GLC products varied from 66%–80%. The accuracy result of the DISCover was calculated by using a confusion matrix. The validation result showed that an overall area-weighted accuracy was determined to be 66.9%, with an average class accuracy of 73.5%, calculated from majority

Introduction

interpretation (Scepan 1999). A confusion matrix was also used for the accuracy assessment of the MODIS product. The overall area-weighted accuracy of the MODIS product was 78.3% (Strahler et al. 2006). A matrix of thematic distance was produced to act as a weight matrix for assessing the accuracy of GLC2000. The GLC 2000 product has 23 classes and an overall accuracy of 68.6% (Mayaux et al. 2006). An overall accuracy weighted by class area reached 79.25% for Globcover2005, and 70.7% for Globcover2009 (Strahler et al. 2006). An overall accuracy reached 74% for CCI land cover. The overall accuracy of GLC-SHARE was determined by dividing the total number of correctly classified points (the sum of the main diagonal elements) by the total number of points, yielding an accuracy of approximately 80%. For MCD12Q1, the overall accuracies of the LCCS layers were slightly higher than those of the IGBP scheme across all years. The results for 2007 (i.e., the midpoint of the current C6 time series) are representative of the results across all years. In 2007, the overall accuracy was 87% for the LCCS3 layer, 81% for LCCS2, 74% for LCCS1, and 67% for the IGBP layer.

1.2.3 CHALLENGES FACING FINER-RESOLUTION GLC VALIDATION

The validation schemes and methods of coarse-resolution GLC data products, as discussed in the previous section, provide us with useful references and experiences for the validation of finer-resolution GLC products. The accuracy of GlobeLand30 has been evaluated by third-party researchers from more than ten countries for either all classes or a single class via sample-based validation or comparison with existing land cover products (Brovelli et al. 2015; Kussul et al. 2015; Manakos et al. 2015; Arsanjani et al. 2016a and 2016b). However, a number of special problems or challenges arise when validating the 30 m resolution GLC data products.

First, there is a significant spatial heterogeneity in land cover across the globe, which can be clearly observed from the 30 m resolution GLC data products. Applying traditional sampling approaches to the 30 m resolution GLC data products might cause several problems, and credible spatial samples may not be obtained. These issues include inappropriate sample sizes for target regions, underrepresented sample numbers for rare classes, and irrational sample distribution in geographical space (Chen et al. 2016). A random stratified sampling method has commonly been adopted in existing GLC products, which does not account for complicated land surface heterogeneity. Considering the spatial heterogeneity of land cover in the sampling design and allowing for higher sample densities or larger sample sizes in more heterogeneous regions is the first challenge in the collaborative validation of 30 m resolution GLC products (Chen et al. 2015).

Second, the collection of sample data is highly labor-intensive, costly, and time-consuming, and becomes even more challenging when sample data need to be collected over large areas or across the entire globe. With the increasing application of the internet and web-service technologies, online tools have been developed to enable people from anywhere in the world to provide and access reference data and other sample information online (Fritz et al. 2012; Han et al. 2015). The information includes geo-tagged photos, visual interpretation of images, and thematic maps. However, the validation tools developed and used in the past were mainly offline, stand-alone operations, and were difficult to support distributed collaborative validation. Additionally, these tools were mostly dedicated to some single specific step of the validation process rather than supporting the entire procedure. Therefore, it is necessary to develop comprehensive online validation systems (or tools) to facilitate the major steps of collaborative validation. These steps include the integration and access of reference materials, estimation of total sample sizes and interclass distribution, geospatial layout of samples, expert interactive checking, generation of accuracy reports, online marking of errors, and the uploading and downloading of samples, among others.

In the past few years, internet-based spatial information resource sharing (such as free remote sensing resources provided by Google, Sky Map) and online mapping tools (such as Google Map Maker, OpenStreet Map) have developed rapidly and provide favorable conditions for online validation (Fritz et al. 2012; Yu and Gong 2012). The Austrian Institute of International Systems (IIASA) has developed a land cover online package tool, Geo-Wiki, to support volunteer-conducted online reporting (Fritz et al. 2012). However, it has not yet provided sample size calculations, sample automatic layout, accuracy calculations, and other functions. Therefore, it is a trend to develop online validation platforms that could support the entire procedure of collaborative GLC validation, including data upload, sample generation, product validation, and accuracy analysis. Such online systems or tools will enable experts and volunteers from different parts of the world to access and use reference information, upload sample data, make sample judgments and labeling, and annotate error messages.

Third, the collaborative accuracy assessment of finer-resolution GLC data products has now become a necessity. Its major motivation comes from the increasing demands of sustainable development monitoring, environmental change studies, and various other application areas, where accuracy assessment reports for large areas or even global scales are required. For instance, the accuracy of GlobeLand30-2000 and GlobeLand30-2010 has been evaluated by third-party researchers from over ten countries. These evaluations were conducted for all classes or a single class through sample-based validation or by comparison with existing land cover products (Brovelli et al. 2015; Manakos et al. 2015; Arsanjani et al. 2016a). However, these accuracy assessments were performed either on 10% of the map sheets selected on a global scale or within specific individual countries or regions. There are still areas where the uncertainty of GlobeLand30 and other finer-resolution GLC data products has yet to be validated and documented. By mobilizing resources and expertise from different countries and organizations around the world for a cooperative validation of the global scale, finer-resolution GLC data products are becoming one of the major concerns for land cover mapping and user communities. On the other side, the number of individuals involved in land cover validation, including users and researchers, has grown rapidly, creating a large community of voluntary validators. They share common interests and have access to their own validation resources. It is therefore possible to move from individual validation practices to an international collaborative validation effort.

1.3 GEO-LED COLLABORATIVE VALIDATION OF 30 m GLC DATA PRODUCTS

To address the challenges facing global-scale accuracy assessment of 30 m resolution GLC data products, a collaborative validation initiative was launched by the Global Land Cover Community Activity of the GEO in 2014 (Chen et al. 2021). Figure 1.5 lists the conceptual framework and major tasks of this GEO-led collaborative validation project.

(1) **Formation of a task force**: A task force was set up by inviting GEO members and participating organizations with their own validation resources and expertise. One of the major tasks of the task force is to develop appropriate approaches and procedures for sampling design, response, and analysis protocols. The second task is to design and develop web-based validation tools to enable online collaborative validation processes by experts from different parts of the world. Then a technical specification was formulated and used for supporting the validation practice at global, regional, and national levels.

(2) **Methodological development**: From the technical point of view, there are a number of methodological issues to be examined in consideration of the complexity and difficulties of the validation of 30 m resolution GLC data products. First of all, both sample-based accuracy assessment and comparison with existing reference datasets should be integrated for global-scale collaborative validation. Appropriate approaches should be developed to perform operational sampling design, reference data collection, sample judgment and labeling, and accuracy assessment of 30 m resolution GLC data products. For instance, the spatial heterogeneity of land cover at the global scale should be taken into consideration in sampling design for obtaining credible sample sizes and their

Introduction

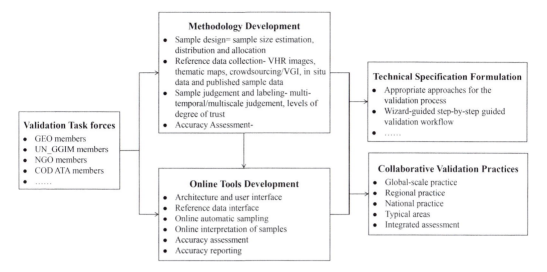

FIGURE 1.5 Major tasks of the GEO-led collaborative GLC validation.

spatial distributions. The utilization of landscape shape index (LSI) for characterizing the spatial heterogeneity of land cover was proposed to solve the critical problems facing sampling design, such as inappropriate sample sizes for target regions, underrepresented sample numbers for rare classes, and irrational sample distribution in the geographical space (Chen et al. 2016).

(3) **Development of online tools**: The development and utilization of an online validation system (or tools) aims to facilitate experts or volunteers from different parts of the world to participate in collaborative validation. On one side, it should be able to support the entire validation processes, from the total sample estimation and interclass distribution, sample space layout, sample judgment and labeling, accuracy assessment, and other online processing. Validation experts or volunteers should be able to make a judgment of given sample points through the online interpretation of high-resolution images or comparison with large-scale land cover maps, as well as in situ measurement data. They may also upload additional supporting information, such as text commentaries and photos of the samples. On the other side, efficient communication and interaction among participants and coordinators of the validation processes in a distributed working environment are essential for collaborative validation. This can be achieved through modeling the relationships among the users and the scopes of their tasks, and their interactions with the help of web-based publish/subscribe and service computing, as shown in Figure 1.6 (Chen et al. 2018).

(4) **Formulation of technical specification**: In order to facilitate the collaborative validation activities, a technical specification was developed to describe the approaches and procedures used for sampling design, response, and analysis protocols at 30 m resolution and global scale. For instance, the criteria and methods for the collection and integration of various reference data and resources are specified and described. In addition to the Very High Resolution (VHR) images and existing high-accuracy thematic maps, a variety of data and information can be used as reference data for GLC validation, such as in situ measurements, crowdsourced, and VGI-generated text commentaries and photos. The quality and usefulness of these potential reference data should be evaluated in terms of their completeness, logical consistency, positional accuracy, thematic accuracy, temporal quality, and usability. In addition, the principles and criteria for sample judgment and labeling should also be defined, such as harmonization of classification systems, multi-scale interpretation strategy, and use of degree of trust in sample judgment.

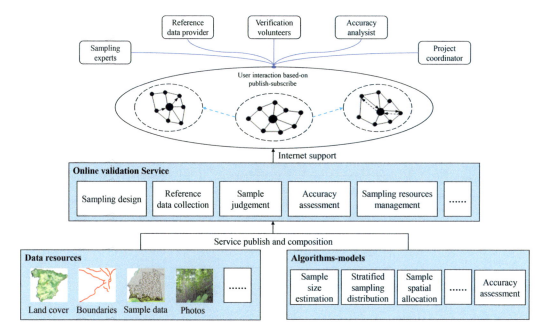

FIGURE 1.6 Service-supported online validation.

(5) **Collaborative validation practices**: From the organizational point of view, the success of such a collaborative initiative depends largely on whether international resources and expertise could be mobilized to form a task force. With the support of the GEO Secretariat and the International Cooperation Program of the Ministry of Science and Technology of China, about 30 GEO members and participating organizations joined this collaborative validation project. A number of workshops, training courses, and other business meetings were organized. A series of validation practices were carried out at national, regional and global scales with the guidance of the technical specification and the support of the developed online tools.

1.4 CHAPTER ORGANIZATION

This book is a summary of the GEO-led collaborative initiative. It presents the validation methodology and practice of 30 m GLC data products. The content of and connection between chapters are illustrated in Figure 1.7.

Chapter 2 introduces the overall sampling design by further elaboration on different sampling methods and corresponding sample size estimation and sample allocation that are relevant to GLC data validation.

Chapter 3 aims to provide an overview of different types of data that can be used as sources of reference data for validating GLC data products, ranging from traditional data sources, including high-resolution images and higher-resolution land cover data, to more recent types of data, such as crowdsourcing and social media data.

Chapter 4 looks into how different types of reference data described in Chapter 3 may be used to facilitate reference sample data generation for validating GLC products. Methods of interpretation, remapping, and analyses are described for deriving reference sample data from remote sensing images, land cover maps, and crowdsourcing data. It also discusses the degree of trust in interpretation and the impact of the reliability of reference data sources on reference sample data derived.

Introduction

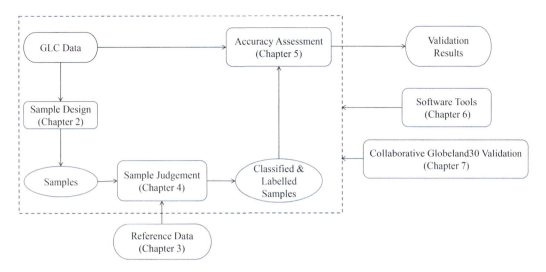

FIGURE 1.7 Chapter organization.

Chapter 5 provides a comprehensive treatment of basic accuracy indicators and other accuracy assessment methods, which are either commonly used for evaluating the accuracy of GLC data or applied to specific land cover data assessments.

Chapter 6 surveys and evaluates the existing desktop and online tools that have been used to support individual and collaborative validation efforts. Based on this brief survey, the requirements and conceptual design are discussed, which are considered in the development of the GLCval tool presented.

Chapter 7 presents the collaborative efforts and results for validating Globeland30 datasets at global, regional, and national scales.

Finally, Chapter 8 concludes the book by summarizing its contents and outlining some future directions that need further developments and investigations.

REFERENCES

Arsanjani, J. J., See, L., Tayyebi, A. 2016a. Assessing the suitability of GlobeLand30 for mapping land cover in Germany. *International Journal of Digital Earth*, 9: 873–891.

Arsanjani, J. J., Tayyebi, A., Vaz, E. 2016b. GlobeLand30 as an alternative fine-scale global land cover map: Challenges, possibilities, and implications for developing countries. *Habitat International*, 55: 25–31.

Bai, Y. C., Wang, J. F. 2003. *Uncertainty of remote sensing-classification and scale effect*. Beijing: Geological Publishing House.

Belward, A.S., Estes, J.E., Kline, K. D. 1999. The IGBP-DIS global 1-km land-cover data set DISCover: A project overview. *Photogrammetric Engineering and Remote Sensing*, 65(9): 1013–1020.

Blakemore, M. 1983. Generalization and error in spatial databases. *Cartographica*, 21: 131–139.

Blanco, P. D., Colditz, R. R., Saldana, G. L., et al. 2013. A land cover map of Latin America and the Caribbean in the framework of the SERENA project. *Remote Sensing of Environment*, 132: 13–31.

Bontemps, S., Defourny, P., Van Bogaert, E., et al. 2011. GLOBCOVER 2009 products description and validation report. *UCLouvain & ESA Team*. https://core.ac.uk/download/pdf/11773712.pdf

Brovelli, M.A., Molinari, M.E., Hussein, E., Chen, J., Li, J. 2015. The first comprehensive accuracy assessment of GlobeLand30 at a national level: Methodology and results. *Remote Sensing*, 7: 4191–4212.

Brown, G. 2013. The relationship between social values for ecosystem services and global land cover: An empirical analysis. *Ecosystem Services*, 5: 58–68.

Burrough, P. A. 1986. *Principles of geographical information systems for land resources assessment (monographs on soil and resources survey)*. New York: Oxford University Press

Buttner, G., Feranec, J., Jaffrain, G., Mari, L., Maucha, G., Soukup, T. 2004. The CORINE land cover 2000 project. *Proceedings of EARSeL eProceedings*, 3: 331–346.

Cao, X., Chen, J., Chen, L. J., et al. 2014. Preliminary analysis of spatiotemporal pattern of global land surface water. *Science China Earth Sciences*, 57: 2330–2339.

Cao, X., Li, A., Lei, G., et al. 2016. Land cover mapping and spatial pattern analysis with remote sensing in Nepal. *Journal of Geo-Information Science*, 18: 1384–1398.

Caspary, W., Scheuring, R. 1993. Positional accuracy in spatial databases. *Computers Environment and Urban Systems*, 17(2): 103–110.

CCI Land Cover, 2017. S2 prototype Land Cover 20 m map of Africa. *CCI Land Cover*. http://2016africa landcover20m.esrin.esa.int/viewer.php

Chen, B., Xu, B., Zhu, Z., Yuan, C., Ping Suen, H., Guo, J., Xu, N., Li, W., Zhao, Y., Yang, J. J. S. B. 2019. Stable classification with limited sample: Transferring a 30-m resolution sample set collected in 2015 to mapping 10-m resolution global land cover in 2017. *Science Bulletin*, 64: 370–373.

Chen, F., Chen, J., Wu, H., et al. 2016. A landscape shape index-based sampling approach for land cover accuracy assessment. *Science China Earth Sciences*, 46: 1413–1425.

Chen, J., Ban, Y. F., Li, S. N. 2014. China: Open access to earth land-cover map. *Nature*, 514: 434.

Chen, J., Chen, F., Wu, H., Chen, L. J., Han, G. 2018. Internet+ land cover validation: Methodology and practice. *Geomatics and Information Science of Wuhan University*, 43(12): 2225–2232.

Chen, J., Chen, J., Liao, A., et al. 2014. Global land cover mapping at 30 m resolution: A POK-based operational approach. *ISPRS Journal of Photogrammetry and Remote Sensing*, 103: 7–27.

Chen, J., Chen, L., Li, R., et al. 2015. Spatial distribution and ten years change of global built-up areas derived from GlobeLand30. *Acta Geodaetica et Vartographica Sinica*, 44(11): 1181–1188.

Chen, J., Chen, L. J., Chen, F., et al. 2021. Collaborative validation of GlobeLand30: Methodology and practices, *Geo-spatial Information Science*, 24(1): 134–144.

Chen, J., Xin, C., Shu, P., Huiru, R. 2017. Analysis and applications of GlobeLand30: A review. *International Journal of Geo-Information*, 6, 230.

Cheng, J. C., Guo, H. D., Shi, W. Z., et al. 2004. *Uncertainty of remote sensing data*. Beijing: Science Press.

Cihlar, J., 2000. Land cover mapping of large areas from satellites: Status and research priorities, *International Journal of Remote Sensing*, 25: 1093–1114.

Corbane, Christina, Martino Pesaresi, Panagiotis Politis, Vasileios Syrris, Aneta J. Florczyk, Pierre Soille, Luca Maffenini et al. 2017. Big earth data analytics on Sentinel-1 and Landsat imagery in support to global human settlements mapping. *Big Earth Data*, 1(1–2): 118–144.

Deng, X., Liu, J. 2012. *Mapping land-cover and land-use changes in China. remote sensing of land use and land cover: Principals and applications*, Boca Raton: CRC Press.

Diogo, V., E. Koomen. 2016. Land cover and land use indicators: Review of Available Data. *OECD Green Growth Papers*. https://www.oecd-ilibrary.org/conte-nt/paper/5jlr2z86r5xw-en

Esch, T., Heldens, W., Hirner, A., Keil, M., Marconcini, M., Roth, A., Zeidler, J., Dech, S., and Strano, E. 2017. Breaking new ground in mapping human settlements from space–The Global Urban Footprint. *ISPRS Journal of Photogrammetry and Remote Sensing*, 134: 30–42.

Esch, T., Felbier, A., Heldens, W., Marconcini, M., Roth, A., and Taubenbock, H. 2013. Spatially detailed mapping of settlement patterns using SAR data of the TanDEM-X mission . In Joint Urban Remote Sensing Event 2013 (April 21 to April 23, 2013, in São Paulo, Brazil).

Feng, M., Sexton, J. O., Channan, S., Townshend, J. R. 2014. A global, high-resolution (30-m) inland water body dataset for 2000: First results of a topographic-spectral classification algorithm. *International Journal of Digital Earth*, 2(9): 113–133.

Florczyk, A. J., Corban, C., Ehrlich, D., et al. 2019. *Ghsl data package 2019*. Luxembourg: Publications Office of the European Union.

Foley, J. A., DeFries, R., Asner, G. P., et al. 2005. Global consequences of land use. *Science*, 309: 570–574.

Friedl, M. A., McIver, D. K., Hodges, J. C. F., et al. 2002. Global land cover from MODIS: Algorithms and early results. *Remote Sensing of Environment*, 83: 135–148.

Fritz, S., et al., 2012. Geo-Wiki: An online platform for improving global land cover. *Environmental Modelling & Software*, 31(7): 110–123.

Ge, Y., Wang, J. F. 2003. *Uncertainty of remote sensing – Error propagation modeling*. Beijing: Geological Publishing House.

Giri, C., Long, J. 2014. Land cover characterization and mapping of south america for the year 2010 using Landsat 30 m satellite data. *Remote Sensing of Environment*, 6(10): 9494–9510.

Giri, C., Pengra, B., Long, J., Loveland, T.R. 2013. Next generation of GLC characterization, mapping, and monitoring. *International Journal of Applied Earth Observation and Geoinformation*, 25: 30–37.

Gong, P. 2009. Assessment of GLC map accuracies using Fluxnet location data. *Progress of Natural Sciences*, 19: 754–759.

Introduction

Gong, P., Li, X. C., Wang, J., et al. 2020. Annual maps of global artificial impervious area (Gaia) between 1985 and 2018. *Remote Sensing of Environment*, 236: January 2020, 111510.

Gong P., Liu, H., Zhang, M. N., et al. 2019. Stable classification with limited sample: Transferring a 30-m resolution sample set collected in 2015 to mapping 10-m resolution global land cover in 2017. *Science Bulletin*, 64(6): 370–373.

Gong, P., Wang, J., Yu, L., et al. 2012. Finer resolution observation and monitoring of global land cover: First mapping results with Landsat Tm and Etm+ data. *International Journal of Remote Sensing*, 34(7): 2607–2654.

Goodchild, M. F. 1992. Geographical information science. *International Journal of Geographical Information Systems*, 6(1): 31–45.

Goodchild, M. F., Gopal, S. 1989. *Accuracy of spatial databases*. Basingstoke: Taylor and Francis.

Grekousis, G., Mountrakis, G., Kavouras, M. 2015. An overview of 21 global and 43 regional land-cover mapping products. *International Journal of Remote Sensing*, 36: 5309–5335.

Guptill, S. C., Morrison, J. L. 1995. *Elements of spatial data quality*. Oxford: Pergamon Press Inc.

Han, G., Chen, J., He, C. Y., et al. 2015. A web-based system for supporting global land cover data production. *ISPRS Journal of Photogrammetry and Remote Sensing*, 103: 66–80.

Hansen, M. C., Defries, R. S., Townshend, J. R. G. 2000. Global land cover classification at 1km spatial resolution using a classification tree approach. *International Journal of Remote Sensing*, 21(6–7): 1331–1364.

Hansen, M.C., Loveland, T.R., 2012. A review of large area monitoring of land cover change using Landsat data, *Remote Sensing of Environment*, 122: 66–74.

Hansen, M. C., Potapov, P. V., Moore, R., et al. 2013. High-resolution global maps of 21st-century forest cover change, *Science*, 342(6160): 850–853.

Hansen, M.C., Reed, B. 2000. A comparison of the IGBP DISCover and University of Maryland 1km global land cover products. *International Journal of Remote Sensing*, 21(6–7): 1365–1373.

Herold, M., Mayaux, P., Woodcock, C., Baccini, A., Schmullius, C. 2008. Some challenges in GLC mapping: An assessment of agreement and accuracy in existing 1km datasets. *Remote Sensing of Environment*, 112: 2538–2556.

Heuvelink, G. B. M. 1998. *Error propagation in environmental modeling in GIS*. London: Taylor and Francis.

Huang, Y. C., Liu, W. B., Li, Z. H., et al. 1995. *GIS spatial data errors analysis and processing*. China Wuhan: University of Geosciences Press.

ISO 19157: 2013(E). ISO. 2023. 19157-1:2023 Geographic information — Data quality. https://www.iso.org/standard/78900.html

Iwao, K., Nishida, K., Kinoshita, T., Yamagata, Y. 2006. Validating land cover maps with degree confluence project information. *Geophysical Research Letters*, 33: L23404.

Jin, S. M., Yang, L. M., Danielson, P., et al. 2013. A comprehensive change detection method for updating the National Land Cover Database to circa 2011. *Remote Sensing of Environment*, 132: 159–175.

Kussul, N., Shelestov, A., Basarab, R., Skakun, S., Kussul, O., Lavreniuk, M. 2015.*Geospatial intelligence and data fusion techniques for sustainable development problems*. ICTERI. http://ceur-ws.org/Vol-1356/paper_48.pdf

Lambin, E. F., Helmut, J.G. 2008. *Land-use and land-cover change: local processes and global impacts*. Berlin: Springer Science & Business Media.

Latifovic, R., Poiliot, D., Dillabaugh, C. 2012. Identification and correction of systematic error in NOAA AVHRR long-term satellite data record. *Remote Sensing of Environment*, 127: 84–97.

Li, D. R., Wang, S. L., Li, D. Y. 2013. *Spatial data mining theories and applications*. Beijing: Science Press.

Li, R., Kuang, W. H., Chen, J., et al. 2016. Spatio-temporal pattern analysis of artificial surface use efficiency based on GlobeLand30. *Scientia Sinica Terrae*, 46: 1436–1445.

Liu, D. J., Shi, W. Z., Tong, X. H., et al. 1996. *GIS spatial data accuracy analysis and quality control*. Shanghai: Shanghai Scientific and Technological Literature Press.

Liu J. Y., Liu M. L., Deng X. Z., et al. 1999. The landuse and land cover database and its relative studies in China, *Journal of Geographic Sciences*, 12(3): 275–282.

Liu, Q. H., Xin, X. Z., Tang, P., et al. 2010. *Quantitative remote sensing model, application and uncertainty*. Beijing: Science Press.

Liu, X. P., Hu, G. H., Chen, Y. M., et al. 2018. High-resolution multitemporal mapping of global urban land using Landsat images based on the Google Earth Engine platform. *Remote Sensing of Environment*, 227–239.

Loveland, T.R., Reed, B.C., Brown, J.F., et al. 2000. Development of a global land cover characteristics database and IGBP DISCover from 1 km AVHRR data. *International Journal of Remote Sensing*, 21: 1303–1330.

Lu, D., Mausel, P., Brondizio, E. 2004. Change detection techniques. *International Journal of Remote Sensing*, 28(5): 823–870.

Lu, D., Weng, Q. 2007. A survey of image classification methods and techniques for improving classification performance. *International Journal of Remote Sensing*, 28(5): 823–870.

Lu, X. H., Jiang H., Zhang X. Y., Jin J. X. 2016. Relationship between nitrogen deposition and LUCC and its impact on terrestrial ecosystem carbon budgets in China. *Science China Earth Sciences*, 59: 2285–2294.

Lymburner, L., Tan, P., McIntyre, A., Lewis, A., Thankappan, M. 2013. Dynamic land cover dataset version 2: 2001–now… a land cover odyssey. *IEEE International Geoscience and Remote Sensing Symposium (IGARSS)*. https://ieeexplore.ieee.org/abstract/document/6723532

Macedo, R. D. C., Moreira, M. Z., Domingues, E., et al. 2013. LUCC (land use and cover change) and the environmental-economic accounts system in Brazil. *Journal of Earth Science and Engineering*, 4: 840–844.

Manakos, I., Chatzopoulos-Vouzoglanis, K., Petrou, Z.I., Filchev, L., Apostolakis, A. 2015. Globalland30 mapping capacity of land surface water in Thessaly, Greece. *Land*, 4: 1–18.

Mayaux, P., Eva, H., Gallego, J., et al. 2006. Validation of the global land cover 2000 map. *IEEE Transactions on Geoscience and Remote Sensing*, 44: 1728–1739.

MODIS land cover team. 2003. Validation of the consistent year 2003. V003.

Nagendra, H., Reyers, B., Lavorel, S. 2013. Impacts of land change on biodiversity: Making the link to ecosystem services. *Current Opinion in Environmental Sustainability*, 5: 503–508.

NCGIA. 1989. The research plan of the National Center for Geographic Information and Analysis. *International Journal of Geographical Information Systems*, 3(2): 117–136.

NGCC. 2014. Research report on the development of global land cover remote sensing product, 863 Plan.

Oliphant, A. J., Thenkabail, P. S., Teluguntla, P., et al. 2017. *NASA making earth system data records for use in research environments (measures) global food security-support analysis data (Gfsad) cropland extent 2015 Southeast Asia 30 mV001*. Monograph. South Dakota, USA: NASA EOSDIS Land Processes DAAC. http://oar.icrisat.org/10979/

Olofsson, P., Foody, G. M., Herold, M., Stehman, S. V., Woodcock, C. E., Wulder, M. A. 2014. Good practices for estimating area and assessing accuracy of land change. *Remote Sensing Environment*, 148: 42–57.

Olthof, I., Latifovic, R., Pouliot, D. 2009. Development of a circa 2000 land cover map of northern Canada at 30 m resolution from Landsat. *Canadian Journal of Remote Sensing*, 35(2): 152–165.

Pearson, E. S., Hartley, H. O. 1966. *Biometrika tables for statisticians*. Cambridge: Cambridge University Press.

Pekel, J. F., Cottam, A., Gorelick, N., Belward, A. S. 2016. High-resolution mapping of global surface water and its long-term changes. *Nature*, 540: 418–422.

Perkal, J. 1956. On epsilon length. *Bulletin de l'académie Polonaise des Sciences*, 4(3): 399–403.

Pickens, A. H., Hansen, M. C., Hancher, M., et al. 2020. Mapping and sampling to characterize global inland water dynamics from 1999 to 2018 with full Landsat time-series. *Remote Sensing of Environment*, 243: 111792.

Pielke, R. A. 2005. Land use and climate change. *Science*, 310: 1625–1626.

Raymond, P. A., Hartmann, J., Lauerwald, R., et al. 2013. Global carbon dioxide emissions from inland waters. *Nature*, 503: 355–359.

Reid, W. V., Chen, D., Goldfarb, L., et al. 2010. Earth system science for global sustainability: Grand challenges. *Science*, 330: 916–917.

Roger, A., Pielke, Sr. 2005. Land use and climate change. *Science*, 310: 1625–1626.

Running, S. W. 2008. Ecosystem disturbance, carbon, and climate. *Science*, 321: 652–653.

Rwanga, S. S., Ndambuki, J. M. 2017. Accuracy assessment of land use/land cover classification using remote sensing and Gis. *International Journal of Geosciences*, 8(4): 611–622.

Scepan, J. 1999. Thematic validation of high-resolution global land-cover data sets. *Photogrammetric Engineering & Remote Sensing*, 65: 1051–1060.

Scoones, I., Hall, R., Borras, Jr., S.M., White, B., Wolford, W. 2013 The politics of evidence: Methodologies for understanding the global land rush. *The Journal of Peasant Studies*, 40: 469–483.

Sexton, Joseph O., Xiao-Peng Song, Min Feng, Praveen Noojipady, Anupam Anand, Chengquan Huang, Do-Hyung Kim et al. 2013. Global, 30-m resolution continuous fields of tree cover: Landsat-based rescaling of MODIS vegetation continuous fields with lidar-based estimates of error. *International Journal of Digital Earth*, 6(5): 427–448.

Sexton, J. O., Feng, M., Channan, S., et al. 2016. Earth science data records of global forest cover and change. *User Guide*, 38.

Shi, W. Z. 1994. *Modeling positional and thematic uncertainties in integration of remote sensing and geographic information systems*. Enschede: ITC Publication.

Shi, W. Z. 2005. *Principles of modeling uncertainties in spatial data and spatial analyses*. Beijing: Science Press

Shi, W. Z., Chen, P. F., Zhang, X. K., 2017. Reliability analysis in geographical conditions monitoring. *Acta Geodaetica et Cartographica Sinica*, 46(10): 1620–1626.

Shi, X. L., Nie, S. P., Ju, W. M., Yu, L. 2016. Climate effects of the GlobeLand30 land cover dataset on the Beijing Climate Center climate model simulations. *Science China Earth Sciences*, 59: 1754–1764.

Shimada, Masanobu, Takuya Itoh, Takeshi Motooka, Manabu Watanabe, Tomohiro Shiraishi, Rajesh Thapa, and Richard Lucas. 2014. New global forest/non-forest maps from ALOS PALSAR data (2007–2010). *Remote Sensing of environment*, 155: 13–31.

Stehman, S. V., Czaplewski, R. L. 1998. Design and analysis for thematic map accuracy assessment: Fundamental principles. *Remote Sensing of Environment*, 64(3): 331–344.

Stehman, S. V., Olofsson, P., Woodcock, C. E., Herold, M., Friedl, M. A. 2012. A global land-cover validation data set, II: Augmenting a stratified sampling design to estimate accuracy by region and land-cover class. *International Journal of Remote Sensing*, 33: 6975–6993.

Sterling, S., Ducharne, A. 2008. Comprehensive data set of global land cover change for land surface model applications. *Global Biogeochemical Cycles*, 22.

Sterling, S. M., Ducharne, A., Polcher, J. 2013. The impact of global land-cover change on the terrestrial water cycle, *Nature Climate Change*, 3: 385–390.

Strahler, A. H., Boschetti, L., Foody, G. M., et al. 2006. *Global land cover validation: recommendations for evaluation and accuracy assessment of global land cover maps*. Luxembourg: European Communities.

Strahler, A. H., Boschetti, L. G., Foody, G. M., et al. 2008. *Global land cover validation: recommendations for evaluation and accuracy assessment of global land cover maps*. Luxembourg: European Communities.

Sulla-Menashe, D., Gray, J. M., Abercrombie, S. P., Friedl, M. A. 2019. Hierarchical mapping of annual global land cover 2001 to present: The MODIS Collection 6 land cover product. *Remote Sensing of Environment*, 222: 183–194.

Tang, P., Zhang, H.W., Zhao, Y., Niu, Z., Zhong, B., Huang, C., Shan, X. 2014. Practice and thoughts of the autoamatic processing of mispectral images with 30m spatial resolution on the global scale, *Journal of Remote Sensing*, 18(2): 231–241.

Thenkabail, P. 2012. Global croplands and their water use for food security in the twenty-first century. *Photogrammetric Engineering and Remote Sensing*, 78: 797–798.

Tong, X. H., Sun, T., Fan, J. Y., et al. 2013. A statistical simulation model for positional error of line features in geographic information systems. *International Journal of Applied Earth Observation and Geoinformation*, 21: 136–148.

Tong, X. H., Wang, Z. H., Xie, H., et al. 2011. Designing a two-rank acceptance sampling plan for quality inspection of geospatial data products. *Computers & Geosciences*, 37: 1570–1583.

Tong, X. H., Xie, H., Liu, S., , et al. 2017. Uncertainty of spatial information and spatial analysis. In: Leng, S. Y., Gao, X. Z., Pei, T., et al. *Geographical science during 1986–2015: From the classics to the frontiers*. Singapore: Springer-Verlag, 511–522.

Tong, X. H., Zhou, L., Jin, Y. M. 2023. Positional error model of line segments with modeling and mearing errors using Brownian Bridge. *Journal of Geodesy and Geoinformation Science*, 6(2):1–10.

Townshend, J. R., Masek, J.G., Huang, C., et al. 2012. Global characterization and monitoring of forest cover using Landsat data: Opportunities and challenges. *International Journal of Digital Earth*, 5: 373–393.

Townshend, J. R. G., Justice, C. O., Li, W. 1991. Global land cover classification by remote sensing: present capabilities and future possibilities. *Remote Sensing of Environment*, 35: 243–256.

Verburg, P. H., Kok, K., Pontius, R. G., Veldkamp, A. 2006. *Modeling land-use and land-cover change*. Berlin: Springer Science & Business Media.

Verburg, P. H., Neumann, W., Linda, N. L. 2011. Challenges in using land use and land cover data for global change studies. *Global Change Biology*, 17: 974–989.

Wang, P., C. Huang, E. C. Brown de Colstoun, J. C. Tilton, and B. Tan. 2017. Global Human Built-up and Settlement Extent (HBASE) Dataset From Landsat. New York: NASA Socioeconomic Data and Applications Center (SEDAC).

Yamazaki, D., Trigg, M. A., Ikeshima, D. K. 2015. Development of a global ~90 m water body map using multi-temporal Landsat images. *Remote Sensing of Environment*, 337–351.

Yang, Y., Xiao, P., Feng, X., Li, H. 2017. Accuracy assessment of seven global land cover datasets over China. *ISPRS Journal of Photogrammetry and Remote Sensing*, 125: 156–173.

Yu, L., P. Gong. 2012. Google Earth as a virtual globe tool for earth science applications at the global scale: Progress and perspectives. *International Journal of Remote Sensing*, 33: 3966–3986.

Yu, L., Wang, J., Clinton, N., et al. 2013. From-Gc: 30 m global cropland extent derived through multisource data integration. *International Journal of Digital Earth*, 6(6): 521–533.

Zanaga, D., Van De Kerchove, R., Daems, D., De Keersmaecker, W., Brockmann, C., Kirches, G., Wevers, J., Cartus, O., Santoro, M., Fritz, S., Lesiv, M., Herold, M., Tsendbazar, N.E., Xu, P., Ramoino, F., Arino, O., 2022. ESA WorldCover 10 m 2021 v200, https://worldcover2021.esa.int/

Zanaga, D., Van De Kerchove, R., De Keersmaecker, W., Souverijns, N., Brockmann, C., Quast, R., Wevers, J., Grosu, A., Paccini, A., Vergnaud, S., Cartus, O., Santoro, M., Fritz, S., Georgieva, I., Lesiv, M., Carter, S., Herold, M., Li, Linlin, Tsendbazar, N.E., Ramoino, F., Arino, O., 2021. ESA WorldCover 10 m 2020 v100, https://worldcover2020.esa.int/

Zell, E., Huff, A. K., Carpenter, A. T., Friedl, L. A. 2012. A user-driven approach to determining critical earth observation priorities for societal benefit. *IEEE Journal of Selected Topics in Applied Earth Observations and Remote Sensing*, 5(6): 1594–1602.

Zhang, J., Li, W., Zhai, L. 2015. Understanding geographical conditions monitoring: A perspective from China. *International Journal of Digital Earth*, 8(1): 38–57.

Zhang, J., Liu, J., Zhai, L., et al. 2016. Implementation of geographical conditions monitoring in Beijing-Tianjin-Hebei, China. *ISPRS International Journal of Geo-Information*, 5(6): 89.

Zhang, J. X., Goodchild, M. F. 2002 *Uncertainty in geographical information*. New York: CRC Press.

Zhang, X., Liu, L. Y., Chen, X. D., Gao, Y., Xie, S., Mi, J. 2020. GlC_FCS30: Global land-cover product with fine classification system at 30 m using time-series Landsat imagery. *Earth System Science Data Discussions*, 13: 2753–2776.

Zhang, X., Liu, L. Y., Wu, C. S., et al. 2020. Development of a global 30 m impervious surface map using multisource and multitemporal remote sensing datasets with the Google Earth Engine platform, *Earth System Science Data*, 12(3): 1625–1648.

Zhang, X. M., Long, T. F., He, G.J., et al. 2020. Rapid generation of global forest cover map using Landsat based on the forest ecological zones. *Journal of Applied Remote Sensing*, 14(2): 022211.

Zimmerman, P. L., Housman, I. W., Perry, C. H., Chastain, R. A., Webb, J. B., Finco, M. V. 2013. An accuracy assessment of forest disturbance mapping in the western Great Lakes. *Remote Sensing Environment*, 128: 176–185.

2 Sampling Design

Huan Xie and Xiaohua Tong
Tongji University, Shanghai, China

Fei Chen
East China University of Technology, Nanchang, China

Chao Wei
Tongji University, Shanghai, China

Zhenhua Wang
Shanghai Ocean University, Shanghai, China

Shicheng Liao
Tongji University, Shanghai, China

Jun Chen
National Geomatics Center of China, Beijing, China

2.1 DESIGN PRINCIPLES AND CHALLENGES

Sampling involves examining just a subset of the data to infer the quality of the entire dataset rather than conducting a comprehensive 100% inspection. Various acceptance sampling plans have been developed with the goal of verifying whether the data product conforms to established quality standards (ISO 2013 2020). An accuracy assessment begins with the definition of population that all the map need to be inspected (Strahler et al. 2006; Congalton 1993; Congalton 1991; Congalton et al. 1983; Czaplewski 1992; Rosenfield and Fitzpatrick-Lins 1986; Hay 1988; Pontius Jr and Millones 2011; Stehman and Czaplewski 1998; Xie et al. 2015; Aronoff 1982; Story and Congalton 1986; Anderson 1976). The sampling plan, defined by its sample size and acceptance number, offers a method to ascertain the sample requirements and quality standards for product inspection. Based on this plan, the user of the sampling design must select a specified number of samples (Stehman 2009; Stehman and Czaplewski 1998; Congalton and Green 2019; Stehman 1997; Smits et al. 1999; Zhuang et al. 1995; Agresti et al. 1995; Fahsi et al. 2000; Ustin et al. 1996; Elias and Abramson 1963; Strehl and Ghosh 2002; Nelson 1983; Pebesma and Wesseling 1998). The sampling process for finer-resolution global land cover (GLC) products has grown increasingly complex due to a substantial rise in both the volume and diversity of data. Identifying an appropriate method for selecting a representative sample from these finer-resolution GLC products to accurately estimate an unbiased quality variable is now a critical aspect of quality assessment. The reference data may be obtained for each sampling unit based on visual interpretation or on-site validation. The method used to determine the reference data classification is called the response design. Consistency comparison between the land cover classification of the sample units and reference data is accuracy validation.

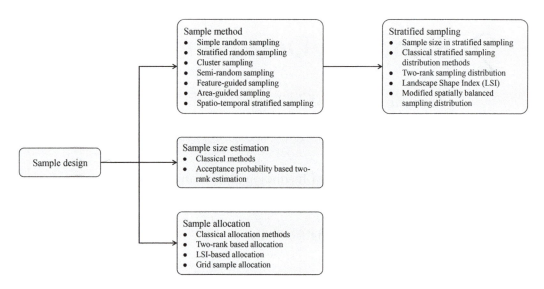

FIGURE 2.1 A comparison of sampling strategies of existing GLC products.

Generally, accuracy analyses rely on an error matrix. The sample design is shown in Figure 2.1, in which stratified sampling is widely used because of its flexibility in sample adjustment.

2.1.1 BASIC PRINCIPLE OF GLC VALIDATION

The sampling principles and methods for the seven earlier GLC products listed in Table 2.1 are different. The differences lie in source imagery, classification system, imagery resolution, mapping objective, and validation principle.

The process that verifies the accuracy of the land cover information provided in as independent a means as practical was first defined as a validation scheme in the DISCover product (Schultz et al. 2017; Loveland and Belward 1997; Helldén 1980; Short 1982; Bishop et al. 1975; Light 1971; Turk 1979; Türk 2002; Finn 1993; Pontius Jr et al. 2004). Considering the key principle of "practical," the

TABLE 2.1
A Comparison of Sampling Strategies of Existing GLC Products

	DISCover	UMD	GLC2000	MOD12	CCI-LC 2015	GLC-SHARE	GlobCover
Resolution	1 km	1 km	1 km	1 km	300 m	30 arc-second	0.3 km
Number of classes	17	14	23	17	22	11	22
Sample size	379	None	253/1,265	1,370 (39,472)	2,400	1,000	3,167
Sampling method	Stratified random sampling	None	Stratified clustered sampling, different sampling rates for each stratum	Stratified random sampling	A two-rank stratified random sampling	Stratified random sampling	Stratified random sampling
Stratified method	class	None	Proportion of priority classes and the landscape complexity	class	class	class	class

stratified random sampling was adopted as the core sampling strategy of the DISCover, and the sample size of each class was calculated to be a minimum of 25 by the standard charts for binomial confidence intervals (Hartley and Pearson 1966) based upon an expected accuracy of 0.85% at 95% confidence level (Belward et al. 1999). Due to the lack of reference data such as TM and SPOT remote sensing images, only 379 samples were selected from the 15 classes of all 17 classes (excluding snow, ice, and water).

The University of Maryland (UMD) product only used a comparison approach to analyze the agreement with the DISCover product, and no sampling was used in validating the UMD product (Hansen et al. 2000).

The principle used for validating the MODIS product was cost control. Since no sufficient reference data was available, part of the training samples for image classification were used as validation samples. The sampling design of the MODIS product adopted random sampling stratified from the training samples. In order to avoid confusion between validation samples and classification training samples, all the training samples were randomly divided into ten groups of nearly equal size, and one of the groups was used for validation, with the remaining used for classification. One thousand three hundred seventy sites (i.e., 39,472 pixels) were selected as the validation samples (Stehman et al. 2003).

Sampling of the GLC 2000 product was the most complex among the seven GLC products. A two-stage stratified clustered sampling strategy was adopted for 23 land cover classes. In the first stage, nonequal probability systematic sampling was used in global-scale sampling. In its implementation, Landsat WRS-2 (World Reference System-2) based Thiessen polygons were introduced in the sampling design. The Shannon index and priority rule were then separately calculated for each equal-area hexagon. According to these rules, the world is divided into four stratifications—namely, homogenous and priority, homogenous and nonpriority, heterogeneous and priority, and heterogeneous and nonpriority,and each strata has different importance in distributing samples reasonably. Then, a sample grid of block (cell of 1,800 km by 1,200 km) was overlaid on the four stratifications, and each sample was assigned a random number. The samples falling in the diverse and priority areas had a large chance to be chosen; thus, nonequal probability systematic sampling was completed. In the secondary stage, five fixed points from each primary sampling unit were selected as the final samples (Mayaux et al. 2006).

The creation of an international expert network was the key element of the sampling of GlobCover 2005 and 2009. The samples were selected using stratified random sampling from a validation dataset. Each sample unit corresponded to 5×5 pixels. Since the size of the sample corresponds to a large area, the area of dominance was used to distinguish whether a sample was homogeneous or heterogeneous. If a land cover type would cover more than 75% of the observational unit, it was considered to be the sole dominant land cover class. Google Earth was first used as the reference data for ground truth (Yu and Gong 2012).

The validation of the GLC-SHARE product selected samples following a stratified random sampling method. It was estimated that around 1000 samples were necessary to conduct the quality assessment (Latham et al. 2014).

In an initial validation effort, the 2015 Climate Change Initiative – Land Cover (CCI-LC) map's accuracy was evaluated with the help of the GlobCover 2009 validation dataset. This reference database for validation was constructed from 2,400 samples, each measuring 150 m $\times$ 150 m, utilizing a biased twofold stratified random sampling approach that focuses on regions most likely to contain errors (Santoro et al. 2017).

Choosing a sampling design for accuracy assessment should be guided by basic principles. The major principles are proposed from the earlier GLC products' validation practices.

(1) Be practical: The sampling design must fit with the practical purposes. The design protocol outlining the conditions and constraints associated with reference data collection is expected to be practical so that the sample data can be collected and analyzed in actual situations. Before the accuracy evaluation, the data products can be examined, and an

appropriate sampling design plan can be formulated to ensure the operability of the accuracy evaluation. When formulating the sampling plan, both the scientific nature of the method and the practical feasibility should be considered.

(2) Be cost-effective: The sampling design must be cost-effective. Considering the financial constraints and the strength of related investigation techniques, each investigation will have requirements in terms of cost, time, etc. For example, a large-scale validation requires more survey costs, but the actual budget costs are obviously insufficient. In this case, the scope of the validation must be narrowed down to accommodate the financial resources.

(3) Be spatially representative: The sample distribution is spatially representative and well-distributed.

The application of the traditional sampling method to GLC maps might cause irrational sample distribution in the geographical space. Considering the spatial complexity, surface coverage correlation, and heterogeneity distribution characteristics of samples at the global scale, the extracted samples should have a spatially representative distribution. Spatial balance is compelling due to its distribution of sample locations evenly across the study area, minimizing large gaps. Typically, this approach yields more precise estimates compared to the designs that lack spatial balance.

(4) Be reliable and precise: The sample size can be determined by the optimal sample size estimation model discussed in Section 2.2. The determination of sample size is a very important issue in accuracy assessment, which is directly related to the cost limitation of sampling and the precision requirement of validation. The sample size calculated by the optimal sample size estimation model under the control of product reliability constraint can increase the reliability of product accuracy results.

2.1.2 Challenges for 30 m Resolution GLC Data Validation

As a new generation of GLC products with 30 m-like resolution emerges, some challenges can be found in the sampling design of these fine-resolution GLC data products, as follows.

1) Lack of a unified multiscale optimal sampling scheme: At present, there is a lack of a unified multiscale optimal sampling scheme with different scales (global, regional, and factor scales). There are many different unit divisions around the global scale, such as regular geographic regions represented by rectangular grids and Thiessen polygonal grids, or irregular geographic regions with reference to geo-ecological data and climate data related to land cover zoning, and geographic divisions that are directly based on administrative divisions (continents, countries, etc.). It is necessary to optimize sampling schemes at different scales (global, regional, and factor scales) on a global scale.

2) It is difficult to evaluate finer-resolution data on a global scale: At the global scale, finer-resolution (30 m or higher) land cover data has larger data volume and richer spatial features, properties, and information compared with 300 m resolution data. How many samples should be selected from the data set is difficult to determine. A rigorous sampling model should be developed to calculate sample size scientifically, considering cost, efficiency, sampling error, confidence, and so on.

3) It is difficult for samples to be representative: Within finer-resolution GLC data, complex spatial relationships exist between each object at different spatial scales—i.e., global scale and region scale. At the global scale, usually areas with dense land types and large mapping areas are more complex, and the characteristics of these areas have a great impact on the accuracy of mapping. According to the first law of geography, things that are close are more relevant than those that are far away. The spatial relationship between land cover at the regional scale and the spatial correlation is reduced to improve the heterogeneity, thereby ensuring the representativeness of the element sample for spatial sampling. How to select samples from the area of spatial correlation region is one of the challenges.

Although the aforementioned validation schemes of existing GLC products provided a useful reference for the validation of high-resolution GLC products, there are some issues that need to be further investigated in the sampling plans and sampling allocation methods: (1) Spatial multilevel sampling frame and sampling unit construction for 30 m-like resolution GLC validation at a different scale. (2) Establishing a global-scale multiscale sampling optimization scheme and sample size estimation for land cover data. (3) Allocating samples by fully considering the spatial heterogeneity and correlation of land cover at high resolution.

2.2 SAMPLING METHODS

The sampling design involves the rational selection of reference sample units (Stehman and Czaplewski 1998). Probability sampling is essential for conducting a thorough statistical accuracy assessment. In probability sampling, each unit within the population has an equal chance of being chosen for the sample during the sampling process. This category includes methods such as simple random sampling, systematic sampling, stratified sampling, and cluster sampling. A key benefit of probability sampling is the ability to compute the sampling error, which indicates the potential divergence of a sample from the population. When extrapolating results to the population, they are presented with an allowance for the sampling error. Sampling methods that are often used in land cover validation are described by (Comber et al. 2012; Lyons et al. 2018; Rwanga and Ndambuki 2017; Shao and Wu 2008; Koukoulas and Blackburn 2001; Liu et al. 2007; Groves et al. 2009; Xie et al. 2015). The four probability sampling methods, together with some alternative methods, are briefly summarized as follows.

2.2.1 SIMPLE RANDOM SAMPLING

Simple random sampling, a probability-based method, entails the random selection of samples. In this method, each individual is selected entirely by chance, ensuring an equal probability of being chosen at any point in the sampling process. Simple random sampling proves effective when the target population exhibits relatively uniform characteristics, lacking significant patterns or clusters. However, this method might not ensure representative coverage across an area, potentially leading to samples that only represent a portion of the area. Figure 2.2 illustrates how samples are designed using the simple random sampling method to validate GlobeLand30 data.

2.2.2 STRATIFIED RANDOM SAMPLING

Stratified sampling involves dividing the population into distinct, nonoverlapping groups, known as strata, where members within each group are more homogeneous compared to those in other groups. This method often achieves greater precision in estimating means and variances than unstratified approaches for the same population. Typically, stratified sampling offers higher accuracy than simple random sampling, except when strata are inappropriately divided such that the variance within strata exceeds that of simple random sampling. For thematic map inspections, incorporating auxiliary data based on prior knowledge about the product characteristics—such as the remote sensing data source, mapping methods, or producer—can facilitate effective stratification. This approach not only reduces variance in the estimates but also enhances estimation accuracy without necessitating larger sample sizes (Wang et al. 2010).

Additionally, a multilevel stratified sampling strategy can be developed for remote sensing-derived products. This approach involves defining multiple strata and detailing the sampling procedures (refer to Figure 2.3). The objective of multilevel stratified sampling is to construct a representative sample subset of remote-sensing-derived products that provide unbiased insights into the overall product. The ultimate goal is to facilitate evaluations based on statistical inferences (Xie et al. 2015).

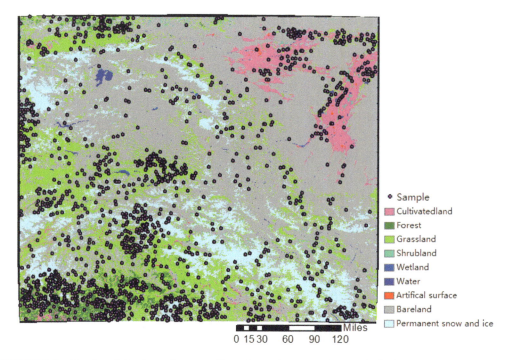

FIGURE 2.2 Results of simple random sampling of one of the sample maps in GlobeLand30 2010.

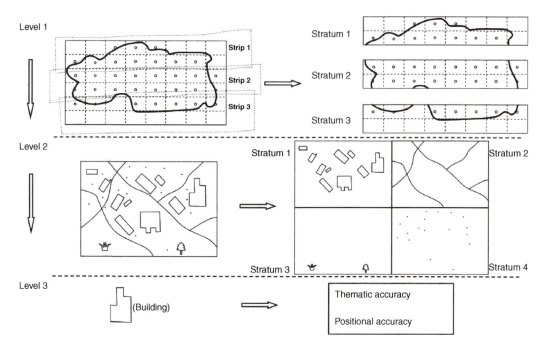

FIGURE 2.3 The process of multilevel stratified sampling for remote-sensing-derived products.

Sampling Design

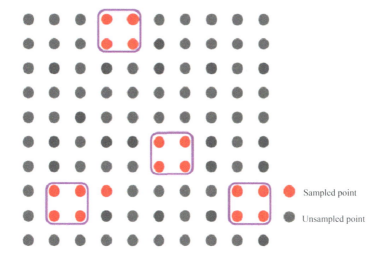

FIGURE 2.4 Cluster sampling.

2.2.3 CLUSTER SAMPLING

Cluster sampling requires clusters (see the boxed areas in Figure 2.4) to be selected randomly from the population, and then all samples of these clusters are inspected (Figure. 2.4). Cluster sampling helps save on the cost and time needed for collecting reference data (Stehman 2008). The disadvantage is that the accuracy of the estimation is poor because the units in the same subgroup are more or less similar, and the sampling error of cluster sampling is usually larger under the condition of the same sample size.

2.2.4 SEMI-RANDOM SAMPLING (SYSTEMATIC SAMPLING)

Semi-random sampling involves randomly selecting the initial sample items, such as location, time, and features, and then following specific rules for selecting the remaining items. For instance, grid sampling exemplifies semi-random sampling, where the initial grid position is determined randomly, and subsequent samples are collected at evenly spaced intervals (grid cells) across the area. Systematic grid sampling is employed to identify clusters and estimate means, percentiles, or other statistical parameters, proving particularly effective for assessing spatial trends or patterns. This method offers a straightforward and efficient means to achieve comprehensive area coverage.

2.2.5 FEATURE-GUIDED SAMPLING (NONSPATIAL SAMPLING)

A feature-guided sampling strategy involves selecting sample items based on the nonspatial attributes of features rather than their spatial locations. Within a defined data quality scope, samples can be randomly selected, assuming that the production characteristics across the scope are uniform. However, if homogeneity is only present within specific subsets and a homogeneous distribution of samples is necessary—for instance, when significant patterns or clusters appear in the sampled characteristics—simple random sampling might not yield a satisfactory sample. In such situations, employing a stratified or semi-random sampling approach could lead to more accurate outcomes.

2.2.6 AREA-GUIDED SAMPLING (SPATIAL SAMPLING)

The sampling units could be established from geographic regions (such as political or statistical divisions) or another segmentation of the domain under investigation. This approach to sampling might serve as an initial phase, succeeded by feature-oriented sampling within each subregion.

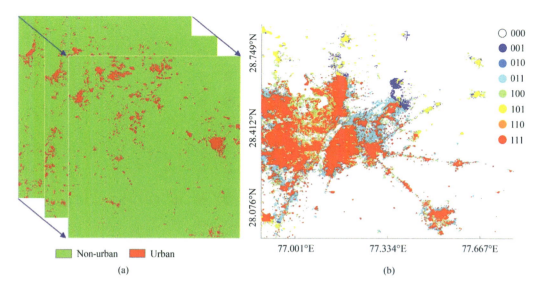

FIGURE 2.5 Spatiotemporal stratified sampling of multitemporal urban land cover: (a) three temporal urban land cover, (b) spatiotemporal stratified sampling.

2.2.7 Spatiotemporal Stratified Sampling

Multitemporal land cover data are located in the same spatial position, with temporal attribute changes. They are distributed in multidimensional space according to space (two-dimensional) + time (three-dimensional), and the sampling unit is a pixel body. The stratified sampling design of multitemporal land cover data products is mainly based on the changes and unchanged types of single elements in different temporal data—for example, the three temporal urban land cover (Figure 2.5a) given the label 1 for urban class and 0 for nonurban class. The combination of standard changes and unchanged types of attributes of the three phases of pixels is used as the basis for stratification, namely, spatiotemporal stratified sampling (Figure 2.5b). By using the combination of characteristics of temporal and spatial changes and unchanged types of land cover as the basis of stratified sampling, the problem of multitemporal data stratification can be solved (Gong et al. 2022).

2.3 SAMPLE SIZE ESTIMATION

Sample size refers to the number of items included in a sample. It is typically determined by balancing the cost of data collection with the necessity for adequate statistical power to conduct meaningful analysis.

In the context of sampling, rather than inspecting 100% of the data, only a subset is evaluated to provide insights into the overall quality of the dataset. The basic method to estimate sample size can be divided into two types (Fauvel et al. 2012; Myint et al. 2011; Stehman 2004; Cohen 1960; Czaplewski 1994; Everitt 1968; Hudson 1987; Landis and Koch 1977; Grzegozewski et al. 2016; Klemenjak et al. 2012; Vorovencii 2014; Yu and Shang 2017; Aickin 1990; Tsendbazar et al. 2015; Kempen et al. 2009; Fraser et al. 2006). Empirical methods are mainly based on available resource, budget, and experience, and statistical methods include those based on calculating relative error and variance, and operating characteristics (OC)–curve-based acceptance sampling method (Poisson distribution, hypergeometric distribution).

It is crucial to collect an adequate number of samples to ensure that the assessment provides a statistically valid representation of the land cover map's accuracy. However, due to the high costs

Sampling Design **31**

associated with gathering reference data at each sample unit, it is necessary to minimize sample size to maintain affordability. Among the various considerations in accuracy assessment, estimating the appropriate sample size should be the primary focus. We recommend employing a global-to-regional sampling approach for general GLC products. How to determine the sample size of the map sheet and the subsequent sample size of land cover samples in each selected map sheet are presented in this section.

Statistical acceptance sampling tables and process-control sampling tables have been tailored to address the requirements of different sampling methods (ISO 2020; ISO 2013). There are two primary types of sampling inspection: by variables and by attributes. In attribute-based inspections, any deviation from specified requirements is classified as a defect (ISO 1999; ISO 1985; ISO 2002). Acceptance sampling employs statistical methods to decide whether to accept or reject a production lot. An acceptance sampling plan specifies a sample size (n) and an acceptance number (c). The count of nonconformances is then compared to the acceptance number to determine the outcome for the lot. The acceptance quality level (AQL) serves as a metric in lot-by-lot sampling inspection by attributes, indicating a quality standard routinely accepted by the sampling plan (ISO 1999). It is usually defined as the percentage of defects (calculated as defects per hundred units × 100%) that the plan will accept 95% of the time. Consequently, lots that meet or exceed the AQL are accepted at least 95% of the time, while those that do not are rejected up to 5% of the time.

2.3.1 CLASSICAL METHODS

(1) Empirical Determination of Sample Size

Widely used in validating GLC products, this kind of method determines sample sizes based on human experience or the situation with regard to available validation sources. For example, when validating DISCover (1 Km resolution GLC product by US Geological Survey (USGS)), a minimum of 25 samples per DISCover class were collected, and a total of 379 samples were used for the validation (Scepan 1999). In the validation of the GLC2000 product, the sample size was determined by the available budget, and a total of 1,265 sample sites were interpreted (Mayaux et al. 2006). For validating GlobCover 2005 and 2009 products, the sample size was determined by the available validation datasets and the expert's confidence in the information they provided. In total, 3,167 samples were used for the validation of GlobCover 2005 (Bicheron et al. 2008; UCLouvain Team 2011).

(2) Simple Sample Size Estimation Model

Early research proposed using equations based on the binomial distribution or its normal approximation to compute the required sample size. These methods are statistically robust for determining the sample size needed to evaluate the overall accuracy of a classification or the accuracy of a single category. The calculations depend on the specified accuracy level (the proportion of correctly classified sample units) and an allowable margin of error.

Determining the appropriate sample size n from the binomial or normal distribution approximation depends on two key parameters: the given acceptable error and the desired level of confidence:

$$n = \frac{pq}{\left[E / Z_\alpha\right]^2}, \tag{2.1}$$

where p is the required accuracy, q is $(1-p)$, E is the acceptable error, and α represents the confidence level. $Z\alpha$ is the critical vale at significance level α. n is the size.

Wulder et al. (2006) provided the following look-up table of appropriate sample sizes for a given level of acceptable error and level of confidence (see Table 2.2).

TABLE 2.2

A Comparison of Sampling Strategies of Existing GLC Products

	80% Accuracy	
Level of Acceptable Error	90%	95%
0.01	2,663	4,356
0.03	296	484
0.05	107	174
0.10	27	44
0.20	7	11

(3) Cost Limitation Model

Assuming that the inspection cost of each sample unit is equal, the total inspection cost function is

$$C = c_0 + c_1 n, \tag{2.2}$$

where C is the total cost, c_0 is the fixed charge, c_1 is the unit inspection cost, and n is the sample size. Then

$$n = \frac{C - c_0}{c_1} \tag{2.3}$$

When designing sampling schemes, there is a trade-off between cost and accuracy. Generally, the larger the sample size, the higher the accuracy, but the cost is bound to increase. In complex sampling design, what should be considered is how to maximize the accuracy with a certain total cost or minimize the total cost with a certain precision requirement. The previous model considers both accuracy and cost.

(4) Search Sampling Table Method

In the process of product quality inspection, the first-level sample size and the second-level sample size can be determined according to the provisions of Table 2.3. The specific look-up table method to determine the sample size can query the geological standard (DZ/T 2014).

2.3.2 Acceptance Probability-Based Two-Rank Estimation

(1) ISO-Designed Acceptance Sampling Plan

ISO 2859 is a series of standards for sampling for inspection by attribute, which can also be used for GLC product sampling. In the quality inspection and evaluation of GLC products, the AQL can be utilized as a metric, with its definition varying across different stages of the evaluation process.

Initially, the sample number n is presumed to be a known quantity. The optimization problem is then formulated to determine the optimal acceptance number c, with the rate of nonconforming items modeled using a Poisson distribution. Concurrently, the optimal sample number n is calculated using the interval estimation method. This approach controls the probability of the relative difference between the proportions of nonconforming items (represented by the AQL) within the lot, where N represents the total population of the lot.

$$n = \frac{\left(\mu_{1-(\alpha/2)}^2 \left(1 - AQL\right)\right) / \left(r^2 AQL\right)}{1 + \frac{1}{N}\left(\left(\mu_{1-(\alpha/2)}^2 \left(1 - AQL\right)\right) / \left(r^2 AQL\right) - 1\right)} \tag{2.4}$$

Sampling Design

TABLE 2.3

Sample Size

The population N	Sample size n
<3	All inspection
3~11	3
12~15	4
16~19	5
20~25	6
26~37	8
38~50	10
51~70	13
71~90	16
91~120	20
121~150	25
151~215	32
216~280	40
281~390	50
391~500	63
501~850	80
851~1,200	100
...	...
36,000,001~400,000,000	5,000
>400,000,001	6,300

In this context, $\mu_{1-(\alpha/2)}$ represents the critical value of the standard normal distribution at the confidence level $1 - (\alpha/2)$, which is essential for determining the threshold for statistical decisions. Additionally, r denotes the limit value of the relative difference. This limit is defined as follows:

$$r = \mu_{1-(\alpha/2)}\sqrt{\frac{AQL(1-AQL)}{n}\frac{N-n}{N-1}} \,/\, AQL \qquad (2.5)$$

After determining the sample number n, the acceptance number c is optimized by minimizing the squared error of the acceptance probability equation at the point $(AQL, 1-\alpha)$ on the OC curve. This sampling plan is tailored for cases where N, the population size, exceeds 250, making it appropriate for the first and second levels of sampling in the evaluation of most remote sensing products. For the third level of sampling, applicable when N is less than 250, the acceptance number c is set to 0, 1, or 2. Subsequently, the sample size n is computed by minimizing the squared error in the acceptance probability equation at the point $(AQL, 1-\alpha)$ on the OC curve, adjusting for the smaller population size.

(2) A Two-Rank Acceptance Sampling Plan

For the inspection of geospatial data products, including land cover data, a Two-Rank Acceptance Sampling Plan (TRASP) has been developed. This plan utilizes a mixed integer optimization and nonlinear programming model to ensure precise and efficient sampling strategies (Tong and Wang 2012; Tong et al. 2011). The extension of the proposed TRASP incorporates adjustments for different lot sizes, as follows: (1) When dealing with small lots, the acceptance number c is fixed. The optimization problem is then formulated to determine the optimal sample number n, with the nonconforming items modeled using a hypergeometric distribution; (2) in scenarios involving larger lots, the sample number n is treated as a known quantity. The optimization problem is focused on obtaining the optimal acceptance number c, with the nonconforming items being modeled by a

Poisson distribution. Concurrently, the optimal sample number n is recalculated using the interval estimation method. This method controls for the probability of the relative difference between the expected proportion of nonconforming items p in the lot and its observed proportion in the sample.

For small lot sizes ($N < 250$), where the acceptance number c is fixed, the OC curve of the acceptance sampling plan is designed to pass through the point (p_0, $1 - a$). An optimization problem is formulated to minimize the squared error of the probability of the acceptance equation at this point on the OC curve with respect to the sample size n. This approach ensures that the sampling plan is optimized for accuracy and efficiency in scenarios involving smaller lot sizes.

$$\min_{n} \varepsilon^2$$

$$s.t. \sum_{d=0}^{c} \frac{\binom{Np_0}{d}\binom{N-Np_0}{n-d}}{\binom{N}{n}} - \left(1-\alpha\right) = \varepsilon \tag{2.6}$$

$$n \geq round\left(Np_0\right)$$

When the lot size increases with respect to the acceptance number c and sample size n, addressing the optimization problem through a Poisson distribution becomes more suitable. In such cases, the optimization equation is initially configured to minimize the sum of squared errors of the acceptance probability at the point (p_0, $1 - a$) on the OC curve, targeting the acceptance number c. However, since the sample size n is not predetermined in this scenario, the TRASP further refines the sample size n. This refinement is achieved by controlling the probability of the relative difference between the actual proportion of nonconforming items p in the lot and its observed proportion in the sample. This method ensures that both c and n are optimally adjusted to accurately reflect the quality of larger lots.

$$\min_{n} \varepsilon^2$$

$$s.t. \sum_{d=0}^{c} \frac{\left(np_0\right)^d}{d!} e^{-(np_0)} - \left(1-\alpha\right) = \varepsilon \tag{2.7}$$

$$0 \leq c \leq n-1$$

The proposed TRASP is articulated as an optimization problem that minimizes the sum of squared errors of the acceptance probability at a specific point on the OC curve. This formulation is based on the AQL, with the modeling of nonconformities tailored to the size of the lot. For small-size lots, nonconformities are modeled using a hypergeometric distribution function. Conversely, for larger lots, a Poisson distribution function is employed. This dual approach ensures that the sampling plan is optimally configured for different lot sizes, providing precise control over quality assessment based on the lot size characteristics.

The fuzzy acceptance sampling plan is tailored for scenarios where sampling parameters are expressed as fuzzy numbers, especially suitable for the quality inspection of geospatial data characterized by ambiguous quality traits. According to Tong and Wang (2012), this method encompasses three scenarios with varying fuzzy sampling parameters: the fuzzy fraction of nonconforming items, the fuzzy sample rate, and a combination of both the fuzzy fraction of nonconforming items and the fuzzy sample rate. The design of this sampling plan is conceptualized as a fuzzy optimization problem. It addresses the variability of lot sizes by utilizing fuzzy versions of the hypergeometric and Poisson distributions, enabling adaptive responses to the uncertainty inherent in the parameters of geospatial data quality inspection.

Sampling Design   **35**

(3) Relative Error-Based Optimized Sample Model

A two-rank sampling contains global sampling and regional sampling, and in each rank, the core issue is to calculate the sample size. The optimized sampling model is derived by the use of the principle of probability and statistics (Tong et al. 2011).

In sampling, errors arise when assessing product quality through a sample estimator. Therefore, managing these errors is critical from the perspectives of both data producers and consumers. This is accomplished by regulating the probability that the relative difference between the proportion of nonconforming items $1-p$ in the lot and the observed proportion of nonconforming item $\hat{p}$ in the sample stays within acceptable limits. Under a confidence level of $1-\alpha$, the probability of the relative difference between $1-p$ and $\hat{p}$ being less than a limit value of the relative difference r is given by

$$\Pr(\frac{p-\hat{p}}{p} < r) = 1-\alpha, \tag{2.8}$$

where $\Pr()$ denotes the probability of the relative difference between $1-p$ and $\hat{p}$ being less than r.

The sample size n is assumed to be sufficiently large; thus, the proportion of nonconforming items follows a normal distribution $N\left(p, V\left(\hat{p}\right)\right)$—that is, $\dfrac{p-p}{\sqrt{V\left(\hat{p}\right)}}$ follows a standard normal distribution, where $V\left(\hat{p}\right)$ the variance is calculated by (DeGroot, 1986):

$$V\left(\hat{p}\right) = \frac{p\left(1-p\right)}{n}\frac{N-n}{N-1} \tag{2.9}$$

Based on Eqs. (2.8) and (2.9), the limit value of the relative difference r is given by

$$r = \left. \mu_{1-\frac{\alpha}{2}}\sqrt{V\left(\hat{p}\right)} \middle/ p \right. = \left. \mu_{1-\frac{\alpha}{2}}\sqrt{\frac{p\left(1-p\right)}{n}\frac{N-n}{N-1}} \middle/ p \right. , \tag{2.10}$$

where $\mu_{1-\frac{\alpha}{2}}$ denotes the critical value of the standard normal distribution at the confidence level of $1-\dfrac{\alpha}{2}$.

Therefore, based on Eq. (2.10), the sample size n can be derived by

$$n = \frac{\dfrac{\mu^2_{1-\frac{\alpha}{2}}\left(1-p\right)}{r^2 p}}{1+\dfrac{1}{N}\left(\dfrac{\mu^2_{1-\frac{\alpha}{2}}\left(1-p\right)}{r^2 p}-1\right)}. \tag{2.11}$$

In Eq. (2.11), the proportion p of nonconforming items in a lot is considered an unknown parameter. Assuming a continuous production process, the AQL, represented by p_0) considered acceptable, is used to determine the sample size in Eq. (2.11). The number of samples required is then calculated using the method proposed by (Tong et al. 2011):

$$\hat{n} = \frac{\dfrac{\mu^2_{1-\frac{\alpha}{2}}\left(1-p_0\right)}{r^2 p_0}}{1+\dfrac{1}{N}\left(\dfrac{\mu^2_{1-\frac{\alpha}{2}}\left(1-p_0\right)}{r^2 p_0}-1\right)} \tag{2.12}$$

The derivation process is rigorous and based on statistical theory. However, in practice, it can be adjusted according to different objects. For GLC maps, the parameter of Expected Classification Accuracy (ECA) is more appropriate than the proportion of conforming items $1 - p$. So, based on Eq. (2.12), replacing the parameter of $(1 - p_0)$ with a parameter of ql (as ECA), the sample size can be calculated by Eq. (2.13) (Tong et al. 2011):

$$\hat{n} = \frac{\dfrac{\mu^2_{1-\alpha/2} \cdot ql}{r^2 (1-ql)}}{1 + \dfrac{1}{N} \left(\dfrac{\mu^2_{1-\alpha/2} \cdot ql}{r^2 (1-ql)} - 1 \right)} \tag{2.13}$$

2.3.3 STRATA SAMPLING

Stratified sampling involves dividing the population into distinct, nonoverlapping groups called strata, where members within each stratum are more similar to each other than to members in other strata. Typically, this method offers greater accuracy than simple random sampling, provided the division into strata is reasonable—i.e., the variance within each stratum is less than that observed in simple random sampling. For thematic map inspections, utilizing auxiliary data reflecting prior knowledge about the product features—such as remote sensing sources, mapping techniques, and producers—can aid in effectively segmenting the maps into appropriate strata. This segmentation helps in reducing the variance of the estimates and enhances accuracy without necessitating an increase in sample size (Xie et al. 2015; Brennan and Prediger 1981; Foody 1992; Ma and Redmond 1995; Allouche et al. 2006; Di Eugenio and Glass 2004; Foody 2004; Foody 2008; Jiang et al. 2011; Kraemer 1979; Gwet 2014).

Spatial stratification varies a lot according to different stratified methods (Aguilar et al. 2013; Duro et al. 2012; Martha et al. 2012; Rozenstein and Karnieli 2011; Sharma et al. 2013; Wickham et al. 2013; Anderson 1971; Thomlinson et al. 1999; Chen et al. 2015; Yu et al. 2013). Strata by land cover type is one of the most popular strategies that can be found in many validation practices, such as DISCover (Scepan 1999), GLCNMO (Tateishi et al. 2011), and GLC-SHARE (Latham et al. 2014). In addition, spatial stratification by regular polygons, such as rectangular or hexagonal, is another choice. For irregular stratification, administration maps or other auxiliary data are often used. For example, Olofsson et al. (2012) used the modified Köppen climate/vegetation classification map as the basis for spatial stratification basis based on the assumption that land cover distribution is closely related to climate.

In addition to spatial stratification, multistage stratified sampling is also a common strategy used in sampling design. Multistage sampling divides the whole sampling process into two or more stages, and at each stage, a different sampling method is applied. In the course of implementation, larger sampling units, also called primary sampling units (PSU), are first chosen from the population. Then, in each PSU, the secondary sampling units are drawn, and so on. In this section, a few examples will be illustrated to demonstrate how existing methods address common practical problems in the accuracy assessment of GLC products.

In the validation of GlobeLand30, a multiscale spatial sampling was used, which consists of global and regional scales. The map sheet sample size was first determined based on an optimized sample model at the global scale (Tong et al. 2011). Map sheet samples are placed according to the spatial complexity. Regional sampling was further performed on each map sheet to calculate the feature sample size with the optimized sample model, and these feature samples were placed based on the spatial correlation. Multiple stratified sampling is the extension of a two-rank sampling. It is a process of dividing data into different strata according to multiple factors, such as spatial scale, data attributes, administrative division, and so on (see Figure 2.6; Xie et al. 2015). Spatial location, information about producers, the type of raw data source, and additional auxiliary data gleaned from

Sampling Design

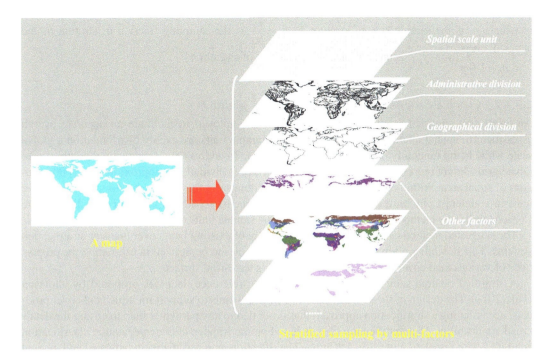

FIGURE 2.6 Multiple stratified sampling.

metadata and common knowledge can all be utilized to design a multilevel stratified sampling framework. This stratified sampling approach is both repeatable and easily adaptable for the validation of GLC products.

2.3.3.1 Sample Size in Stratified Sampling

Multiple stratified sampling is a general method applicable to different datasets. In each stratification, the sample size is calculated by the sampling model presented in Section 2.3, as well as other estimation models such as the percent sampling model (10%, 20%, …), expert experience, and existing sampling tables.

In stratified sampling, it is common to divide the sample into multiple subsamples corresponding to different strata. Assuming there are H strata, each subsample from a stratum will have a sample size n_h, $h = 1, 2, …, H$. The sizes of these subsamples must adhere to the constraint that $n_1 + n_2 + … + n_H = n$, meaning the total sample size is the aggregate of all individual subsample sizes. The selection of these n_h values can be optimized in various manners. Typically, the sample size for each stratum is set proportionally to the stratum's size, a method known as proportional allocation (referenced in Section 2.4.1). Additionally, Neyman's optimal allocation method may also be applied to determine the sizes of samples across different strata.

If the sample size of each stratum is proportionated to the population size of the stratum, the sample size of each stratum is:

$$n_h = n \cdot \omega_h \quad \omega_h = W_h = \frac{N_h}{N} \tag{2.14}$$

If the sample size is estimated by Neyman allocation, the sample size of each stratum is

$$n_h = n \cdot \omega_h \quad \omega_h = \frac{W_h S_h}{\sum W_h S_h}, \tag{2.15}$$

in which, n_h is the sample size of stratum h, $f_h = \frac{n_h}{N_h}$ is the sampling fraction of stratum h, and $s_h^2 = \frac{1}{n_h - 1} \sum_{i=1}^{n_h} (y_{hi} - \bar{y}_h)^2$ is the variance of the sample of stratum h.

2.3.3.2 Classical Stratified Sampling Distribution Methods

An equal-area stratified random sampling method is employed to facilitate an accurate overall assessment of spatial characteristics, ensuring that sample units are evenly distributed on a global scale while being randomly distributed on a local scale. To address the issue that geographic coordinate grids do not represent equal areas, Earth's surface is divided into approximately 7,000 hexagons of equal area. Within each hexagon, a simple random sampling approach is used, selecting five units to counteract any biases that might arise from repetitive patterns in land cover distribution. Consequently, this method results in a total of 38,664 sample units (Zhao et al. 2014).

The suggested method can be applied independently of existing land cover maps or stratification systems. This flexibility allows the results from the samples to be utilized in estimating the proportions of various land cover types that make up Earth's terrestrial surface.

A stratified random sampling design for collecting reference data was proposed by Olofsson et al. (2012). The design of the global validation database aims to support multiple land cover products, necessitating a stratification approach that is not tied to any particular map to ensure the data's broad applicability. The stratification used is based on the Köppen climate/vegetation classification combined with population density. A Köppen classification map was manually adjusted and merged with two layers, indicating population density and a land-water mask. Consequently, 21 strata were established, and an initial global sample of 500 reference sites was chosen, with each site encompassing a 5×5 km block.

A two-stage stratified clustered sampling strategy was proposed by (Mayaux et al. 2006). The stratified sampling was first used to determine the sample unit, which is the Voronoï polygons computed from the WRS-2 centroids. Meanwhile, each unit will be thought to be a "Priority" if the class proportion satisfies one of the following conditions: >30% forest, >10% cropland, >10% wetland. Then the entropy of each unit H can be estimated by Eq. (2.16), where p_k is the proportion of land cover class k. This unit will be identified as "Homogenous" if H is smaller than 0.5; otherwise, it was "Heterogeneous."

$$H = -\sum_{k=1}^{m} P_k \cdot \log(P_k) \tag{2.16}$$

Subsequently, four strata can be defined as follows: (1) Homogenous and Priority, (2) Homogenous and Nonpriority, (3) Heterogeneous and Priority, (4) Heterogeneous and Nonpriority. The second stage is cluster sampling, which means five fixed-position samples were selected from the selected polygon samples, and each sample size was a block of 3×3 pixels. Finally, 253 Voronoï polygons were selected by stratified sampling, and 1,265 samples were collected by cluster sampling given that 5 samples (3×3 pixels) can be assigned in each polygon.

The proposed method was used in the validation of GLC2000. One thousand two hundred sixty-five samples were drawn from the Landsat TM images, aerial photographs, and thematic maps, and NDVI profiles were used as reference data within each sample, and the final accuracy of GLC2000 was 68.6%.

2.3.3.3 Two-Rank Sampling Distribution

In the validation of high-resolution GLC products, a two-rank sampling strategy is suggested (Tong et al. 2011). The first-rank sampling is called global scale sampling (i.e., determining map sheet

Sampling Design

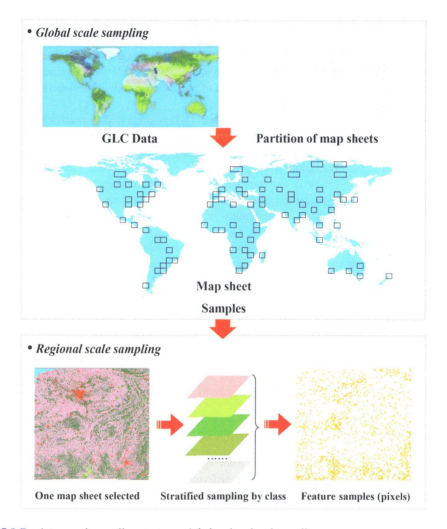

FIGURE 2.7 A two-rank sampling strategy: global and regional sampling.

samples from global map sheets), by which the sample unit is a map sheet of GlobeLand30, and the second-rank is called regional sampling (i.e., determining features from the selected map sheet), by which the sample unit is a land cover feature or pixel.

Figure 2.7 shows the whole process of global and regional sampling. In global sampling, the first step is to determine the sample size of the map sheet, which was discussed in Section 2.3.2, and the second task is to allocate the spatial locations of map sheet samples, which will be described in Section 2.4.5. After that, a regional sampling is adopted in each selected map sheet sample. A stratified sampling is adopted by class, and in each class, the sample size is calculated, and the allocation of samples is determined by calculating regional spatial correlations (Section 2.4.3).

2.3.3.4 Landscape Shape Index–Based Sample Distribution

In principle, regions with greater heterogeneity should be allocated larger sample sizes or higher sample densities. This is because such areas are typically more challenging to map and validate accurately (Herold et al. 2008; Olofsson et al. 2012). Traditional sampling methods determine sample sizes based on several factors, including the desired level of precision, the necessary confidence interval, the geographical extent of the target regions, or the constraints of the available budget (Foody 2002; Foody 2009; Mayaux et al. 2006; Tong et al. 2011; Olofsson et al. 2014). However, many of these traditional methods struggle to manage the high spatial heterogeneity of land cover effectively across extensive areas, often failing to produce reliable spatial samplings (Tung and LeDrew 1988; Koukoulas and Blackburn 2001; Smith 2002; Wang et al. 2013; Olofsson et al. 2014). The primary challenges associated with traditional sampling methods include inadequate sample sizes for specific target regions, underrepresentation of rare classes, and illogical sample distribution across geographical areas.

To address these issues, the landscape shape index (LSI) is employed to characterize the spatial heterogeneity of land cover and to refine the approach for estimating sample sizes. The LSI is calculated as a constant multiplied by the ratio of a polygon's perimeter to the square root of its area. This constant is set at 0.25 for square patches and 0.28 for circular patches. Consequently, sample sizes and their distribution can be determined using the LSI as a quantitative measure across three levels. Figure 2.8 shows the general framework of the LSI-based approach.

The estimation of sample size is significantly influenced by the spatial heterogeneity of land cover within a region. Thus, calculating the LSI for the target area is essential. This calculation involves determining a regional LSI (rLSI), which aids in estimating the necessary sample size and establishing sample density ratios among various regions. These ratios, when adjusted for the areas of the regions, help in determining the sample sizes required. Additionally, for each land cover class, a class-level LSI (cLSI) is calculated to assess their spatial heterogeneity. Incorporating cLSIs into the traditional Neyman optimal allocation formula allows for adjusting the standard errors associated with each land cover class. Given that classes with rarity exhibit higher cLSIs, there is a justified increase in their sample allocation. Furthermore, samples for each class are to be geographically distributed based on their heterogeneity, necessitating the computation of a per-class LSI (uLSI) for each geographic unit. From these uLSIs, a curve (referred to as the uLSI curve, depicted in Figure 2.21c) is constructed to facilitate the selection of optimal geographical units for locating sample sites.

(1) Spatial Stratified Sampling Distribution

These methodologies do not consider spatial heterogeneity, potentially leading to identical sample size estimates for regions of similar areas but differing spatial characteristics. Such oversight in sample sizing can result in an inaccurate assessment of the overall accuracy of land cover data sets, as the variations in spatial heterogeneity are crucial determinants of sample requirements (Stehman 2009).

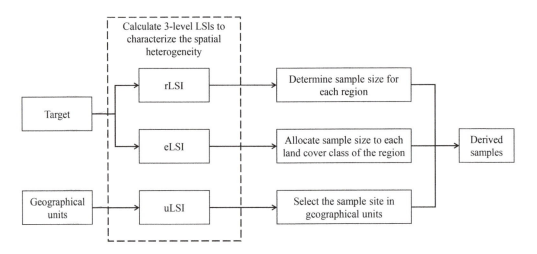

FIGURE 2.8 Three-level LSIs-based sampling approach.

Regions with greater spatial heterogeneity should allocate larger sample sizes per unit area. Consequently, the sample density ratio between two regions ought to mirror the ratio of their spatial heterogeneities. Regional landscape shape indexes (rLSIs) are computed to quantify the spatial heterogeneities of targeted regions and establish the sample density ratios among them. The sample size for each region is then determined by multiplying the sample density by the region's area. By associating the sample densities with their respective rLSIs, a direct correlation between sample size and rLSI can be established, as outlined next.

Assume that N represents the total sample size across all regions, which may be determined by expert analysts or through the application of optimal sampling strategies, such as a two-stage acceptance sampling plan (Wang et al. 2010; Tong et al. 2011).

Ideally, regions with higher rLSI values should allocate larger sample sizes per unit area. Let n denote the number of regions, with $rLSI_i$ and $rLSI_j$ representing the rLSIs of regions i and j, respectively. Additionally, let A_i and A_j indicate the areas of regions i and j. The sample sizes for region i can be calculated as follows:

$$N_i = \frac{rLSI_i \times A_i}{\sum_{j=1}^{j=n} rLSI_j \times A_j} \times N, \left(i = 1, 2, \ldots, n\right). \tag{2.17}$$

Eq. (2.17) delineates the correlation between sample sizes and the rLSIs. For any specified region i, the sample size n_i can be calculated by employing its rLSI and area in conjunction with those of other regions. Consequently, regions exhibiting greater heterogeneity possess increased sample densities and, subsequently, larger sample sizes.

(2) Classical Stratified Sampling Distribution

In a designated target region, the calculated sample size is distributed among various land cover classes to achieve class-specific accuracy levels. This allocation method may lead to the underrepresentation of samples for rare classes covering smaller geographical areas. Given their relatively minor areal proportions, the number of samples assigned to these classes could be minimal, potentially approaching zero (Stehman et al. 2012; Olofsson et al. 2014).

Neyman optimal allocation is a fundamental sampling formula utilized to distribute sample sizes across various classes:

$$cN_{i,k} = N_i \times \frac{S_{i,k} \times W_{i,k}}{\sum S_{i,k} \times W_{i,k}}, \left(k = 1, 2, \ldots, m\right), \tag{2.18}$$

the variables $cN_{i,k}$, $S_{i,k}$ and $W_{i,k}$ represent the sample size, standard error, and areal proportion of class k in region i, respectively. The variable n_i denotes the total sample size for region i, while m represents the total number of classes. The term $S_{i,k}$ quantifies the spatial variability within land cover classes and is crucial for determining how sample sizes are allocated. A class with a larger $S_{i,k}$ has a more complex spatial distribution and higher spatial heterogeneity; thus, more samples are allocated. For practical applications, the cLSIs serve as proxies for $S_{i,k}$:

$$cN_{i,k} = N_i \times \frac{cLSI_{i,k} \times W_{i,k}}{\sum_{k=1}^{k=M} cLSI_{i,k} \times W_{i,k}}, \left(k = 1, 2, \ldots, m\right), \tag{2.19}$$

where $cLSI_{i,k}$ denotes the cLSI of class k in region i. In a given region i, the sample number of class can be derived according to the sample size N_i, the areal proportion, and the cLSI of class k. The $cLSI_{i,k}$ is computed as the average cLSI of class k per unit area, thereby providing a metric that quantifies the spatial complexity of each class within the region.

2.3.3.5 Modified Spatially Balanced Sampling Distribution

(1) Spatially Balanced Sampling

Currently, two algorithms are predominantly utilized for spatially balanced sampling: (1) Reversed Randomized Quadrant-Recursive Raster (RRQRR) and (2) Generalized Random Tessellation Stratified (GRTS). The RRQRR algorithm employs a hierarchical quadrant-recursive ordering based on Morton order. Morton order arranges data into N- or Z-shaped patterns, each consisting of four cells labeled sequentially as lower-left, upper-left, lower-right, and upper-right (numbered 1, 2, 3, and 4, respectively, in the N pattern). These cells are capable of nesting at multiple hierarchical levels, forming a recursive, space-filling sequence. This quadrant-recursive ordering is specifically designed to preserve proximal relationships in two-dimensional (2D) space when data is translated into one-dimensional (1D) space, ensuring that addresses that are close in 1D order are also proximal in 2D space. Although the RRQRR and GRTS algorithms bear general similarities, there are some minor distinctions. To illustrate the primary steps involved, the RRQRR algorithm is described as an example:

1) Morton Code

The Morton codes are generated recursively from the quadtree hierarchy. First, the study area is divided into four units (quadrants), and a sequencing number of 1, 2, 3, or 4 is assigned to each unit, which follows the "N" or "Z" order, thus obtaining the first layer, L_1. Each unit is then further divided into four parts, each with an assigned number, to form the second layer, L_2. This division and assignment process is repeated until the entire study area is exhausted. From the bottom to the top, each unit has a series of layer numbers that make a unique code—i.e., the Morton code.

2) Linear Address

Beginning at the top-left corner, the Morton code for each unit is transformed into a one-dimensional linear address, adhering to the same "N" or "Z" order. This linearization process effectively translates the 2D spatial configuration into a sequenced 1D format while preserving the spatial proximity and arrangement inherent to the original 2D space.

3) Reversed Morton Code

To generate the reversed Morton code, the layer numbers of each unit are inverted. For instance, if the original Morton code is M_{1234}, the reversed code becomes M'_{4321}. Concurrently, the linear address corresponding to each unit is modified to reflect this reversal. This process ensures that the sequence in the 1D space mirrors the reversed order, maintaining spatial relationships while altering their hierarchical representation.

4) Hierarchical Random Sorting

Proceed with randomly sorting the four parts of L_1, followed by a random sorting of the four parts of L_2, and continue this process sequentially through L_k. Concurrently, with each level of sorting, the linear address of each unit is updated to reflect these changes. This stepwise randomization introduces variability while ensuring that the linear addresses remain consistent with the new, randomized order of the parts at each hierarchical level (see Figure 2.9).

5) Sampling

To guarantee that samples of varying significance are selected based on different probabilities, an inclusion probability raster is implemented. This raster assigns a value to each cell that represents the likelihood of that cell being sampled relative to others. During the filtering process, a comparison is made between the inclusion probability and a random raster generated from a uniform

Sampling Design

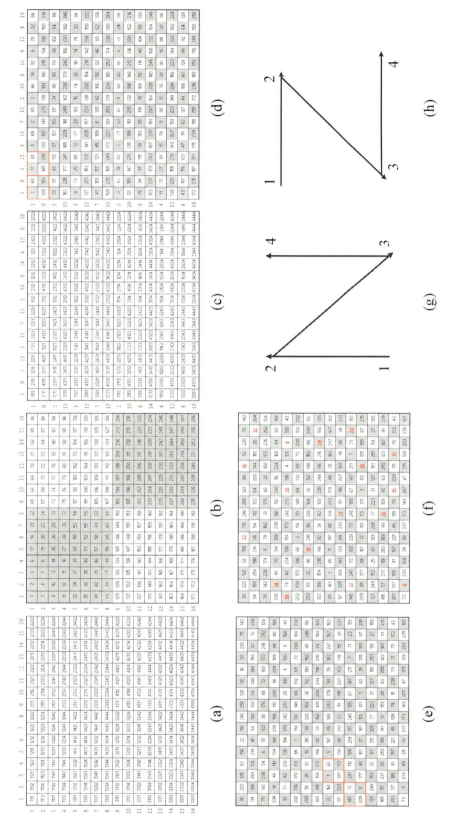

FIGURE 2.9 RRQRR algorithm of spatially balanced sampling. (a) Morton code, (b) linear address, (c) reversed Morton code, (d) reversed linear address, (e) hierarchical random sorting, (f) spatially balanced sampling, (g) the "N" order, (h) the "N" or "Z" order.

distribution within the range [0,1]. Cells with an inclusion probability exceeding their corresponding random probability are retained. Subsequently, following the newly generated linear address from Step 4, a random starting point is selected, and n consecutive cells are extracted from this point. This procedure ensures that the sampling is spatially balanced (see Figure 2.9).

The inclusion probability raster is crucial for achieving spatially balanced sampling. The factors considered in its design vary across different applications. For instance, in field surveys, accessibility significantly influences the sampling method design. Road network data are commonly incorporated to facilitate this. Accessibility is typically quantified within various ranges around road buffers, leading to decreased selection likelihood for points distant from roads, thereby mitigating the risk of no-response sampling units. In contrast, for land cover products with a 30 m resolution, the landscape in transitional areas is often highly fragmented. Assigning equal inclusion probabilities to patches with varying degrees of fragmentation can lead to a spatial distribution of sample points that fails to reflect the influence of patch characteristics, such as size and shape. In spatially balanced sampling, the distribution of the sample points is illustrated as depicted in Figure 2.10. In contrast, if the patches are given different inclusion probabilities according to the complexity. Although the characteristics of the patch are considered, the overall distribution is still relatively uniform. The distribution of the sample points in spatially balanced sampling will be as shown in Figure 2.11. These spatial balanced

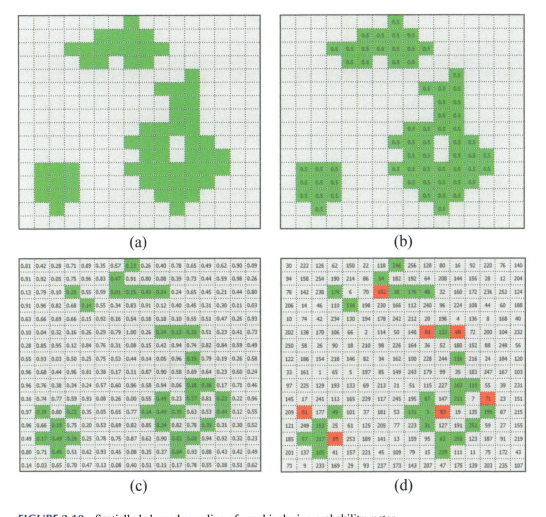

FIGURE 2.10 Spatially balanced sampling of equal inclusion probability raster.

Sampling Design

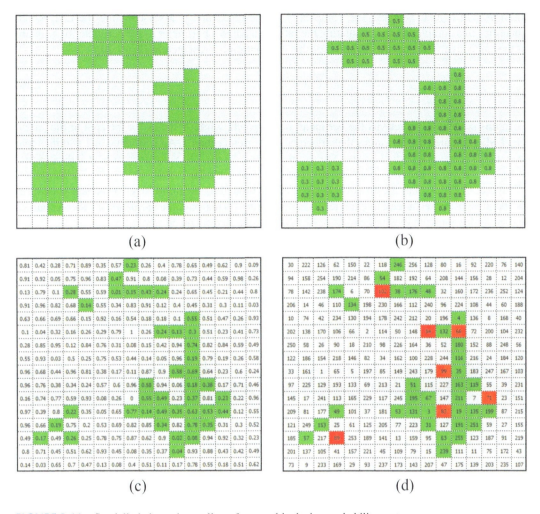

FIGURE 2.11 Spatially balanced sampling of unequal inclusion probability raster.

sampling methods are still difficult to obtain representative samples in the face of complex terrain. Therefore, how to consider the data characteristics and design the appropriate inclusion probability raster are key to obtaining representative samples. In order to evaluate the accuracy of land cover classification, LSI is regarded as an important parameter to define the probability layer of balanced sampling. Complex, heterogeneous, and broken land cover will have a greater probability of being sampled, while simple, uniform, and regular land cover sample points have a relatively small probability of being sampled, thus ensuring the representativeness of regional sample points.

(2) Using the Landscape Pattern–Based Inclusion Probability to Improve Spatially Balanced Sampling

It is widely acknowledged that classification errors in remotely sensed images are not uniformly distributed but adhere to certain identifiable patterns (Pontius 2001; Fleiss et al. 1969; Næsset 1996). Classification accuracy tends to be higher for spatial objects characterized by regular structures and extensive areas, whereas those with fragmented and irregular forms are more susceptible to misclassification errors. Thus, acknowledging the correlation between landscape complexity and the likelihood of classification errors, landscapes comprising complex, heterogeneous, and fragmented patches are assigned a higher inclusion probability compared to those consisting of simpler, more uniform, and regular patches. The complexity of the landscape is assessed using the LSI, which quantifies this complexity by calculating the ratio of a patch's perimeter to the perimeter of a square of the same area. The specific formula for this calculation is presented in Eq. (2.20):

$$\text{LSI} = \frac{0.25E}{\sqrt{A}} \tag{2.20}$$

In this context, E represents the perimeter of the patch, while A stands for the area of the patch. The LSI is calculated with values that are equal to or greater than 1. An LSI value of 1 signifies that the patch's shape is a perfect square, the simplest form. As the LSI value increases, it indicates that the patch's shape is becoming increasingly complex and irregular.

Figure 2.12 illustrates the city of Suzhou, China, situated in the Taihu River Basin and known for its abundant water resources, as depicted in Figure 2.12(a). This example of Suzhou's water resources is used to demonstrate the application of LSI-based spatially balanced sampling.

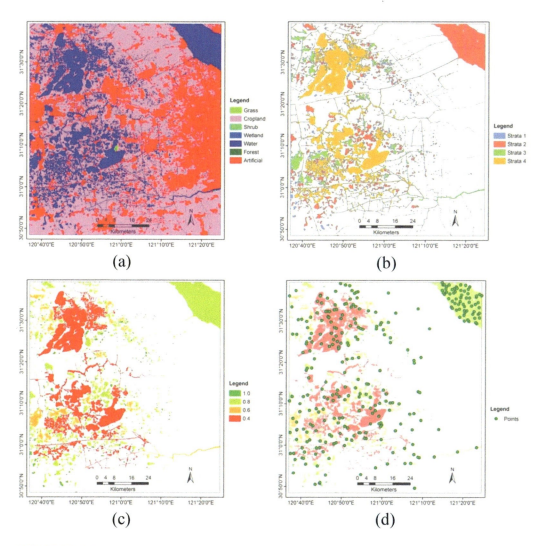

FIGURE 2.12 Example of using the landscape pattern–based inclusion probability to improve spatially balanced sampling: (a) test area, (b) spatial clustering based on the LSI, (c) inclusion probability raster based on the LSI, (d) spatially balanced sampling based on the LSI.

Sampling Design

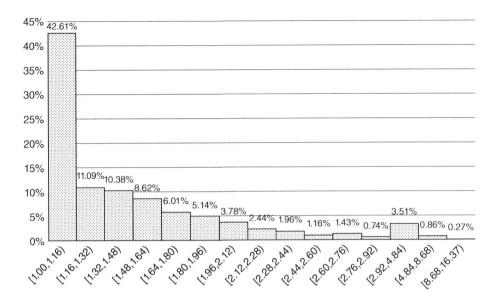

FIGURE 2.13 The frequency distribution histogram of the LSI.

Given that the inclusion probability reflects the likelihood of a unit being sampled relative to other units in the population, it is essential to normalize the LSI to ensure appropriate scaling and comparability. The normalization process involves the following steps:

1) The LSI values for all water patches within the area are computed, followed by the creation of a frequency distribution histogram, as illustrated in Figure 2.13. Influenced by the spatial distribution of surface land cover, there is a high count of fragmented patches with small area proportions, whereas complex patches are fewer but cover larger areas. Consequently, the frequency histogram typically exhibits a left-skewed distribution, with statistics such as the mean, median, and mode being low, rendering them unsuitable for stratification purposes.
2) Consequently, to minimize disparities among the patches, they are clustered into four groups based on their LSI values. This grouping is depicted in Figure 2.12(b).
3) The greater the landscape fragmentation, the higher the likelihood of misclassification. To address this, an unequal inclusion probability is applied based on the fragmentation level of the patches. As illustrated in Figure 2.12(c), patches are categorized into four layers. Inclusion probabilities of 1, 0.8, 0.6, and 0.4 are respectively assigned based on their LSI. Patches characterized by significant heterogeneity and extensive fragmentation receive more samples, thereby enhancing the representativeness of the sample points.
4) As depicted in Figure 2.12(d), an inclusion probability raster is generated, and utilizing the RRQRR algorithm, 300 spatially balanced sample points are extracted.

2.3.3.6 Spatiotemporal Stratified Sampling

The paramount criterion for a statistically rigorous sampling design is adherence to the principles of probability sampling. This ensures that the derived estimators consistently and accurately represent the parameters of interest (Stehman 2001). The inclusion probability must be both known and greater than zero (Olofsson et al. 2014). Stratified sampling is an effective method for accurately estimating rare categories. For a thorough accuracy assessment of three periods of urban land cover data on a global scale, stratified sampling is utilized to collect reference samples. The multitemporal global urban land maps employed in this analysis distinguish between two classes: urban and nonurban (Liu et al. 2018). Evaluating accuracy epoch by epoch demands significant effort and time, while single-epoch assessments only provide insight into the data quality of that specific epoch. In contrast, evaluating multitemporal data not only involves estimating accuracy for a single epoch but also entails identifying various combinations of temporal changes and what is unchanged. Therefore, it is crucial to assess accuracy using a broad set of spatial and temporal samples. The stratification in this approach ensures that all potential scenarios evident in the product are represented (Boschetti et al. 2016). The sampling units were delineated spatially using pixels with a resolution of 30 m and temporally based on the acquisition dates of the multitemporal land cover images.

In contrast to the methodology used by Xie et al. (2015), which applied stratified random sampling based on classification labels, the temporal stratification in this study is derived from the combination of land cover types observed on three distinct dates. Considering the changes and what is unchanged in the three periods, there are eight types of temporal changes. An example of the spatiotemporal stratified sampling is shown in Figure 2.14. Here, "1" means that the land cover is urban, and "0" refers to other land cover except urban.

Thus, spatiotemporal stratification involves two steps. Initially, a global urban land cover map serves to establish spatial stratification, which is based on the urban ecoregions outlined by Schneider et al. (2010). This spatial stratification, as noted by Boschetti et al. (2016), ensured adequate representation of the ecoregions within the reference data samples. Following this, temporal stratification was conducted to complement the spatial bases. The intent of the proposed spatiotemporal stratification is to guarantee sufficient sample sizes at the same spatial locations while capturing changes in time-series attributes.

2.3.3.7 Modified Spatial Sampling Based on Gray-Level Co-occurrence Matrix

Spatial sampling based on the Gray-Level Co-occurrence Matrix (GLCM-sampling model) considers both sampling size calculation and sample point distribution, permitting users and producers to determine the sampling rate using spatial autocorrelation and spatial heterogeneity. Moreover, it could ensure that sample points are uniformly distributed in the spatial region and proportionally distributed in different types of land cover.

(1) Gray-Level Co-occurrence Matrix

In the GLCM-sampling model, each pixel is considered an accuracy-assessed item, referring to an element in the dataset. The remote sensing image is assumed to be rectangular with a certain number of columns N_x and rows N_y. The lot size (N) of accuracy-assessed items is denoted as $N = N_x \times N_y$.

Suppose that the gray level at each pixel is quantified into N_g levels, and let G be the set of quantified gray levels, $G_x = \{0, 1, \ldots, N_g - 1\}$. The remote sensing image H is represented as a function that assigns some gray levels in G to each pixel or pair of coordinates in $N = N_x \times N_y$.

The texture-context information is specified by the matrix of relative frequencies (P_{ij}) with two neighboring pixels separated by optimal distance d in the remote sensing image, where one pixel is with gray level i and the other pixel is with gray level j ($i, j \in G_x$).

The matrices of gray-level co-occurrence frequency (P_{ij}) are represented as a function of the angular relationship (θ) and optimal distance (d) among the neighboring pixels as,

$$
\begin{aligned}
p(i,j,d,\theta) &= \#\left\{\left[(k,l),(m,n)\right] \in \left(N_y \times N_x\right) \times \left(N_y \times N_x\right)\right\} \\
&\left\{\left(k-m=0, |l-t|=d\right) \times \left(|k-m|=d, l-t=0\right), H(k,l)=i, H(m,t)=j\right\}
\end{aligned}
$$

(2.21)

Sampling Design

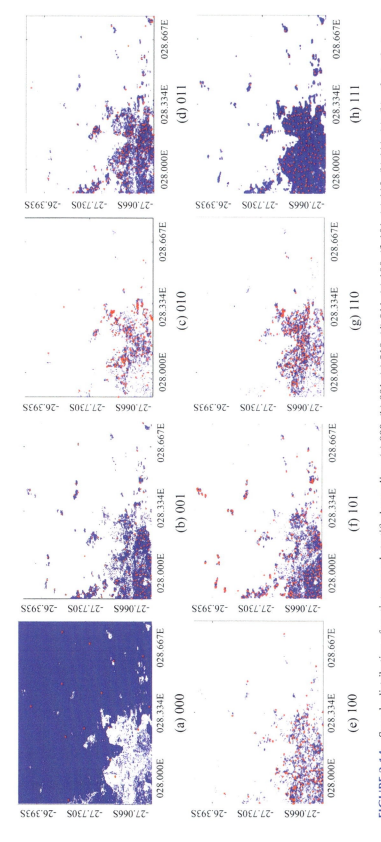

FIGURE 2.14 Sample distribution of spatiotemporal stratified sampling: (a) 000, (b) 001, (c) 010, (d) 011, (e) 100, (f) 101, (g) 110, (h) 111. (a)–(h) refer to the transformation of urban and other land features between 2000, 2005, and 2010. For example, 111 means that the pixel has always been urban in these three periods; 001 means that other features have become urban during the period of 2005–2010.

where # is the item number; (k, l) and (m, t) are the rows and columns information of the pixel with i and j gray, respectively; and d is the number of interval pixels between (k, l) and (m, t) on angular (θ) in the practical calculation.

The GLCM-correlation parameter (r) of each pixel is calculated as

$$r = \frac{\sum_i \sum_j (ij) p(i,j,d,\theta) - \mu_x \mu_y}{\sigma_x \sigma_y},\qquad(2.22)$$

where $P_{(i,j,d,\theta)}$ is the entry in a normalized GLCM. The mean (μ) and standard deviations (σ) for the rows and columns of the matrix are calculated as follows:

$$\mu_x = \sum_i \sum_j i \cdot p(i,j,d,\theta)\qquad(2.23)$$

$$\mu_y = \sum_i \sum_j j \cdot p(i,j,d,\theta)\qquad(2.24)$$

$$\sigma_x = \sum_i \sum_j (i - \mu_x)^2 \cdot p(i,j,d,\theta)\qquad(2.25)$$

$$\sigma_y = \sum_i \sum_j (j - \mu_y)^2 \cdot p(i,j,d,\theta)\qquad(2.26)$$

The GLCM-correlation parameter (r) ranges from -1 to 1. When r is close to 1, the pixels have a strong spatial correlation, which is at coordinates (k, l) and (m, t). Or the pixels have a weak spatial correlation.

(2) GLCM-Sampling Model

In the GLCM-sampling model, the GLCM-correlation parameter (r), the sampling size (n), and optimal distance (d) are deduced as

$$\begin{cases} \min_n \varepsilon^2 \\ s.t. \left(\dfrac{\sum_i \sum_j (ij) p(i,j,d,\theta) - \mu x \mu y}{\sigma x \sigma y} - r0 \right) = \varepsilon, \\ n = \left| \dfrac{N_x \cdot N_y}{n0^\circ \cdot n90^\circ} \right| \end{cases}\qquad(2.27)$$

where ε is an arbitrarily small value, r_0 is the critical value of the GLCM-correlation parameter (r) provided by the users and producers to balance data redundancy and accuracy, θ is defined as the value with four different orientation information, including $0°$ and $90°$. Here for simplified calculation, only two different orientations are considered. n is the optimal sample size. $n_{0°}$ and $n_{90°}$ are the number of the interval pixels on $0°$ and $90°$, respectively.

Sampling Design

2.4 SAMPLE ALLOCATION

The purpose of the allocation of samples is to determine the spatial locations of samples by considering spatial relations based on the designed sampling plan.

2.4.1 CLASSICAL ALLOCATION

Simple random sampling, cluster sampling, stratified random sampling, and semi-random sampling are four fundamental sample methods, each of which has its own unique sample allocation. They were widely used in the validation of GLC products. In the validation of DISCover, simple stratified random sampling was used to collect samples in each land cover class (Scepan 1999).

Allocating sample sizes to strata is a vital aspect of designing stratified sampling. There are three main options for allocation: equal, optimal, and proportional. Equal allocation is often employed when the goal is to accurately estimate the user's accuracy for each class. This method is appropriate when all classes are deemed equally significant and possess identical user accuracy. Conversely, optimal allocation is suitable when the primary aim is to reduce the standard error of a singular estimate, like overall accuracy or the area of a specific class. Proportional allocation, on the other hand, offers the advantage of straightforward analysis due to its nature as an equal probability sampling design. Original plans called for the global DISCover 1.0 data set to be produced by a number of participating laboratories processing portions of the global data set on a continental basis. In an effort to document possible variations in classification accuracy resulting from this system of distributed processing, a random sample stratified by continent was considered. The sample distribution is shown in Figure 2.15.

Systematic sampling can obtain samples with uniform spatial distribution. Systematic sample, that is, to begin with, element i, and select elements $i, i + k, …, i + (n − 1)k$, as the sample, the starting point i being chosen at random and $N = kn$ approximately. In September 2017, the ESA CCI Land Cover Team released a prototype land cover map at 20 m resolution over Africa for the year 2016. This is the first land cover map produced at such a high resolution covering an entire continent

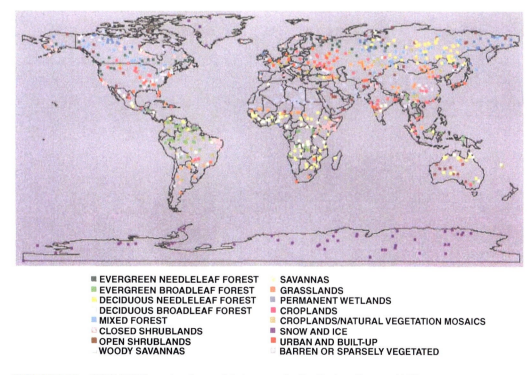

FIGURE 2.15 IGBP DISCover by class validation sample distribution (Scepan 1999).

for the year 2016. Evaluation based on the Copernicus Global Land Services (CGLS) reference dataset at 10 m resolution, the sample design of reference data has been systematic (with the same distance between sample sites – 35 km) in order to represent well the African landscapes (Lesiv et al. 2017). Sample points for systematic sampling are shown in Figure 2.16.

While one-stage cluster sampling might be the simplest design, its practical use is limited due to the high cost of acquiring reference data. As an alternative, various combinations of sampling designs can be employed in two-stage cluster sampling. A common approach involves implementing simple random sampling without replacement (SRSWOR) at each stage, as noted by Stehman (2009). Initially, PSUs are selected randomly without replacement from the total pool of PSUs. Subsequently, within each chosen PSU, a simple random sample of pixels is drawn without replacement from all available pixels. This process defines the second stage of the sampling design, making two-stage cluster sampling a more prevalent choice in practical scenarios.

Two-stage cluster sampling was applied in the sampling design for assessing the US National Land Cover Database 2001 (NLCD2001). It divided each area into 120 km × 120 km frame units. Two-stage sampling was implemented from these frame units. The first-stage samples of 12 km × 12 km PSU were randomly selected in each sampling area. The target number of PSUs per region was 55. This number of samples takes into account both the cost and the distribution of samples throughout the region. In the two-stage cluster design, where a pixel serves as the secondary sampling unit (SSU), 100 sample pixels from each class were selected through stratified random sampling from the PSUs chosen in the first stage within the region. The sample distribution of this sampling method is shown in Figure 2.17.

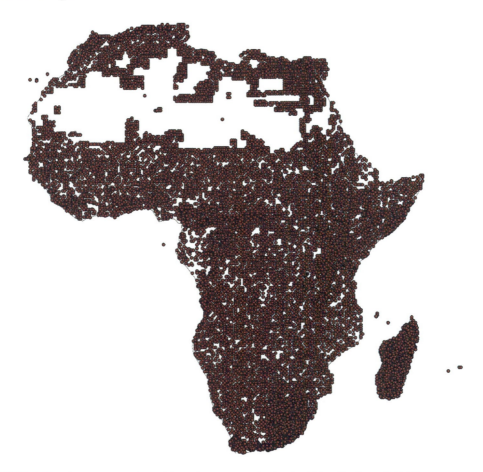

FIGURE 2.16 Spatial distribution of the reference sample sites in Africa (Lesiv et al. 2017).

Sampling Design

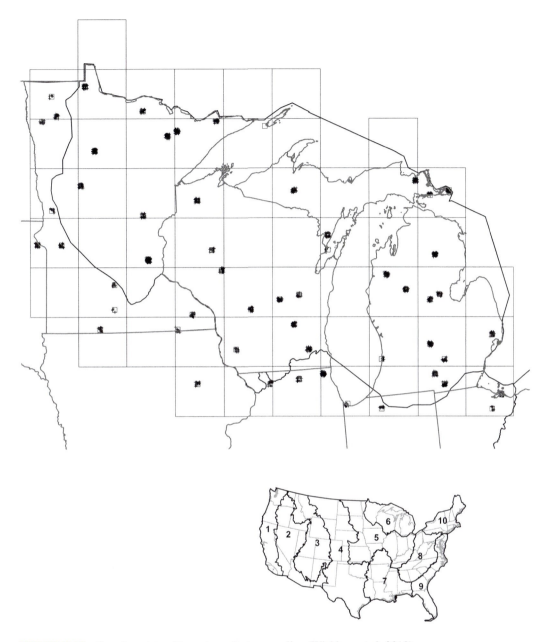

FIGURE 2.17 Sample points of two-stage cluster sampling (Wickham et al. 2010).

2.4.2 Two-Rank-Based Allocation

At high resolution, areas with richer land cover information are more representative of the overall quality of the product. Generally, the maps with more land cover types or larger land areas are preferred in map sheet sampling, while those with single class or small areas are not. Therefore, two criteria are adopted to measure the land cover complexity: global kernel density estimation and effective land area. This multiscale spatial sampling method was proposed by Xie et al. in Xie et al. 2015. The proposed method consists of two stages—i.e., global scale sampling and regional scale sampling. In the first stage, stratified sampling was used and the sample size was firstly determined based on an optimized sample model at the global scale (Tong et al. 2011), and then primary samples were placed according to the spatial complexity. The spatial complexity for each PSU was determined by two factors—i.e., Kernel density and effective area ratio. With the proposed method, 80 PSUs were selected from the world, and the overall accuracy of GlobeLand30 was determined as 83.5%.

Global kernel density estimation is used to measure the abundance of land cover complexity, and the Rosenblatt-Parzen estimator is expressed as

$$f_n(x) = \frac{1}{nh} \sum_{i=1}^{n} k\left(\frac{x - X_i}{h}\right), \quad (2.28)$$

where $f_n(x)$ is the density value within a window of h, $k()$ denotes a kernel function, and n is the total number within the window. On a global scale, the spatial distribution of land cover types is expressed by the kernel density estimator, and it is one of the indicators to determine the sample location.

The index of effective land area can be calculated by land area divided by the map sheet area as follows:

$$R_{area} = \frac{A_{land}}{A_{mapsheet}} \quad (2.29)$$

Each map sheet has two normalized indices, i.e., kernel density estimator (KDE-low: 0–0.5; KDE-high: 0.5–1) and effective land area ratio (ER-minority: 0%–50%; ER-majority: 50%–100%). Therefore, all the map sheets are stratified into four strata based on the two indices, and the priority of map sheet samples is determined by the four strata (see Figure 2.18 and Table 2.4). Four strata are developed by two parameters: KDE and ER.

As a result, the allocation of map sheet samples is conducted by the use of a spatial complexity–based stratified sampling method, in which the selected map sheet samples are representative in both characteristics and space, see Figure 2.19.

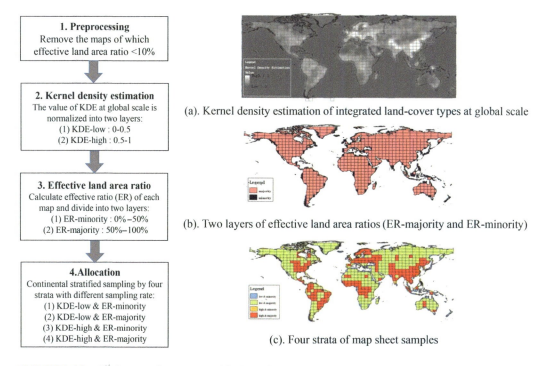

FIGURE 2.18 All the map sheets are stratified into four strata.

Sampling Design

TABLE 2.4
Stratified Sampling by Four Strata With Different Sampling Rates

Layer	Strata	Sampling rate (%)
1	KDE-low and ER-minority	10
2	KDE-low and ER-majority	20
3	KDE-high and ER-minority	30
4	KDE-high and ER-majority	40

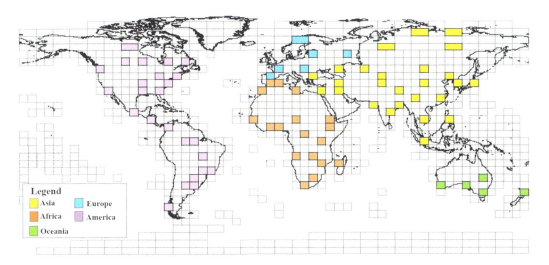

FIGURE 2.19 Result of the selected map sheet samples.

2.4.3 SPATIAL CORRELATION–BASED SAMPLE ALLOCATION

Within the scope of regional areas, spatial correlation has a significant influence on sample layout. The closer the distance between the two objects is, the more similar their spatial properties are. Spatial correlation is the inherent attribute, and the core idea of feature sample allocation is to reduce the correlation among samples by analyzing local spatial relations.

Local Morton's *I*, which is a local spatial correlation statistic, can figure out the space pattern of the sample's property:

$$I_i = \frac{x_i - \bar{X}}{S_i^2} \sum_{j=1, j \neq i}^{n} w_{i,j}\left(x_j - \bar{X}\right), \qquad (2.30)$$

where *i* is the *i*-th land cover item, I_i is the value of local Morton's *I*, x_i is the characteristic value, $\bar{X}$ is the average characteristic, $w_{i,j}$ denotes the spatial weight between *i* and *j*, *n* is the total of items, and *S* is calculated by

$$S_i^2 = \frac{\sum_{j=1, j \neq i}^{n}\left(x_j - \bar{X}\right)^2}{n-1} - \bar{X}^2 I_i = \frac{x_i - \bar{X}}{S_i^2} \sum_{j=1, j \neq i}^{n} w_{i,j}\left(x_j - \bar{X}\right) \qquad (2.31)$$

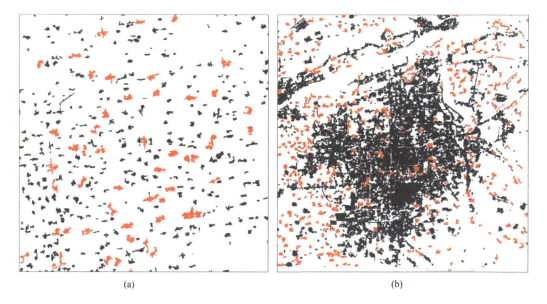

FIGURE 2.20 The effect of spatial correlation analysis: (a) discrete distribution of original items (in dark shade of gray) and random distribution of samples (in light shade of gray) after spatial correlation analysis and (b) clustered distribution of original items (in black) and random distribution of samples (in gray) after spatial correlation analysis.

Each land cover item within a map sheet obtains a spatial correlation index by Eqs. (2.30) and (2.31). This index reflects the relationship between the two objects on a local scale: if the index I is positive, the item is similar to the adjacent one, and if the index I is negative, the item is different from the adjacent one. Filtering the items whose index values are less than 0, the rest of the items conform to the requirement of feature samples, which have low spatial correlation and high heterogeneity (see Figure 2.20).

2.4.4 LSI-Based Allocation

Once the sample sizes have been determined for both region and class types, it is necessary to choose their geographical locations. Preferably, a greater number of sample sites should be allocated in areas characterized by heterogeneous landscapes (Wang et al. 2009; Olofsson et al. 2012). Traditional methods, such as stratified random sampling and the two-stage cluster sampling approach, often involve the random or systematic selection of sample locations. However, these methods do not account for the spatial heterogeneity of land cover (Scepan 1999; Friedl et al. 2002; Stehman et al. 2012; Congalton et al. 2014; Zhao et al. 2014). Since one of the major problems in the sampling is irrational sample distribution in the geographical space, taking into consideration the spatial heterogeneity of land cover is crucial for accuracy assessments in large areas.

An LSI-based sampling distribution approach is proposed. LSI can be calculated in each geographical unit (called uLSI) in order to distribute the derived sample size of each target region or each class into geographical sampling units according to their spatial heterogeneities. A curve, referred to as the uLSI curve, is derived from the uLSIs to determine the optimal geographical units for locating sample sites.

Suppose that region i can be divided into $R \times L$ geographical units. The uLSI of class k in row r and column l can be calculated and called $_k^i\text{uLSI}_{r,l}$. The uLSI of each unit is then calculated for all land cover classes. Figure 2.21 shows an example sample distribution of cultivated land, where geographical units are shown in Figure 2.21(a), and the result for cultivated land is shown in Figure 2.21(b).

Sampling Design

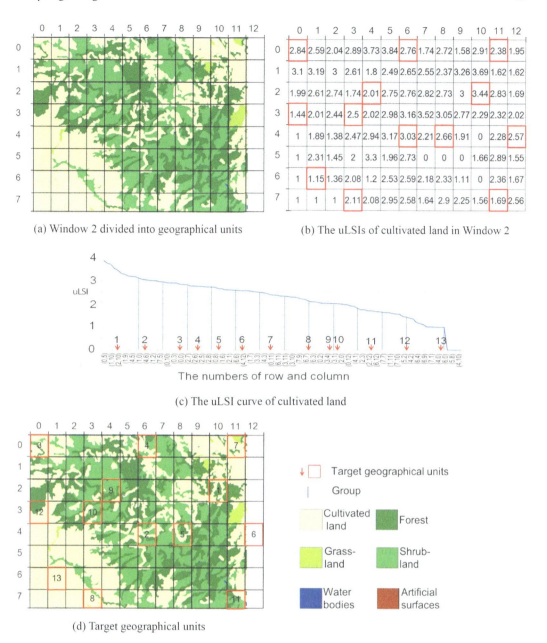

FIGURE 2.21 An illustration of the distribution of cultivated land samples.

Geographical units can be ranked in descending order according to the uLSIs, from which a specific uLSI curve for cultivated land can be derived (see igure 2.21[c]). The x-axis represents the numerical values of rows and columns for each geographic unit, while the y-axis represents the uLSI of the geographic unit. The x-axis evenly divides all geographical units into groups, with the number of groups equal to the sample size of the land cover classes. Subsequently, a geographical unit is randomly selected from each group. In each selected geographical unit, a pixel sample could be generated, employing a random sampling methodology. The final sampling outcome is an even distribution across geographical units exhibiting diverse landscape heterogeneity rather than being concentrated in areas characterized by high or low levels of heterogeneity.

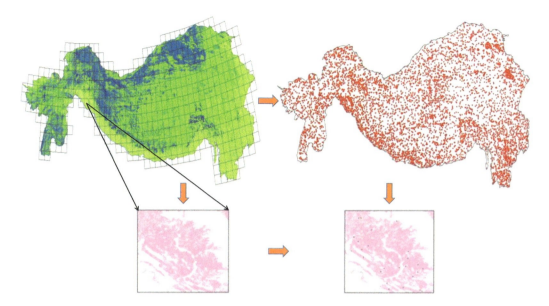

FIGURE 2.22 Grid sampling in a region (the left is a single feature map, and the right is the result of spatial allocation in the grid).

Suppose that the sample number of class k is $cN_{i,k}$. The geographical units would be divided into $cN_{i,k}$ groups along the x-axis of the uLSI curve. One target geographical unit is selected randomly from each group. Figure 2.21(c) illustrates the selection of target geographical units. The target geographical units are shown in Figure 2.21(c) using red arrows. The spatial distribution of geographic units indicated by the red arrows is depicted in Figure 2.21(d). Ultimately, within a target geographical unit, one pixel of cultivated land is randomly chosen as a sample site. Experimental findings indicate that the LSI approach designates more sample sites in areas with high heterogeneity.

2.4.5 Grid Sample Allocation

Because of the complex spatial distribution of high-resolution land cover remote sensing data, a strategy of uniformly pattering spatial elements is introduced, ensuring that rare classes and samples are spread out uniformly. . As shown in Figure 2.22, the grid is $1 \times 1°$ to ensure the sample is uniformly distributed and data easily downloaded. In the process of sampling, the stratified random sampling method is adopted; that is, samples are distributed according to the features of each 1×1 grid, and the number of sample points is the number of samples of the features × the number of pixels of the features/the total number of pixels of the grid. The whole roof area of the world is divided into 475 1×1 grids. The workflow of the sample spatial assigned is shown in Figure 2.22.

We calculated the total of assigned elements in the entire region and allocated the sample number based on the proportion of each element, to ensure there is a corresponding sample of each feature type.

2.5 SUMMARY

In order to organize an efficient validation for GLC data products, a reasonable sampling design is required to guarantee samples with good spatial representativeness and even spatial distribution. Because GLC products are extensively utilized across various application fields and scenarios, there is no single sampling method that universally applies to all GLC products. This chapter provides a set of sampling design methods:

Sampling Design

1) Sample size estimation methods: including classical methods, sample size based on acceptance probability, and sample size based on LSI
2) Design of sampling methods: including how to define strata and sample size in stratified sampling, classical stratified methods, two-rank sampling method, LSI-based sampling method, and modified spatially balanced sampling distribution methods
3) Samples allocation methods: including classical allocation methods, two-rank based allocation, LSI-based allocation, and grid allocation methods

Systematic sampling is generally motivated because of its ease of implementation in the field and because it achieves the criterion of spatial balance, therefore also tending to produce better precision than simple random sampling. When class-specific accuracy is a priority objective, the stratified design becomes a prime candidate. If cost is a dominant design criterion, then cluster sampling is optional. The rationale for geographic stratification on a global scale and stratifying by map class can be used to control sample allocation and class-specific accuracy. When it is necessary to consider the characteristics of sample representativeness or spatial heterogeneity, the two-rank sample distribution and allocation and LSI-based sample distribution and allocation can be good choices.

In summary, this chapter tries to tell how an efficient sampling strategy can be designed for developing a cost-effective accuracy assessment protocol. The key is to choose an adequate design, not necessarily the perfect design; recognize the strengths and weaknesses of different designs; and understand the trade-offs among objectives and desirable design principles.

REFERENCES

Agresti, Alan, Atalanta Ghosh, and Matilde Bini. 1995. "Raking kappa: Describing potential impact of marginal distributions on measures of agreement." *Biometrical Journal* 37(7): 811–820.

Aguilar, MA, MM Saldaña, and FJ Aguilar. 2013. "GeoEye-1 and WorldView-2 pan-sharpened imagery for object-based classification in urban environments." *International Journal of Remote Sensing* 34(7): 2583–2606.

Aickin, Mikel. 1990. "Maximum likelihood estimation of agreement in the constant predictive probability model, and its relation to Cohen's kappa." *Biometrics* 46(2): 293–302.

Allouche, Omri, Asaf Tsoar, and Ronen Kadmon. 2006. "Assessing the accuracy of species distribution models: prevalence, kappa and the true skill statistic (TSS)." *Journal of Applied Ecology* 43(6): 1223–1232.

Anderson, James R. 1971. "Land-use classification schemes." *Photogrammetric Engineering* 37(4): 379–387.

Anderson, James Richard. 1976. *A land use and land cover classification system for use with remote sensor data.* Vol. 964: US Government Printing Office.

Aronoff, Stan. 1982. "Classification accuracy: a user approach." *Photogrammetric Engineering and Remote Sensing* 48(8): 1299–1307.

Belward, Alan S, John E Estes, and Karen D Kline. 1999. "The IGBP-DIS global 1-km land cover data set DISCover: A project overview." *Photogrammetric Engineering and Remote Sensing* 65(9): 1013–1020.

Bicheron, Patrice, Pierre Defourny, Carsten Brockmann, Leon Schouten, Christelle Vancutsem, Mireille HUC, Sophie Bontemps, Marc Leroy, Frederic Achard, and Martin Herold. 2008. GLOBCOVER-products description and validation report.

Bishop, Yvonne MM, Stephen E Fienberg, and W Paul. 1975. Holland. *Discrete multivariate analysis: Theory and practice.* The MIT Press.

Boschetti, Luigi, Stephen V Stehman, and David P Roy. 2016. "A stratified random sampling design in space and time for regional to global scale burned area product validation." *Remote Sensing of Environment* 186: 465–478.

Brennan, Robert L, and Dale J Prediger. 1981. "Coefficient kappa: Some uses, misuses, and alternatives." *Educational and Psychological Measurement* 41(3): 687–699.

Chen, Jun, Jin Chen, Anping Liao, Xin Cao, Lijun Chen, Xuehong Chen, Chaoying He, Gang Han, Shu Peng, and Miao Lu. 2015. "Global land cover mapping at 30 m resolution: A POK-based operational approach." *ISPRS Journal of Photogrammetry and Remote Sensing* 103: 7–27.

Cohen, Jacob. 1960. "A coefficient of agreement for nominal scales." *Educational and Psychological Measurement* 20(1): 37–46.

Comber, Alexis, Peter Fisher, Chris Brunsdon, and Abdulhakim Khmag. 2012. "Spatial analysis of remote sensing image classification accuracy." *Remote Sensing of Environment* 127: 237–246.

Congalton, Russell G, and Kass Green. 2019. *Assessing the accuracy of remotely sensed data: Principles and practices*: CRC press.

Congalton, Russell G, Jianyu Gu, Kamini Yadav, Prasad Thenkabail, and Mutlu Ozdogan. 2014. "Global land cover mapping: A review and uncertainty analysis." *Remote Sensing* 6(12): 12070–12093.

Congalton, Russell G, Richard G Oderwald, and Roy A Mead. 1983. "Assessing Landsat classification accuracy using discrete multivariate analysis statistical techniques." *Photogrammetric Climate Change Initiative – Land Cover (CCI-LC) Engineering and Remote Sensing* 49(12): 1671–1678.

Congalton, Russell G. 1991. "A review of assessing the accuracy of classifications of remotely sensed data." *Remote Sensing of Environment* 37(1): 35–46.

Congalton, Russell G. 1993. "Accuracy assessment of remotely sensed data: Future needs and directions." *Proceedings of Pecora 12 Symposium*, Sioux Falls, SD. August 1993.

Czaplewski, Raymond L. 1992. "Misclassification bias in areal estimates." *Photogrammetric Engineering and Remote Sensing* 58(2): 189–192.

Czaplewski, Raymond L. 1994. *Variance approximations for assessments of classification accuracy*. Vol. 316: US Department of Agriculture, Forest Service, Rocky Mountain Forest and Range Experiment Station.

DeGroot, Morris H. 1986. Boston. Probability and Statistics. 2nd ed. Addison-Wesley.

Duro, Dennis C, Steven E Franklin, and Monique G Dubé. 2012. "A comparison of pixel-based and object-based image analysis with selected machine learning algorithms for the classification of agricultural landscapes using SPOT-5 HRG imagery." *Remote Sensing of Environment* 118: 259–272.

DZ/T. 2014. *Inspection and evaluation of quality for geological data*: Ministry of Land and Resources in China.

Elias, P, and N Abramson. 1963. *Information theory and coding*: Mc2Graw2Hill.

Eugenio, Barbara Di, and Michael Glass. 2004. "The kappa statistic: A second look." *Computational Linguistics* 30(1): 95–101.

Everitt, BS. 1968. "Moments of the statistics kappa and weighted kappa." *British Journal of Mathematical and Statistical Psychology* 21(1): 97–103.

Fahsi, Ahmed, T Tsegaye, W Tadesse, and T Coleman. 2000. "Incorporation of digital elevation models with Landsat-TM data to improve land cover classification accuracy." *Forest Ecology and Management* 128(1–2): 57–64.

Fauvel, Mathieu, David Sheeren, Jocelyn Chanussot, and Jon Atli Benediktsson. 2012. "Hedges detection using local directional features and support vector data description." *2012 IEEE International Geoscience and Remote Sensing Symposium*, 22-27 July 2012, Munich, Germany.

Finn, John T. 1993. "Use of the average mutual information index in evaluating classification error and consistency." *International Journal of Geographical Information Science* 7(4): 349–366.

Fleiss, Joseph L, Jacob Cohen, and Brian S Everitt. 1969. "Large sample standard errors of kappa and weighted kappa." *Psychological Bulletin* 72(5): 323.

Foody, Giles M. 1992. "On the compensation for chance agreement in image classification accuracy assessment." *Photogrammetric Engineering and Remote Sensing* 58(10): 1459–1460.

Foody, Giles M. 2002. "Status of land cover classification accuracy assessment." *Remote Sensing of Environment* 80(1): 185–201.

Foody, Giles M. 2004. "Thematic map comparison." *Photogrammetric Engineering & Remote Sensing* 70(5): 627–633.

Foody, Giles M. 2008. "Harshness in image classification accuracy assessment." *International Journal of Remote Sensing* 29(11): 3137–3158.

Foody, Giles M. 2009. "Sample size determination for image classification accuracy assessment and comparison." *International Journal of Remote Sensing* 30(20): 5273–5291.

Fraser, Clive S, Gene Dial, and Jacek Grodecki. 2006. "Sensor orientation via RPCs." *ISPRS Journal of Photogrammetry and Remote Sensing* 60(3): 182–194.

Friedl, Mark A, Douglas K McIver, John CF Hodges, Xiaoyang Y Zhang, D Muchoney, Alan H Strahler, Curtis E Woodcock, Sucharita Gopal, Annemarie Schneider, and Amanda Cooper. 2002. "Global land cover mapping from MODIS: Algorithms and early results." *Remote Sensing of Environment* 83(1–2): 287–302.

Gong, Yali, Huan Xie, Yanmin Jin, and Xiaohua Tong. 2022. "Assessing multi-temporal global urban land-cover products using spatio-temporal stratified sampling." *ISPRS International Journal of Geo-Information* 11(8): 451.

Groves, Robert M, Floyd J Fowler Jr, Mick P Couper, James M Lepkowski, Eleanor Singer, and Roger Tourangeau. 2009. *Survey methodology*. Vol. 561: John Wiley & Sons.

Grzegozewski, Denise Maria, Jerry Adriani Johann, Miguel Angel Uribe-Opazo, Erivelto Mercante, and Alexandre Camargo Coutinho. 2016. "Mapping soya bean and corn crops in the State of Paraná, Brazil, using EVI images from the MODIS sensor." *International Journal of Remote Sensing* 37(6): 1257–1275.

Gwet, Kilem L. 2014. *Handbook of inter-rater reliability: The definitive guide to measuring the extent of agreement among raters*: Advanced Analytics, LLC.

Hansen, Matthew C, Ruth S DeFries, John RG Townshend, and Rob Sohlberg. 2000. "Global land cover classification at 1 km spatial resolution using a classification tree approach." *International Journal of Remote Sensing* 21(6–7): 1331–1364.

Hartley, Hermann Otto, and Egon Sharpe Pearson. 1966. *Biometrika tables for statisticians*: Published for the Biometrika Trustees at the University Press.

Hay, AM. 1988. "The derivation of global estimates from a confusion matrix." *International Journal of Remote Sensing* 9(8): 1395–1398.

Helldén, Ulf. 1980. "A test of landsat-2 imagery and digital data for thematic mapping illustrated by an environmental study in northern Kenya, Lund University." *Natural Geography Institute Report No. 47*.

Herold, Martin, Philippe Mayaux, CE Woodcock, A Baccini, and C Schmullius. 2008. "Some challenges in global land cover mapping: An assessment of agreement and accuracy in existing 1 km datasets." *Remote Sensing of Environment* 112(5): 2538–2556.

Hudson, William D. 1987. "Correct formulation of the kappa coefficient of agreement." *Photogrammetric Engineering and Remote Sensing* 53(4): 421–422.

ISO 1985. *Sampling procedures for inspection by attributes – Part 2: Sampling plans indexed by limiting quality (LQ) for isolated lot inspection*: ISO.

ISO 1999. *Sampling Procedures for Inspection by Attributes – Part 1: Sampling Schemes Indexed by Acceptance Quality Limit (AQL) for Lot-by-lot Inspection*: ISO.

ISO 2002. *Sampling procedures for inspection by attributes – Part 4: Procedures for assessment of declared quality levels*: ISO.

ISO 2013. *Sampling Procedures and Charts for Inspection by Variables for Percent Defective*: ISO.

ISO 2020. *Sampling Procedures and Tables for Inspection by Attributes*: ISO.

Jiang, Long, Mo Yu, Ming Zhou, Xiaohua Liu, and Tiejun Zhao. 2011. "Target-dependent twitter sentiment classification." Proceedings of the 49th annual meeting of the association for computational linguistics: human language technologies.

Kempen, Bas, Dick J Brus, Gerard BM Heuvelink, and Jetse J Stoorvogel. 2009. "Updating the 1: 50,000 Dutch soil map using legacy soil data: A multinomial logistic regression approach." *Geoderma* 151(3–4): 311–326.

Klemenjak, Sascha, Björn Waske, Silvia Valero, and Jocelyn Chanussot. 2012. "Automatic detection of rivers in high-resolution SAR data." *IEEE Journal of Selected Topics in Applied Earth Observations and Remote Sensing* 5(5): 1364–1372.

Koukoulas, Sotlrlos, and George Alan Blackburn. 2001. "Introducing new indices for accuracy evaluation of classified images representing semi-natural woodland environments." *Photogrammetric Engineering and Remote Sensing* 67(4): 499–510.

Kraemer, Helena Chmura. 1979. "Ramifications of a population model for κ as a coefficient of reliability." *Psychometrika* 44(4): 461–472.

Landis, J Richard, and Gary G Koch. 1977. "The measurement of observer agreement for categorical data." *Biometrics* 33(1): 159–174.

Latham, John, Renato Cumani, Ilaria Rosati, and Mario Bloise. 2014. "Global land cover share (GLC-SHARE) database beta-release version 1.0-2014." *FAO: Rome, Italy* 29.

Lesiv, Myroslava, S Fritz, I McCallum, N Tsendbazar, M Herold, J-F Pekel, M Buchhorn, B Smets, and R Van De Kerchove. 2017. "Evaluation of ESA CCI prototype land cover map at 20m."

Light, Richard J. 1971. "Measures of response agreement for qualitative data: Some generalizations and alternatives." *Psychological Bulletin* 76(5): 365.

Liu, Canran, Paul Frazier, and Lalit Kumar. 2007. "Comparative assessment of the measures of thematic classification accuracy." *Remote Sensing of Environment* 107(4): 606–616.

Liu, Xiaoping, Guohua Hu, Yimin Chen, Xia Li, Xiaocong Xu, Shaoying Li, Fengsong Pei, and Shaojian Wang. 2018. "High-resolution multi-temporal mapping of global urban land using Landsat images based on the Google Earth Engine Platform." *Remote Sensing of Environment* 209: 227–239.

Loveland, Thomas R, and AS Belward. 1997. "The IGBP-DIS global 1km land cover data set, DISCover: First results." *International Journal of Remote Sensing* 18(15): 3289–3295.

Lyons, Mitchell B, David A Keith, Stuart R Phinn, Tanya J Mason, and Jane Elith. 2018. "A comparison of resampling methods for remote sensing classification and accuracy assessment." *Remote Sensing of Environment* 208: 145–153.

Ma, Zhenkui, and Roland L Redmond. 1995. "Tau coefficients for accuracy assessment of classification of remote sensing data." *Photogrammetric Engineering and Remote Sensing* 61(4): 435–439.

Martha, Tapas R, Norman Kerle, Cees J Van Westen, Victor Jetten, and K Vinod Kumar. 2012. "Object-oriented analysis of multi-temporal panchromatic images for creation of historical landslide inventories." *ISPRS Journal of Photogrammetry and Remote Sensing* 67: 105–119.

Mayaux, Philippe, Hugh Eva, Javier Gallego, Alan H Strahler, Martin Herold, Shefali Agrawal, Sergey Naumov, Evaristo Eduardo De Miranda, Carlos M Di Bella, and Callan Ordoyne. 2006. "Validation of the global land cover 2000 map." *IEEE Transactions on Geoscience and Remote Sensing* 44(7): 1728–1739.

Myint, Soe W, Patricia Gober, Anthony Brazel, Susanne Grossman-Clarke, and Qihao Weng. 2011. "Per-pixel vs. object-based classification of urban land cover extraction using high spatial resolution imagery." *Remote Sensing of Environment* 115(5): 1145–1161.

Næsset, Erik. 1996. "Use of the weighted Kappa coefficient in classification error assessment of thematic maps." *International Journal of Geographical Information Systems* 10(5): 591–603.

Nelson, Ross F. 1983. "Detecting forest canopy change due to insect activity using Landsat MSS." *Photogrammetric Engineering and Remote Sensing* 49(9): 1303–1314.

Olofsson, Pontus, Giles M Foody, Martin Herold, Stephen V Stehman, Curtis E Woodcock, and Michael A Wulder. 2014. "Good practices for estimating area and assessing accuracy of land change." *Remote Sensing of Environment* 148: 42–57.

Olofsson, Pontus, Stephen V Stehman, Curtis E Woodcock, Damien Sulla-Menashe, Adam M Sibley, Jared D Newell, Mark A Friedl, and Martin Herold. 2012. "A global land-cover validation data set, part I: Fundamental design principles." *International Journal of Remote Sensing* 33(18): 5768–5788.

Pebesma, Edzer J, and Cees G Wesseling. 1998. "Gstat: A program for geostatistical modelling, prediction and simulation." *Computers & Geosciences* 24(1): 17–31.

Pontius Jr Robert G, Emily Shusas, and Menzie McEachern. 2004. "Detecting important categorical land changes while accounting for persistence." *Agriculture, Ecosystems & Environment* 101(2–3): 251–268.

Pontius Jr Robert Gilmore, and Marco Millones. 2011. "Death to Kappa: Birth of quantity disagreement and allocation disagreement for accuracy assessment." *International Journal of Remote Sensing* 32(15): 4407–4429.

Pontius, RG. 2001. "Quantification error versus location error in comparison of categorical maps (vol 66, pg 1011, 2000)." *Photogrammetric Engineering and Remote Sensing* 67(5): 540–540.

Rosenfield, George H, and Katherine Fitzpatrick-Lins. 1986. "A coefficient of agreement as a measure of thematic classification accuracy." *Photogrammetric Engineering and Remote Sensing* 52(2): 223–227.

Rozenstein, Offer, and Arnon Karnieli. 2011. "Comparison of methods for land-use classification incorporating remote sensing and GIS inputs." *Applied Geography* 31(2): 533–544.

Rwanga, Sophia S, and Julius M Ndambuki. 2017. "Accuracy assessment of land use/land cover classification using remote sensing and GIS." *International Journal of Geosciences* 8(4): 611.

Santoro, Maurizio, G. Kirches, J. Wevers, M. Boettcher, C. Brockmann, C. Lamarche, S. Bontemps, I. Moreau, and P. Defourny. 2017. "Land Cover CCI." *Product User Guide. Version* 2 .

Scepan, Joseph. 1999. "Thematic validation of high-resolution global land-cover data sets." *Photogrammetric Engineering and Remote Sensing* 65: 1051–1060.

Schneider, Annemarie, Mark A Friedl, and David Potere. 2010. "Mapping global urban areas using MODIS 500-m data: New methods and datasets based on 'urban ecoregions.'" *Remote Sensing of Environment* 114(8): 1733–1746.

Schultz, Michael, Janek Voss, Michael Auer, Sarah Carter, and Alexander Zipf. 2017. "Open land cover from OpenStreetMap and remote sensing." *International Journal of Applied Earth Observation and Geoinformation* 63: 206–213.

Shao, Guofan, and Jianguo Wu. 2008. "On the accuracy of landscape pattern analysis using remote sensing data." *Landscape Ecology* 23: 505–511.

Sharma, Laxmi Kant, Mahendra Singh Nathawat, and Suman Sinha. 2013. "Top-down and bottom-up inventory approach for above ground forest biomass and carbon monitoring in REDD framework using multi-resolution satellite data." *Environmental Monitoring and Assessment* 185: 8621–8637.

Short, Nicholas M. 1982. *The Landsat tutorial workbook: Basics of satellite remote sensing*. Vol. 1078: National Aeronautics and Space Administration, Scientific and Technical Information Branch.

Smith, Eric P. 2002. "BACI design." *Encyclopedia of Environmetrics* 1: 141–148.

Smits, PC, SG Dellepiane, and RA Schowengerdt. 1999. "Quality assessment of image classification algorithms for land-cover mapping: A review and a proposal for a cost-based approach." *International Journal of Remote Sensing* 20(8): 1461–1486.

Stehman, Stephen V, and Raymond L Czaplewski. 1998. "Design and analysis for thematic map accuracy assessment: Fundamental principles." *Remote Sensing of Environment* 64(3): 331–344.

Stehman, Stephen V, JD Wickham, JH Smith, and L Yang. 2003. "Thematic accuracy of the 1992 National Land-Cover Data for the eastern United States: Statistical methodology and regional results." *Remote Sensing of Environment* 86(4): 500–516.

Stehman, Stephen V, Pontus Olofsson, Curtis E Woodcock, Martin Herold, and Mark A Friedl. 2012. "A global land-cover validation data set, II: Augmenting a stratified sampling design to estimate accuracy by region and land-cover class." *International Journal of Remote Sensing* 33(22): 6975–6993.

Stehman, Stephen V. 1997. "Selecting and interpreting measures of thematic classification accuracy." *Remote Sensing of Environment* 62(1): 77–89.

Stehman, Stephen V. 2001. "Statistical rigor and practical utility in thematic map accuracy assessment." *Photogrammetric Engineering and Remote Sensing* 67(6): 727–734.

Stehman, Stephen V. 2004. "A critical evaluation of the normalized error matrix in map accuracy assessment." *Photogrammetric Engineering & Remote Sensing* 70(6): 743–751.

Stehman, Stephen V. 2008. "Sampling designs for assessing map accuracy." *Proceedings of the 8th International Symposium on Spatial Accuarcy Assessment in Natural Resources and Environmental Sciences.*

Stehman, Stephen V. 2009. "Sampling designs for accuracy assessment of land cover." *International Journal of Remote Sensing* 30(20): 5243–5272.

Story, Michael, and Russell G Congalton. 1986. "Accuracy assessment: A user's perspective." *Photogrammetric Engineering and Remote Sensing* 52(3): 397–399.

Strahler, Alan H, Luigi Boschetti, Giles M Foody, Mark A Friedl, Matthew C Hansen, Martin Herold, Philippe Mayaux, Jeffrey T Morisette, Stephen V Stehman, and Curtis E Woodcock. 2006. "Global land cover validation: Recommendations for evaluation and accuracy assessment of global land cover maps." *European Communities, Luxembourg* 51(4): 1–60.

Strehl, Alexander, and Joydeep Ghosh. 2002. "Cluster ensembles---a knowledge reuse framework for combining multiple partitions." *Journal of Machine Learning Research* 3(Dec): 583–617.

Tateishi, Ryutaro, Bayaer Uriyangqai, Hussam Al-Bilbisi, Mohamed Aboel Ghar, Javzandulam Tsend-Ayush, Toshiyuki Kobayashi, Alimujiang Kasimu, Nguyen Thanh Hoan, Adel Shalaby, and Bayan Alsaaideh. 2011. "Production of global land cover data–GLCNMO." *International Journal of Digital Earth* 4(1): 22–49.

Thomlinson, John R, Paul V Bolstad, and Warren B Cohen. 1999. "Coordinating methodologies for scaling landcover classifications from site-specific to global: Steps toward validating global map products." *Remote Sensing of Environment* 70(1): 16–28.

Tong, Xiaohua, and Zhenhua Wang. 2012. "Fuzzy acceptance sampling plans for inspection of geospatial data with ambiguity in quality characteristics." *Computers & Geosciences* 48: 256–266.

Tong, Xiaohua, Zhenhua Wang, Huan Xie, Dan Liang, Zuoqin Jiang, Jinchao Li, and Jun Li. 2011. "Designing a two-rank acceptance sampling plan for quality inspection of geospatial data products." *Computers & Geosciences* 37(10): 1570–1583.

Tsendbazar, Nandin-Erdene, Sytze De Bruin, Steffen Fritz, and Martin Herold. 2015. "Spatial accuracy assessment and integration of global land cover datasets." *Remote Sensing* 7(12): 15804–15821.

Tung, F, and E LeDrew. 1988. "The determination of optimal threshold levels for change detection using various accuracy indexes." *Photogrammetric Engineering and Remote Sensing* 54(10): 1449–1454.

Turk, G. 2002. "Map evaluation and "chance correction". *Photogrammetric Engineering and Remote Sensing* 68(2): 123–125+133.

Türk, Goksel. 1979. "Gt index: A measure of the success of prediction." *Remote Sensing of Environment* 8(1): 65–75.

UCLouvain Team: Sophie Bontemps, Pierre Defourny, Eric Van Bogaert, ESA Team, Olivier Arino, Vasileios Kalogirou, and Jose Ramos Perez. 2011. GLOBCOVER 2009 products description and validation report.

Ustin, SL, QJ Hart, L Duan, and G Scheer. 1996. "Vegetation mapping on hardwood rangelands in California." *International Journal of Remote Sensing* 17(15): 3015–3036.

Vorovencii, Iosif. 2014. "Assessment of some remote sensing techniques used to detect land use/land cover changes in South-East Transilvania, Romania." *Environmental Monitoring and Assessment* 186: 2685–2699.

Wang, Jin-Feng, Cheng-Sheng Jiang, Mao-Gui Hu, Zhi-Dong Cao, Yan-Sha Guo, Lian-Fa Li, Tie-Jun Liu, and Bin Meng. 2013. "Design-based spatial sampling: Theory and implementation." *Environmental Modelling & Software* 40: 280–288.

Wang, Jin-Feng, George Christakos, and Mao-Gui Hu. 2009. "Modeling spatial means of surfaces with stratified nonhomogeneity." *IEEE Transactions on Geoscience and Remote Sensing* 47(12): 4167–4174.

Wang, Jinfeng, Robert Haining, and Zhidong Cao. 2010. "Sample surveying to estimate the mean of a heterogeneous surface: Reducing the error variance through zoning." *International Journal of Geographical Information Science* 24(4): 523–543.

Wickham, James D, Stephen V Stehman, Leila Gass, Jon Dewitz, Joyce A Fry, and Timothy G Wade. 2013. "Accuracy assessment of NLCD 2006 land cover and impervious surface." *Remote Sensing of Environment* 130: 294–304.

Wickham, JD, SV Stehman, JA Fry, JH Smith, and Collin G Homer. 2010. "Thematic accuracy of the NLCD 2001 land cover for the conterminous United States." *Remote Sensing of Environment* 114(6): 1286–1296.

Wulder, Michael A, Steven E Franklin, Joanne C White, Julia Linke, and Steen Magnussen. 2006. "An accuracy assessment framework for large-area land cover classification products derived from medium-resolution satellite data." *International Journal of Remote Sensing* 27(4): 663–683.

Xie, Huan, Xiaohua Tong, Wen Meng, Dan Liang, Zhenhua Wang, and Wenzhong Shi. 2015. "A multilevel stratified spatial sampling approach for the quality assessment of remote-sensing-derived products." *IEEE Journal of Selected Topics in Applied Earth Observations and Remote Sensing* 8(10): 4699–4713.

Yu, Bing, and Songhao Shang. 2017. "Multi-year mapping of maize and sunflower in Hetao irrigation district of China with high spatial and temporal resolution vegetation index series." *Remote Sensing* 9(8): 855.

Yu, Le, and Peng Gong. 2012. "Google Earth as a virtual globe tool for Earth science applications at the global scale: progress and perspectives." *International Journal of Remote Sensing* 33(12): 3966–3986.

Yu, Le, Jie Wang, Nicholas Clinton, Qinchuan Xin, Liheng Zhong, Yanlei Chen, and Peng Gong. 2013. "FROM-GC: 30 m global cropland extent derived through multisource data integration." *International Journal of Digital Earth* 6(6): 521–533.

Zhao, Yuanyuan, Peng Gong, Le Yu, Luanyun Hu, Xueyan Li, Congcong Li, Haiying Zhang, Yaomin Zheng, Jie Wang, and Yongchao Zhao. 2014. "Towards a common validation sample set for global land-cover mapping." *International Journal of Remote Sensing* 35(13): 4795–4814.

Zhuang, Xin, Bernard A Engel, Xiaoping Xiong, and Chris J Johannsen. 1995. "Analysis of classification results of remotely sensed data and evaluation of classification algorithms." *Photogrammetric Engineering and Remote Sensing* 61(4): 427–432.

3 Reference Data

Xiaohua Tong and Chao Wei
Tongji University, Shanghai, China

Lijun Chen
National Geomatics Center of China, Beijing, China

Dan Liang
Zhejiang A&F University, Hangzhou, China

Zhengxing Wang
Institute of Geographical Sciences and Natural Resources Research, Chinese Academy of Sciences Beijing, China

Maria Antonia Brovelli
Politecnico di Milano, Milan, Italy

Hanfa Xing
South China Normal University, Foshan, China

3.1 INTRODUCTION

Reference data refers to benchmarks or standards used to verify the accuracy and reliability of land cover products. These reference data are usually high-quality, high-resolution ground observation data, field sampling data, high-resolution remote sensing images, and other geographic information system (GIS) data. Reference data play a vital role, as they are used to verify and analyze land cover products with respect to their accuracy, completeness, and authenticity (Pengra et al. 2015; Fritz et al. 2017; See et al. 2022).

When selecting reference data sources for validating land cover products, considerations should be given to the data quality of the reference data sources themselves, such as positional accuracy, logical consistency, thematic accuracy, temporal quality, and usability (ISO 19157 2013). Positional accuracy refers to the degree of closeness between the geographic location of features on a map and their true positions on Earth's surface. Logical consistency is defined as the extent to which data adhere to logical rules concerning their structure, attributes, and relationships, whether at conceptual, logical, or physical levels. Thematic accuracy refers to the accuracy of quantitative attributes, the correctness of qualitative attributes, and the accuracy of features and their relationships. For global land cover (GLC) data, thematic accuracy refers to classification accuracy. Temporal quality is defined as the quality of temporal attributes of features and relationships. In the context of spatial data, temporal quality refers to the accuracy, reliability, and consistency of time-related information associated with geographic features, such as data acquisition dates, time periods represented by the dataset, and update frequencies. Usability is indeed based on user needs and can be assessed

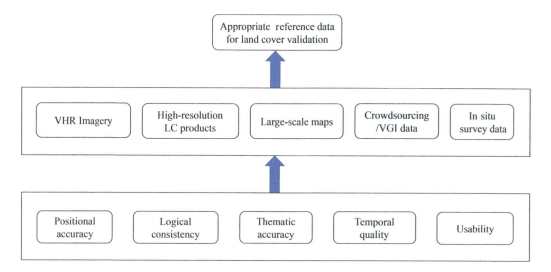

FIGURE 3.1 Reference data selected based on quality indicators.

using various quality elements. These quality elements include accuracy, completeness, consistency, timeliness, and relevance, among others, all of which contribute to the overall usability of the dataset. Usability assessment involves evaluating how well the dataset meets the specific needs and expectations of its intended users. Through comprehensive evaluation, high-quality and reliable reference data can be chosen for validating land cover products (see Figure 3.1).

The remaining part of this chapter proceeds as follows: Starting with the second section, the reference data sources are introduced successively, which are very high resolution (VHR) imagery (encompassing satellite imagery and other imagery) in Section 3.2, finer-resolution land cover data with spatial resolution finer than 30 meters in Section 3.3, large-scale maps in Section 3.4, crowdsourcing/Volunteered Geographic Information (VGI) data (OSM, Mapillary, DCP, etc.) in Section 3.5, and in situ data in Section 3.6, respectively. Finally, a summary is given in Section 3.7.

3.2 VHR IMAGERY

VHR images have the characteristics of possessing highly detailed and spatially finely resolved information about entities and materials on Earth's surface. VHR satellite images can capture detailed features, such as individual structures, vehicles, and small vegetation areas. It provides detailed imagery at submeter resolution, making it relatively easy to manually identify land cover types. Therefore, VHR images are a good source of reference datasets for land cover products.

3.2.1 Satellite Imagery

In view of the great progress of remote sensing technology, many remote sensing platforms, such as QuickBird, GeoEye, Worldview, and IKONOS, can capture features on the ground at VHR. These images can display other spatial features, such as edges, shapes, and textures (Grenier et al. 2007; Costa et al. 2002; Johansen et al. 2010; Grenier et al. 2008). These satellite remote sensing platforms are conducive to evaluating the spatiotemporal changes in large-scale land cover and land use. Using remote sensing images to detect land cover change is an important application of Earth observation data because it provides insights into environmental protection, ecological management, and urban governance (Wang et al. 2017; Dronova 2015; Yang et al. 2012; Shin et al. 2023; Frazier and Hemingway 2021; Wang et al. 2019). In particular, VHR remote sensing images can capture

Reference Data

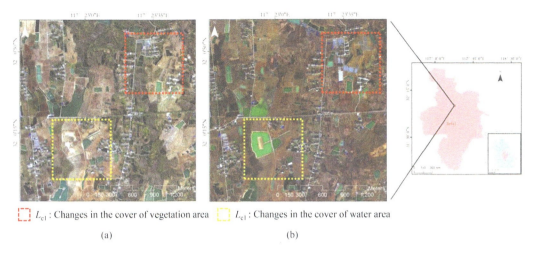

FIGURE 3.2 Multi-temporal GF-1 images used to extract the real change regions: (a) the image acquired on April 14, 2015, and (b) the image acquired on January 26, 2016 (Wang et al. 2017).

the details of ground objects and provide opportunities for detailed detection of land cover changes. The study region, as represented in Figure 3.2, was flown with GF-1 images acquired at two different times. Through these two VHR images, the real change regions were identified (Figure 3.3). Table 3.1 summarizes the submeter resolution commercial satellites.

1) IKONOS
   IKONOS satellite was successfully launched on September 24, 1999. It is the first commercial remote sensing satellite in the world to provide high-resolution satellite images. IKONOS is a commercial satellite that can collect panchromatic images with 1 m resolution and multispectral images with 4 m resolution. At the same time, panchromatic images and multispectral images can be fused into color images with 1m resolution. Up to now, IKONOS has collected images covering more than 250 million square kilometers over continents, and many of them are widely used by governments in national defense, military mapping, sea and air transportation, and other fields. From the orbit at a height of 681 km, IKONOS has a revisit period of 3 days and can transmit data directly from satellites to 12 ground stations around the world. The image dates are from 1999 to 2015.
2) GeoEye-1
   GEOEye-1 was launched in September 2008, with a spatial resolution of 0.5 m, which is one of the highest-resolution commercial satellites at present. GeoEye-1 satellite has the characteristics of finer spatial resolution, extremely strong mapping ability, and a very short revisit period, and it has outstanding advantages for large-scale mapping and for interpreting fine spatial objects.
3) QuickBird
   QuickBird satellite was launched by the US company DigitalGlobe in October 2001. It is one of the commercial satellites that can provide submeter resolution in the world at present. It has the highest geographical positioning accuracy, massive onboard storage, and spatial resolution two to ten times finer than other commercial high-resolution satellites. Moreover, the QuickBird satellite system can collect images covering 75 million square kilometers every year, and the archived data is increasing at an unprecedented rate. There are at least two or three transit tracks in China every day, with archived image data covering about 7 million square kilometers.

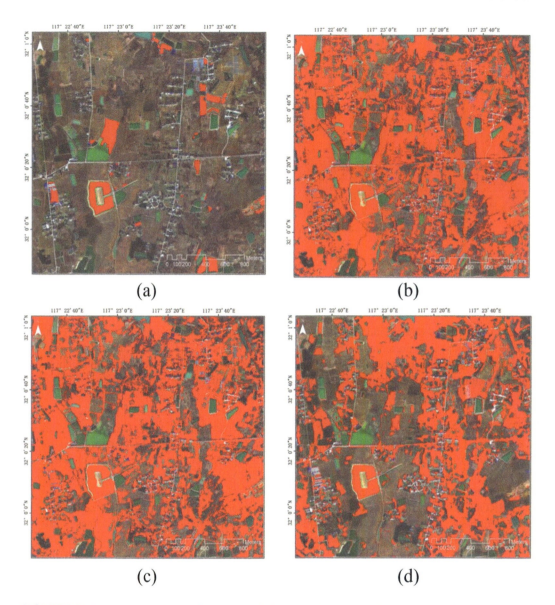

FIGURE 3.3 Unsupervised change detection results by using the tested methods: (a) ground-truth image and (b)–(d) results with different change detection methods (Wang et al. 2017).

4) PlanetScope

It is operated by Planet Labs Inc. in San Francisco, California, United States. It captures daily high-resolution imagery and represents a significant development in the remote sensing community. Since 2016, Planet has maintained the largest fleet of Earth-imaging small satellites, acquiring multispectral imagery for various applications. These compact satellites, roughly the size of a loaf of bread, can be produced and deployed more rapidly than traditional satellites. They do not require specialized launch vehicles; rather, they can hitch a ride as secondary payloads on other space missions. This logistical advantage accelerates the enhancement of Earth observation capabilities as it allows for the quick incorporation of new technologies and swift replacement or expansion of satellite constellations in response to operational failures. Consequently, Planet's PlanetScope satellites are becoming increasingly vital for Earth imaging.

Reference Data

TABLE 3.1
Commercial Remote Sensing Satellites of Submeter Resolution (PAN, Panchromatic, XS, Multispectral) (Wang et al. 2019)

No	Sensor/Instrument	Sensor Type	Spatial Resolution (Nadir)	Agency	Launch Year
1	IKONOS	Optical	PAN 1 m, XS 4 m	Digital Globe, USA	1999
2	QuickBird-2	Optical	PAN 0.61 m, XS 2.62 m	Digital Globe, USA	2001
3	GeoEye-1	Optical	PAN 0.41 m, XS 1.65 m	Digital Globe, USA	2008
4	WorldView-1	Optical	PAN 0.46 m	Digital Globe, USA	2007
5	WorldView-2	Optical	PAN 0.46 m, XS 1.85 m	Digital Globe, USA	2009
6	WorldView-3/4	Optical	PAN 0.31 m, XS 1.24 m	Digital Globe, USA	2014/2016
7	COSMO-SkyMed 1/2/3/4/5/6	SAR	X-Band up to 1 m	Italian Space Agency	2007–2018
8	Pleiades-1/2	Optical	PAN 0.5 m, XS 2 m	French space agency and EADS Astrium	2011/2012
9	TerraSAR-X TanDEM-X	SAR	X-Band up to 1 m	German Aerospace Center and EADS Astrium	2007
10	Resurs-DK1	Optical	PAN 1 m, XS 2–3 m	Russian Space Agency	2006
11	Kompsat-2	Optical	PAN 1 m, XS 4 m	Korean Academy of Aeronautics and Astronautics	2006
12	Kompsat-3	Optical	PAN 0.7 m, XS 2.8 m	Korean Academy of Aeronautics and Astronautics	2012
13	CartoSat-2/2A/2B	Optical	PAN 1 m	Indian Space Research Organization	2007
14	EROS-B	Optical	PAN 0.7 m	Israeli Aircraft Industries Ltd. (built) and ImageSat International N.V. (own)	2006
15	GF-2	Optical	PAN 0.8, XS 3.2 m	State Administration of Science, Technology and Industry for National Defense, China	2014
16	Beijing 2	Optical	PAN 0.8 m, XS 3.2 m	Twenty First Century Aerospace Technology Co., Ltd, China	2015
17	SuperView-1	Optical	PAN 0.5 m, XS 2 m	China Aerospace Science and Technology Corporation	2016

The most important feature of Planet satellite constellation is that it can provide customers with rapidly updated commercial satellite images. This constellation can provide a wide range of Earth images with a resolution of 3 m, and its coverage and update rate are unparalleled. Its brightest label is that corrected orthophoto images can be obtained directly.

Apart from the aforementioned satellites, VHR images can also be obtained from web mapping applications. Traditional GIS tools are expensive, and they are not flexible enough in geographical visualization. Thus, it is difficult to automatically and seamlessly operate and integrate massive data from different sources (Conroy et al. 2008; Renner et al. 2009; Yu and Gong 2012; Wood et al. 2007). The rapid development of virtual Earth technology provides a free and convenient tool that can efficiently communicate and share land cover data with users all over the world (Mccoy 2017; Agapiou 2017). Common virtual tool platforms, such as Google Earth,[1] NASA's World Wind,[2] and ESRI's ArcGIS Explorer, are widely used by many researchers (Craglia et al. 2008; Goodchild 2008; Biradar et al. 2009; Tsendbazar et al. 2017; Esch et al. 2018; Fritz et al. 2012; See et al. 2015; Fritz et al. 2017; Schepaschenko et al. 2015; Bey et al. 2016).

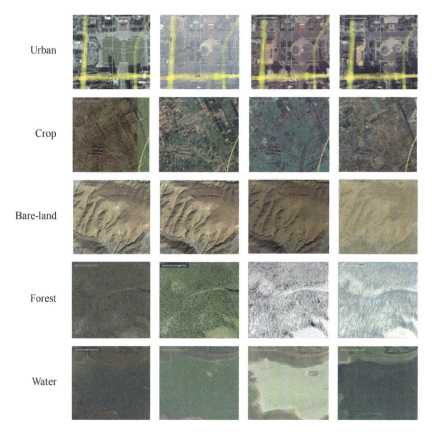

FIGURE 3.4 Images of land cover in different periods on Google Earth.

Through the online platform, a large number of VHR images can be freely obtained. These images are compressed true color (red, green, and blue) composite images, which are suitable for visual interpretation. In the field of remote sensing, a popular use of such imagery involves creating reference datasets through visual interpretation. This process often serves to augment existing training data or to aid in the validation of land cover products (Luo et al. 2014; Hansen et al. 2014; Kobayashi et al. 2014; Carroll et al. 2009; Pekel et al. 2016; Schneider et al. 2010; Melchiorri et al. 2018; Luo et al. 2018). Google Earth is extensively used to produce detailed thematic maps and to verify coarse and medium-resolution products, such as those produced based on MODIS and Landsat images, respectively. These products include global mapping of tree cover, forests, water bodies, and urban areas, thereby providing valuable references. Google Earth includes a time-slider feature for temporal manipulation of images. Figure 3.4 shows images of various periods found by web mapping applications. A regional assessment of the availability of VHR imagery in Google Earth and Bing Maps[3] is presented in Table 3.2.

Lesiv et al. (2018) analyzed the spatial distribution of the most recent VHR satellite imagery accessible through Microsoft Bing Maps and Google Earth. Figures 3.5 and 3.6 illustrate the distributions of VHR images in Bing Maps and Google Earth, respectively. Figure 3.7 shows the similarities and differences between their distributions.

Google Maps uses Web Mercator, which has the advantage of using a spherical model instead of an ellipsoid model for Earth, thus being computationally more efficient. However, errors in projecting latitude and longitude from the WGS84 ellipsoid to a sphere mean that distances and angles should not be estimated using Web Mercator maps, according to Battersby et al. (2014). The research team of Battersby et al. (2014) further elaborated on the effects of employing the Web Mercator

TABLE 3.2
The Availability of VHR Satellite Imagery (<5 m Resolution) From Google Earth and Microsoft Bing Maps as of January 2017 by Region (Lesiv et al. 2018)

Region	Google Earth Coverage with VHR Imagery (%)	Google Earth Most Recent Year, Calculated as the Median	Microsoft Bing Maps Coverage with VHR Imagery (%)	Microsoft Bing Maps Most Recent Year, Calculated as the Median
North America	47	2016	51	2011
Central America	95%	2016	94	2013
Most South America	72	2016	88	2010
Northern Europe	90%	2015	64	2012
Southern Europe	99	2016	96	2011
Western Europe	99	2016	100	2012
Eastern Europe	39	2016	58	2012
Northern Africa	73	2013	67	2013
Western Africa	70	2016	67	2013
Middle Africa	75	2013	65	2013
Eastern Africa	96	2016	90	2013
Southern Africa	10	2016	99	2013
Eastern Asia	89	2013	82	2012
Western Asia	79	2016	93	2013
Southern Asia	91	2016	92	2015
Southeastern Asia	89	2016	58	2013

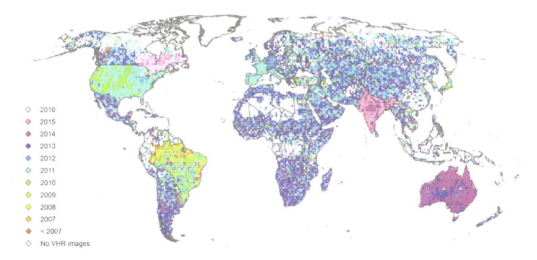

FIGURE 3.5 The dates of the most recent VHR satellite imagery (<5 m resolution) available in Microsoft Bing Maps as of January 2017 (Lesiv et al. 2018).

projection across various online map services. Consequently, while network maps offer a user-friendly interface for displaying geographic information, they do not constitute an accurate GIS system. Data collected using the standard WGS84 ellipsoid model are immediately transferred to the Web Mercator spherical model. Numerous researchers have documented that the positional accuracy of Google Earth imagery varies by location (Goudarzi and Landry 2017; Mohammed et al. 2013;

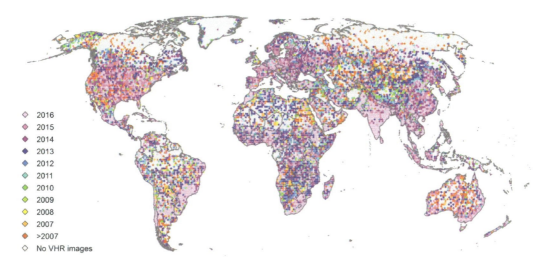

FIGURE 3.6 The dates of the most recent VHR satellite imagery (<5 m resolution) available in Google Earth as of January 2017 (Lesiv et al. 2018).

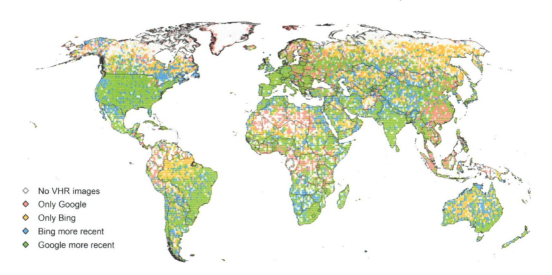

FIGURE 3.7 Comparison of the most recent VHR satellite imagery (<5 m resolution) available in Microsoft Bing Maps and Google Earth as of January 2017 (Lesiv et al. 2018).

Pulighe et al. 2016; Mulu and Derib 2019; Methakullachat and Witchayangkoon 2019). These investigations compare the positions of multiple ground control points (GCPs; of established WGS84 coordinates) against their corresponding positions as displayed on Google Maps. Notably, in these scenarios, greater errors do not correlate with higher latitudes, indicating that the Web Mercator projection may not be the primary issue. Instead, the discrepancies are more likely related to the resolution of the Google Maps imagery available and the user's precision in pinpointing a checkpoint on the maps. For instance, as of April 13, 2023, Google Maps provides satellite imagery of similar finer resolution for both Montreal and Rome but a coarser resolution for Khartoum (Dhonju et al. 2023).

Tianditu[4] is China's inaugural state-sponsored web mapping service. On October 21, 2010, the State Bureau of Surveying and Mapping (SBSM) released the beta version of Tianditu, marking

Reference Data 73

FIGURE 3.8 An example VHR image in Tianditu.

China's first official free web mapping platform. Following the beta testing phase, Tianditu was officially launched on January 18, 2011. In June 2011, a trial version for mobile phones was introduced, and by October 2011, the mobile phone version was officially released.

Google provides more precise geographic information than Tianditu. For example, Tianditu's accuracy for longitude and latitude is limited to only two decimal places, while Google's is accurate to six decimal places. However, for place names within mainland China, Tianditu offers more accurate and current information than Google. Tianditu is outstanding in providing detailed information about China, even in remote mountainous and rural regions. Regarding the user interface, Google and Tianditu are on par. Users familiar with digital maps should have no difficulty using either platform (Chen et al. 2013). The main drawback of Tianditu is its support in Chinese only, whereas Google offers multilingual support. Figure 3.8 displays the Tianditu user interface.

3.2.2 OTHER IMAGERY

Drones, commonly referred to as unmanned aerial vehicles (UAVs), are being increasingly employed to collect data for various applications and execute tasks that would otherwise be difficult or unfeasible for humans (Mohsan et al. 2022; Alsamhi et al. 2022; Iqbal et al. 2023). The advent of UAVs has revolutionized land cover/land use research by providing VHR imagery, ranging from decimeter to millimeter levels (Zhang and Zhu 2023). This allows for detailed study of small objects, such as people, vehicles, animals, and plants, which were previously difficult with images of coarser resolution. Integrating drones with GIS is advantageous because it decreases costs and improves access to geospatial data collection. Traditional aerial photography with aircraft is expensive due to the costs associated with renting planes, employing pilots, and hiring photographers. Drones equipped with sophisticated cameras and AI software systems can replace these older methods, thus making the data collection process more cost-effective and quicker. Currently, some researchers are exploring methods to generate precise geo-referenced datasets from UAV images by collecting GCPs and supporting comparisons with geo-referenced satellite and aerial images. Ground-truth data are collected through the field work of UAV mapping and used for

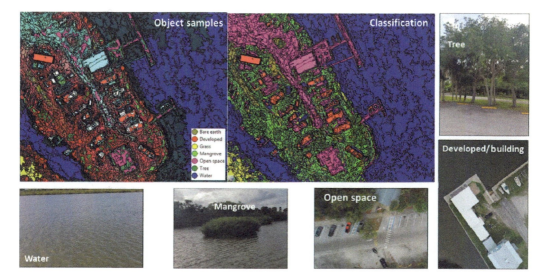

FIGURE 3.9 Selection of training samples for validations (Yang et al. 2019).

training and verification of remote sensing data (Figure 3.9). Figure 3.10 illustrates the comparison of classification results from UAV, satellite, and aerial data. The UAV data produced classifications, which are of higher accuracy and more details than satellite data. In satellite remote sensing classifications, only larger vegetated areas can be extracted from the image.

Although UAVs offer high flexibility in data acquisition, they need ground operators, management of massive data, and data preprocessing. In contrast, satellite data are very convenient to obtain from web-based platforms and directly used for analysis. Moreover, UAVs provide data at resolutions higher than that of satellites, although they do not match satellite data's extensive coverage and are often limited by national and/or international regulations to fly over specific territories.

3.3 FINER-RESOLUTION LAND COVER DATA

Finer-resolution land cover data refers to detailed maps or datasets that provide information about different types of land cover at a spatial resolution finer than a certain threshold. These data typically have a spatial resolution of finer than 30 m, allowing for the identification and mapping of individual features on Earth's surface with high accuracy.

Building upon the groundbreaking tradition of the freely accessible Landsat missions, the European Space Agency (ESA) and Copernicus Programme have provided globally consistent optical and radar data through the Sentinel satellites (with resolutions ranging from 10 to 20 m) since 2014 (Venter et al. 2022). Coupled with advancements in machine learning algorithms and cloud computing platforms tailored for Earth observation, such as Google Earth Engine (Gorelick et al. 2017) and openEO (Schramm et al. 2021), the Sentinel satellites have facilitated large-scale mapping of land use and land cover (LULC) at a 10 m resolution. From 2021 onwards, three sets of global Sentinel-based 10 m resolution LULC maps have been published, including Google's Dynamic World (Brown et al. 2022), ESA's World Cover 2020, and Esri's 2020 Land Cover. With World Cover 2020 and Esri's 2020 Land Cover being annually updated, Dynamic World provides near real-time LULC maps as new Sentinel-2 scenes are released, being updated every five days.

This section briefly reviews some of the widely used finer-resolution land cover data, with brief description about individual products, their classification systems, their data-sharing links, their data preview maps, etc.

Reference Data 75

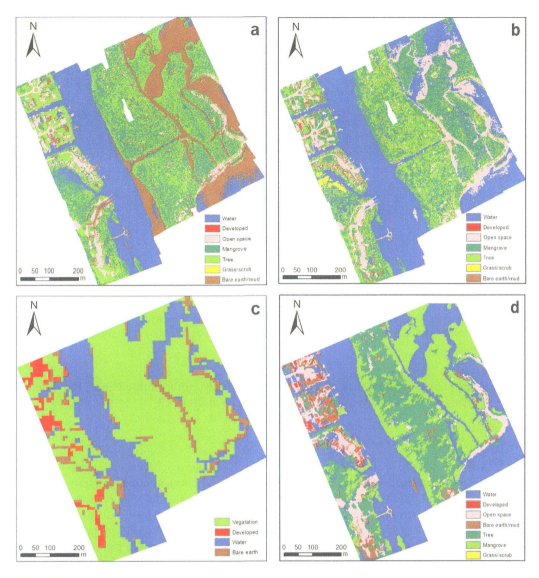

FIGURE 3.10 Comparison of classification results: (a) pixel-based classification using UAV data, (b) object-oriented classification using UAV data, (c) object-oriented classification using Sentinel-2 data, (d) object-oriented classification using aerial data (Yang et al. 2019).

3.3.1 GOOGLE'S DYNAMIC WORLD[5]

Dynamic World is a first-of-its-kind near real-time (NRT) LULC classification product developed using deep learning on 10 m Sentinel-2 imagery (Brown et al. 2022). It leverages a highly scalable cloud-based system to provide globally consistent and high-resolution LULC predictions in parallel with Sentinel-2 acquisitions. Dynamic World offers an open and continuous feed of LULC data, accommodating various user needs, such as up-to-date information and customized global composites for specific date ranges. For example, Figure 3.11 shows the mode composite of all Dynamic World NRT products from April 1, 2021, to May 1, 2021. The continuous nature of the product's outputs allows for refinement, extension, and redefinition of the LULC classification, providing

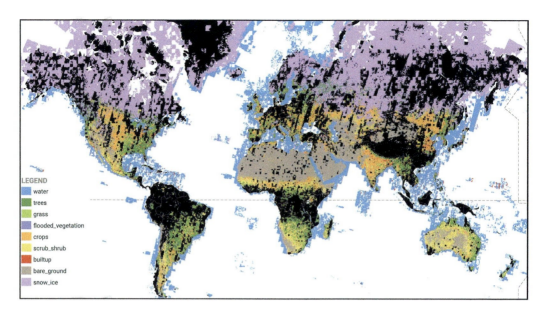

FIGURE 3.11 Mode composite of all dynamic world NRT products from April 1, 2021, to May 1, 2021, with black areas indicating no data over land (due to cloud cover) and white areas indicating no data over water (Brown et al. 2022).

unprecedented flexibility for a diverse user community across different disciplines. The different land cover categories of Dynamic World LULC are shown in Table 3.3.

3.3.2 ESA WorldCover 10 m[6]

The ESA WorldCover 10 m product is a comprehensive GLC map that is generated using data from Sentinel-1 and Sentinel-2 satellites. It provides detailed land cover information at 10 m resolution and includes 11 different land cover categories (see Table 3.4), which are consistent with the land cover classification system (LCCS) of the Food and Agriculture Organization of the United Nations. This product is developed within the framework of the ESA WorldCover project with the goal of offering accurate and up-to-date land cover information for the entire globe.

In response to the 2017 WorldCover conference, the ESA launched the WorldCover project. The primary objective of this project was to develop a comprehensive GLC product with a resolution of 10 m for the year 2020. This product (ESA WorldCover 10 m V100; see Figure 3.12) was created using data from both the Sentinel-1 and Sentinel-2 satellites. It encompasses 11 distinct land cover classes and has undergone independent validation, resulting in a global overall accuracy of 74.4%.

Based on the positive user feedback, ESA made the decision to expand the WorldCover project and tasked the WorldCover consortium with producing an updated version of the product (ESA WorldCover 10 m V200) (see Figure 3.13) for the year 2021, aiming for even higher quality. The new WorldCover map for 2021 was released on October 28, 2022, and demonstrated a global overall accuracy of 76.7%.

It is important to note that the WorldCover maps for 2020 and 2021 were generated using different algorithm versions, namely v100 and v200, respectively. As a result, the changes observed between the two maps encompass both actual changes in land cover and variations attributable to the differences in the algorithms employed. This distinction is crucial in understanding the differences depicted in the maps, as they reflect a combination of real-world land cover changes and algorithmic adjustments. More information is available at the URL: https://esa-worldcover.org/en.

Reference Data

TABLE 3.3
Dynamic World Land Use Land Cover Classification Taxonomy

Class ID	LULC Type	Description
0	Water	• Water present in the image • Contains little-to-no sparse vegetation, no rock outcrop, and no built-up features like docks • Does not include land that can or has previously been covered by water
1	Trees	• Any significant clustering of dense vegetation, typically with a closed or dense canopy • Taller and darker than surrounding vegetation (if surrounded by other vegetation)
2	Glass	• Open areas covered in homogenous grasses with little-to-no taller vegetation • Other homogenous areas of grass-like vegetation (blade-type leaves) that appear different from trees and shrubland • Wild cereals and grasses with no obvious human plotting (i.e., not a structured field)
3	Flooded vegetation	• Areas of any type of vegetation with obvious intermixing of water • Do not assume an area is flooded if flooding is observed in another image • Seasonally flooded areas that are a mix of grass/shrubs/trees/bare ground
4	Crops	• Human planted/plotted cereals, grasses, and crops
5	Shruband scrub	• Mix of small clusters of plants or individual plants dispersed on a landscape that shows exposed soil and rock • Scrub-filled clearings within dense forests that are clearly not taller than trees. Appear grayer/browner due to less dense leaf cover
6	Built-up area	• Clusters of human-made structures or individual very large human-made structures • Contained industrial, commercial, and private buildings, and the associated parking lots • A mixture of residential buildings, streets, lawns, trees, isolated residential structures, or buildings surrounded by vegetative land covers • Major road and rail networks outside of the predominantly residential areas • Large homogeneous impervious surfaces, including parking structures, large office buildings, and residential housing developments containing clusters of cul-de-sacs
7	Bare ground	• Areas of rock or soil containing very sparse to no vegetation • Large areas of sand and deserts with no to little vegetation • Large individual or dense networks of dirt roads
8	Snowand ice	• Large homogeneous areas of thick snow or ice, typically only in mountain areas or highest latitudes • Large homogenous areas of snowfall

Source: Adopted from Brown et al. 2022.

TABLE 3.4
Coding of the Map Layer in ESA WorldCover 10 m Product

Map code	Land cover class
10	Tree cover
20	Shrubland
30	Grassland
40	Cropland
50	Built-up
60	Bare/sparse vegetation
70	Snow and ice
80	Permanent water bodies
90	Herbaceous wetland
95	Mangroves
100	Moss and lichen

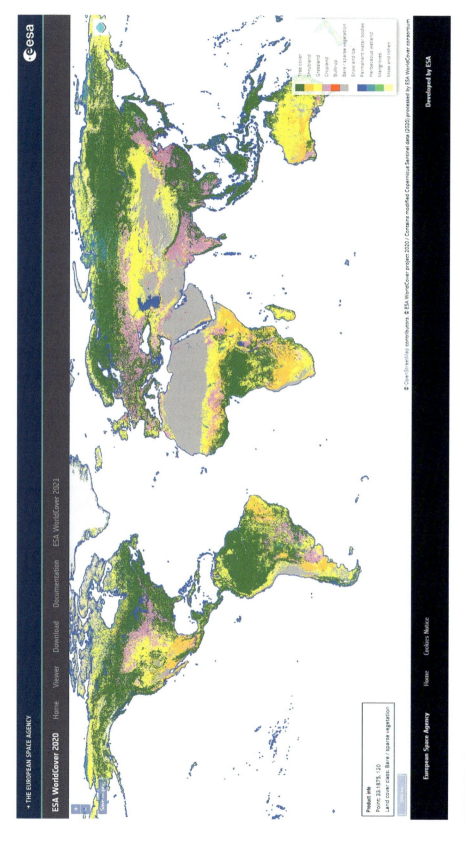

FIGURE 3.12 The WorldCover 2020 map v100.

Reference Data 79

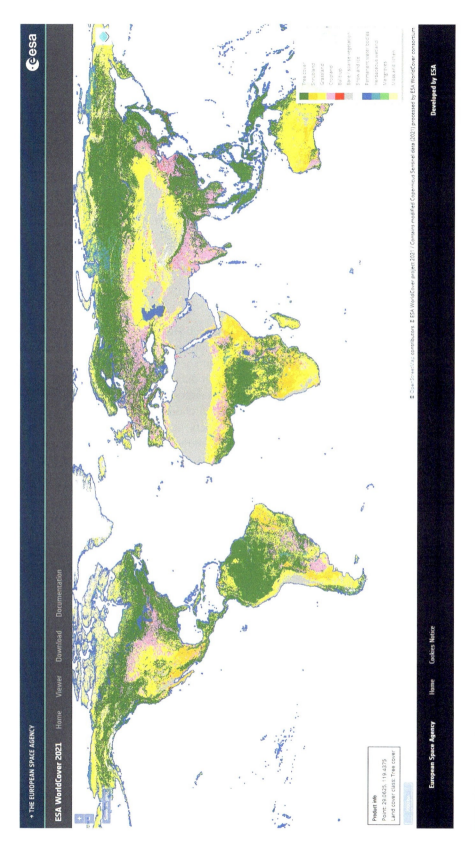

FIGURE 3.13 The WorldCover 2021 map v200.

TABLE 3.5
Types and Codes Included in ESRI 10 m

Category	Land Cover Class	Hex Code
1	No Data	#FFFFFF
2	Water	#1A5BAB
3	Trees	#358221
4	Grass	#A7D282
5	Flooded Vegetation	#87D19E
6	Crops	#FFDB5C
7	Scrub/Shrub	#EECFA8
8	Built Area	#ED022A
9	Bare Ground	#EDE9E4
10	Snow/Ice	#F2FAFF
11	Clouds	#C8C8C8

3.3.3 ESRI 10 M LAND USE LAND COVER[7]

ESRI 10 (2017–2023) is a land cover product developed by ESRI, a leading provider of GIS software and solutions. This product provides information about land cover occurrences and dynamics for the years 2017 to 2023. It offers a comprehensive and consistent view of land cover changes over this period. The ESRI 10 (2017–2023) land cover product is derived from a combination of satellite imagery, aerial photography, and other geospatial data sources. It utilizes advanced image processing techniques and machine learning algorithms to classify and map different land cover types. The maps are derived from ESA Sentinel-2 imagery at 10 m resolution. Each map is a composite of LULC predictions for nine classes throughout the year in order to generate a representative snapshot of each year (see Table 3.5). The datasets, produced by Impact Observatory and licensed by ESRI, sourced from Microsoft Planetary Computer's data catalog and storage (Karra et al. 2021), have each been assessed to have an average accuracy of over 75%.

ESRI 10 m annual land use land cover data are available for preview and download at https://livingatlas.arcgis.com/landcoverexplorer/. A GLC map for 2020 is shown in Figure 3.14.

3.3.4 FROM-GLC10

FROM_GLC10 (2017) data was produced by a research group at China's Tsinghua University based on a 10 m resolution Sentinel-2 image, the first global land cover map with 10 m resolution (Gong et al. 2019). Figure 3.15 displays an overview of this map. Based on the classifier trained by more than 300,000 training sample subsets collected from different seasons at about 93,000 sample pixels around the world, the method of stable classification of finite land cover samples was proposed by FROM_GLC10 developers through in-depth analysis and simulation of sample size and errors. The overall accuracy of FROM-GLC10 map was 72.7%. The dataset is divided into ten categories (see Table 3.6). FROM_GLC10 (2017) data are available for public use at http://data.ess.tsinghua.edu.cn/fromglc10_2017v01.html

3.3.5 ESA-S2-LC20

ESA-S2-LC20 is the Sentinel-2 land cover prototype map of Africa, created for 2016 at a resolution of 20 m by the ESA. Its primary purpose was to gather user feedback for further improvements. It utilizes a geographic coordinate system based on the WGS84 reference ellipsoid. Cloud-free surface

Reference Data

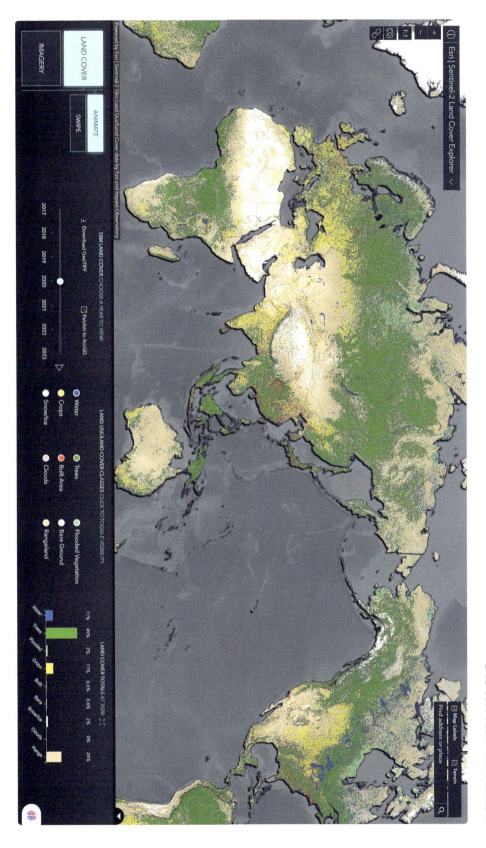

FIGURE 3.14 ESRI 2020 GLC map.

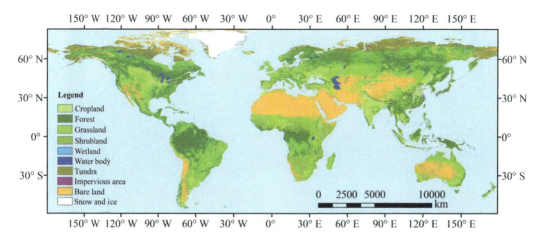

FIGURE 3.15 FROM-GLC10 (2017) land cover map (10 m resolution).

TABLE 3.6
Land Cover Types and Codes in FROM-GLC10

Class Value	Remapped Value	Land Cover Class	Hex Code
1	1	Water	#1A5BAB
2	2	Trees	#358221
4	3	Flooded Vegetation	#87D19E
5	4	Crops	#FFDB5C
7	5	Built-Up Area	#ED022A
8	6	Bare Ground	#EDE9E4
9	7	Snow/Ice	#F2FAFF
10	8	Clouds	#C8C8C8
11	9	Rangeland	#C6AD8D

reflectance composites over Africa were derived from over 30,000 S2A L1C images. The final product, featuring ten land cover classes and one class for cropland, was generated by integrating two machine learning classification algorithms (Defourny et al. 2017). The legend includes ten generic land cover classes that appropriately describe the land surface at 20 m: trees cover areas, shrubs cover areas, grassland, cropland, vegetation aquatic or regularly flooded, lichen and mosses/sparse vegetation, bare areas, built-up areas, snow and/or ice, and open water (see Figure 3.16). The high-resolution land cover map at 20 m over Africa can be downloaded from https://2016africalandcover20m.esrin.esa.int/.

3.3.6 SinoLC-1

SinoLC-1, the first 1 m resolution national scale land cover map of China, was developed using deep learning-based algorithms and open-access datasets, such as GLC products, OSM, and Google Earth imagery. The overall accuracy of SinoLC-1 was determined to be 73.6 %, with a kappa coefficient of 0.660 (Li et al., 2023). SinoLC-1 includes 11 land cover types, including forests, cultivated land, shrubs, grasslands, roads, buildings, water bodies, wetlands, ice and snow, tundra, barren and sparse vegetation (see Figure 3.17). In March 2023, the thematic map was first published on Zenodo, an internationally renowned data-sharing website, with the link https://doi.org/10.5281/zenodo.7707461.

Reference Data

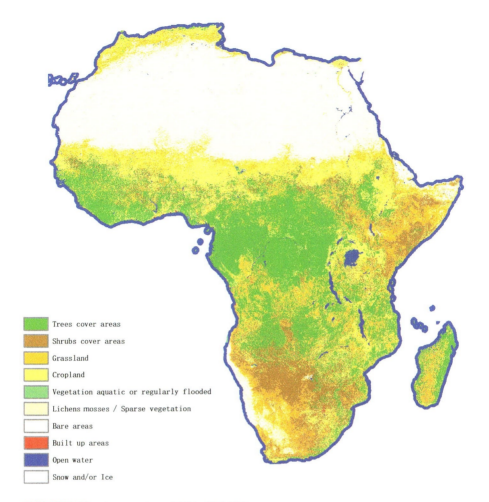

FIGURE 3.16 An overview of ESA-S2-LC20 map.

3.4 LARGE-SCALE MAPS

In addition to finer-resolution land cover data, large-scale maps can also provide valuable information for validation of land cover classifications. In this chapter, large-scale maps refer to maps at relatively large scales (i.e., showing spatial objects in greater detail than smaller-scale maps), such as large-scale topographic maps, large-scale thematic maps, and other detailed representations of Earth's surface features, etc. These maps usually provide detailed representations of land cover types, land use patterns, and terrain characteristics. The following are some of the benefits of using large-scale maps for accuracy assessment and validation of land cover classifications:

- They provide detailed information about land cover types at a local level. This detailed information can be used to verify the accuracy of land cover classifications derived from other sources, such as remote sensing data.
- They can serve as ground-truth data for land cover classifications. Field validation can be conducted using the information provided in large-scale maps to verify the accuracy of land cover classifications and improve the overall quality of map classifications.
- They offer a spatial context that can help in understanding the relationships between different land cover types and their spatial distributions. This spatial context can aid in the interpretation and validation of land cover classifications.

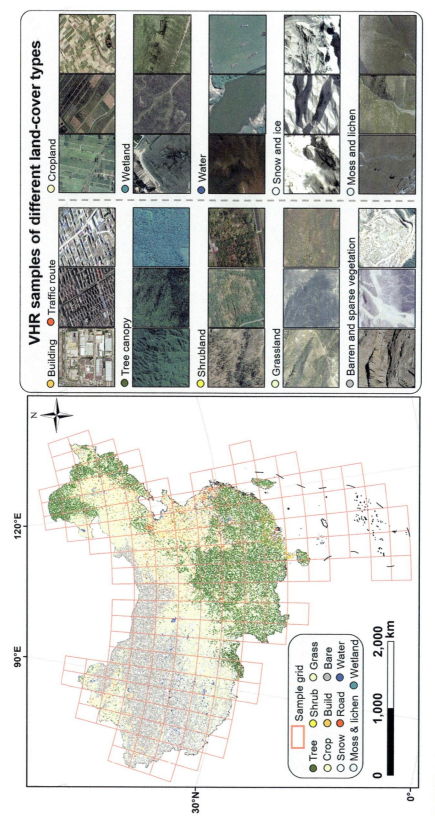

FIGURE 3.17 SinoLC-1 land cover types (Li et al. 2023).

Reference Data

- They can be used to detect changes in land cover over time. By comparing historical large-scale maps with current ones, changes in land cover can be identified and validated, providing insights into land cover dynamics and trends.
- They can be used to check the consistency of land cover classifications across different sources. By comparing land cover information from large-scale maps with classifications derived from other data sources, inconsistencies, and errors can be identified and corrected.

Despite these advantages, there are some issues for consideration when using large-scale maps as reference data for validation purposes. These issues are as follows:

- Different maps may use different classification schemes or terminology for land cover types. It is important to ensure consistency between the land cover categories in the map legends and the land cover products being validated.
- It is crucial to check for temporal consistency between the maps; as land cover can change over time, the maps used for validation should be up to date to avoid discrepancies incurred by using outdated maps as references.
- As official and authoritative maps are generally reliable and accurate, it is necessary to verify the credibility and trustworthiness of the reference maps to ensure the validity of the information used for validation.
- It should be recognized that map interpretations are often subjective, as different individuals may interpret and classify the same land cover types differently based on their understanding and expertise; awareness of potential subjectivity helps to ensure consistency in the interpretation and validation processes.
- It is critical to integrate multiple map sources or reference data for validation. Combining different sources contributes to a more comprehensive and reliable assessment of land cover accuracy.

3.4.1 LARGE-SCALE TOPOGRAPHIC MAPS

Topographic maps provide detailed and accurate representations of natural and human-made features on Earth's surface. These maps use contour lines to show the elevation and shape of the terrain, allowing individuals to visualize the three-dimensional landscape in a two-dimensional format. Topographic maps also display various features, such as rivers, lakes, forests, roads, buildings, and other landmarks. They provide valuable information for navigation, land surveying, urban planning, environmental analysis, and outdoor recreational activities. The detailed information provided by topographic maps aids in the interpretation of land cover classes within their geographical settings.

Topographic maps play a pivotal role in the validation of land cover by furnishing indispensable details about elevation, slope, aspect, and terrain characteristics, being informative covariates for classifying land cover types. National mapping agencies worldwide, such as the US Geological Survey (USGS), Ordnance Survey in the United Kingdom, and China's National Administration of Surveying, Mapping and Geoinformation (NASG), provide detailed topographic maps at various scales. These maps depict not only entities in specific areas but also give an overview across broader regions. In addition to physical entities, such as mountains, valleys, rivers, and natural landscapes, topographic maps also portray cultural entities. These cultural attributes are essential in comprehending the relationship between land cover classes and the surrounding landscape. By integrating data from topographic maps with other large-scale maps, researchers and land cover analysts can gain deeper insights into the landscape context of different land cover categories. This integration provides a comprehensive and detailed view of surface features, aiding in understanding the complex relationships between different land cover types.

Detailed information provided by topographic maps may assist in interpreting land cover classes within their geographical settings. For instance, understanding the undulating terrain helps to comprehend land cover types such as forests, wetlands, and agricultural fields in different regions. The details from topographic maps support in-depth analysis of patterns and trends among land cover types. By observing terrain in relation to land cover, inferences can be made about possible reasons for land cover changes, such as vegetation in valleys being more closely associated with water sources. Topographic maps enable a better understanding of the interactions between land cover categories and terrain patterns. This is crucial for revealing mutual influences and dependencies between different land cover types.

3.4.2 LARGE-SCALE THEMATIC MAPS

In addition to topographic maps, there are also large-scale thematic maps that are valuable for the validation of land cover map classifications:

1) Soil maps

A soil map shows the distribution of different soil types and their properties within a specific geographical area (e.g., Figure 3.18; source: https://www.eea.europa.eu/legal/copyright). Soil maps are created through soil surveys, which involve collecting soil sample data at various locations; analyzing soil properties, such as texture, fertility, pH, and drainage; and then mapping out the spatial distributions of different soil types based on these characteristics. Soil maps provide information on soil types and their properties that are closely related to land cover patterns. By comparing soil types indicated on soil maps with land cover classifications, it is possible to verify the consistency between the land cover types and the underlying soil characteristics.

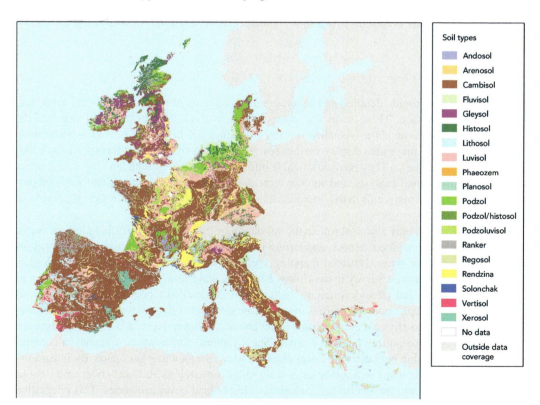

FIGURE 3.18 A map presentation of Corine European soil database version.

Land cover refers to the distributions of natural and man-made covers in different areas on Earth's surface. Soil types directly affect the formation of land cover because different soil types have different adaptability and growth conditions for vegetation. The consistency of land cover data can be verified by comparing soil types on soil maps to land cover classifications. If the soil type in an area does not match the land cover classification therein, it may indicate inaccurate or inconsistent land cover classifications.

2) Water maps

A water map depicts the distribution and characteristics of water bodies within a specific geographic region. Similar to soil maps, water maps are crafted through a comprehensive process that involves surveying, data collection, and analysis of various water bodies, such as rivers, lakes, reservoirs, wetlands, and ponds. Water maps serve multiple purposes, including environmental management, water resource planning, flood risk assessment, and biodiversity conservation. They provide critical information for decision-making related to water management, land use planning, and infrastructure development. Water maps can be used for land cover product validation by comparing the spatial distributions and characteristics of water bodies depicted on the water map with the corresponding land cover classifications.

The China Water Cover Map (CWaC, 2020/10 m; Li and Niu 2022) was created by the wetland remote sensing research team of the Institute of Aerospace Information, Chinese Academy of Sciences. This map is based on Sentinel-1 and Sentinel-2 time-series images that cover the entirety of China in 2020. It was generated by utilizing a remote sensing big data cloud platform with shape-based and inundation frequency-based automatic classification methods. The overall accuracy of CWaC is reported to be 85.6%, with a kappa coefficient of 0.83 (Li and Niu 2022). The classification accuracy for rivers, reservoirs, lakes, agricultural ponds, and paddy fields exceeds 80%, while the classification accuracy for seasonal wetlands is greater than 70% (Li and Niu 2022). CWaC marks the first-of-its-kind map of water bodies with a spatial resolution of 10 m over China, providing crucial information for water resources management and facilitating research on the relationships between changes in water bodies and human activities. Figure 3.19 shows the map of CWaC.

The development and evolution of water cover types is a process that is influenced by both human activities and natural processes. Accurate and timely access to information about water bodies and aquatic land cover is essential to understanding the relationship between changes in the natural environment and sustainable development. Li and Niu (2022) have proposed a two-level classification system of water bodies and aquatic cover types, taking into account water sources, water use, and water body shapes, defining them in conjunction with currently available remote sensing data (see Table 3.7). The first level mainly distinguishes between artificial and natural aquatic cover types, and the second level mainly distinguishes between permanent and seasonal water bodies according to the frequency of inundation. According to this classification system, there are a total of six categories, including lakes, reservoirs, rivers, seasonal wetlands, agricultural ponds, and rice fields.

3) Vegetation maps

Vegetation maps show the distributions and types of vegetation in a particular area. These maps provide valuable information about the plant cover, vegetation types, and ecosystems present in a specific region, and they can be used to validate land cover classifications. For example, vegetation types and their spatial distributions in the Tibetan Plateau with 10 m spatial resolution in 2020 are shown in Figure 3.20. Land cover data often include vegetation categories. To accurately reflect actual vegetation cover, it is necessary to ensure that the vegetation types match the vegetation map. Comparing the two kinds of maps can help confirm the accuracy and consistency of vegetation classes in land cover maps. Validating land cover products with vegetation maps can identify

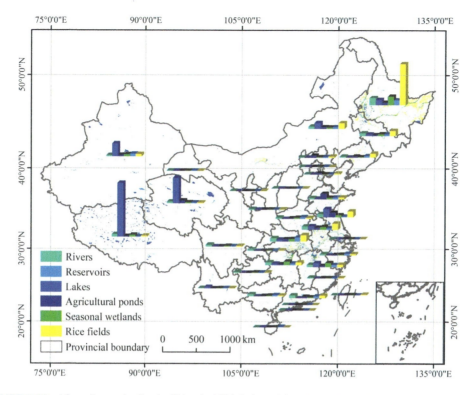

FIGURE 3.19 Map of water bodies in China in 2020 (adopted from Li and Niu. 2022).

TABLE 3.7
Water Bodies and Aquatic Cover Type Classification System

Code	Name
10	Cropland
20	Forest
30	Grassland
40	Shrubland
50	Wetland
60	Water
70	Tundra
80	Impervious surface
90	Breland
100	Snow/ice

Source: Modified from Li and Niu (2022).

potential classification errors or inconsistencies. It is critical in ecological research and environmental monitoring to ensure data reliability and availability.

4) Land use maps

A land use map represents different land use types and human activities within a specific area. It typically depicts different land use categories, such as agricultural land, urban areas, forests, and

Reference Data

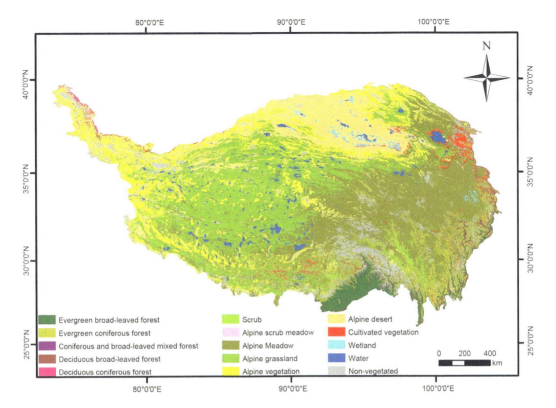

FIGURE 3.20 Vegetation types and their spatial distributions at 10 m spatial resolution on the Qinghai-Tibet Plateau in 2020.

Source: **https://phys.org/news/2023-02-vegetation-qinghai-tibet-plateau-based-terrain-climate-remote. html#google_vignette**

wetlands, through the use of symbols, colors, and markers. Land use maps reveal the impacts of human activities on Earth's surface, such as those reflected by agricultural lands, urban construction, and industrial areas. By comparing land use maps with land cover classifications, the contribution of human activities to land cover can be better understood, thereby improving the interpretability of land cover data and helping to assess the relationships between land cover and actual land use practices in the region. Consistency checking provides insight into the accuracy of classification results. For example, a land use map (shown in Figure 3.21) is used to display the land use of NUAs (non-urbanized areas; La Rosa and Privitera 2013).

3.5 CROWDSOURCING/VGI DATA

Crowdsourcing, also known as VGI, refers to the practice of obtaining data or information from a large group of individuals, typically through online platforms or mobile applications. The emergence of a large number of crowdsourced geographic data, such as OSM, Geo-Wiki, and Mapillary (see Table 3.8), provides a new source of inexpensive, current, and plentiful reference data.

Issues of data quality could significantly restrict and even undermine the use of VGI for validating land cover maps (Fonte et al. 2015). When using crowdsourced data to validate land cover products, there are several matters for consideration: data quality and accuracy, data bias and

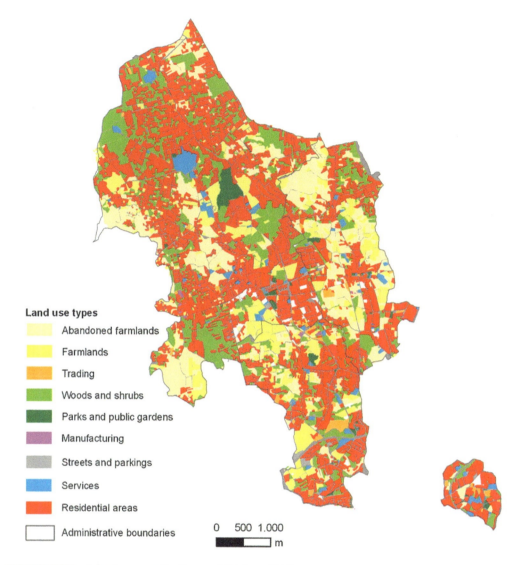

FIGURE 3.21 A land use map (La Rosa and Privitera 2013).

TABLE 3.8
Crowdsourcing/VGI Data Platforms

Crowdsourcing/ VGI Data Platforms	Dates Founded	Main Functions	Data Sources
OSM	2004	Create a free and editable map of the world	Volunteers' data (GPS)
Geo-Wiki and LACO-Wiki	2009, 2016	To get land cover data	Satellites' images
Mapillary	2013	To get street-view data	Volunteers' photos and videos
DCP	1996	To get images at the intersection of longitude and latitude lines	Volunteers' photos
VIEW-IT	2000–2010	Annotate satellite imagery with land cover features	Satellite Imagery
Public webcams	/	To provide live or periodically updated video feeds over the internet	Videos

Reference Data 91

subjectivity, data consistency and standardization, data coverages and spatial distributions, and data updates and timeliness. They are briefly elaborated on next:

- Crowdsourced data quality and accuracy can vary depending on the capabilities and expertise of the contributors. When validating land cover products, it is important to carefully screen and verify the crowdsourced data to ensure its quality and accuracy.
- Crowdsourced data can be influenced by the subjective judgments and personal preferences of the contributors, as different contributors may have different interpretations and labeling approaches for land cover types. During validation, it is important to consider these factors and perform appropriate data processing and standardization.
- Crowdsourced data may exhibit inconsistencies and lack standardization due to differences between multiple individuals' contributions. It is important to be aware of these differences and perform necessary data processing and standardization to ensure consistency in map comparisons and analyses.
- Crowdsourced data may have uneven coverages and spatial distributions, as some areas may have more data contributions while others may have less. It is important to understand the coverages and spatial distributions of the crowdsourced data and consider the impact of these factors on the validation results.
- Crowdsourced data updates and timeliness can be influenced by the activities and frequency of data uploads by contributors. It is important to ensure that the crowdsourced data used is up to date; there may exist possible delays or gaps in data availability.

In summary, when using crowdsourced data to validate land cover products, it is important to consider data quality and accuracy, to account for data bias and subjectivity, to perform data consistency and standardization, to understand data coverage and spatial distribution, and to ensure the use of up-to-date data. Additionally, depending on specific application requirements, it may be necessary to incorporate other data sources and professional expertise for validation and comparison.

3.5.1 OSM[8]

OSM is a collaborative and open-source mapping platform that allows users to create, edit, and share geographic data. It was launched in 2004 and has since grown into a global community-driven project. OSM aims to provide free and editable map data that can be used for various purposes, including navigation, research, and humanitarian efforts. Unlike traditional mapping services, OSM relies on the contributions of its users to collect and update map data. Anyone can become a contributor by creating an account on the platform. Users can add or edit features such as roads, buildings, landmarks, and points of interest, as well as attributes, including names, classifications, and descriptions. Data in OSM are stored in a structured format called OpenStreetMap Data (OSM XML), which includes information about the geometry, attributes, and relationships of map features. This data can be accessed and downloaded by individuals, organizations, and developers for their own use.

OSM provides a range of tools and editors that make it easy for users to contribute maps to it. The primary web-based editor is called iD,[9] which allows users to add and edit features directly on the OSM website. There are also other advanced editors like JOSM[10] (Java OpenStreetMap Editor) that offer more functionality for experienced contributors. One of the key strengths of OSM is its global community of contributors. Thousands of volunteers around the world actively participate in mapping their local areas, resulting in a rich and detailed map dataset. The community also organizes mapping events, workshops, and collaborations to improve the quality and coverage of the maps on the platform.

The data in OSM are licensed under the Open Database License (ODbL), which allows users to freely access, use, and share the data as long as they attribute the sources and share any modifications they make. This open licensing encourages collaboration and innovation, making OSM a valuable resource for a wide range of applications.

3.5.2 Mapillary[11]

Mapillary, a kind of crowdsourcing platform, provides images of people and streets around the world (Neuhold et al. 2017). On this website, images of scenery alongside streets (with image tagging), taken by contributors and uploaded to the platform, allow people around the world to have real-time views of the places.

In 2013, Mapillary was founded by Mapillary AB in Malmö, Sweden. The creators wanted to use crowdsourcing to store the entire world (not just streets) in the form of photographs, resulting in a huge dataset with 25,000 high-resolution tagged images organized in 152 categories. There are a variety of ways to collect data, such as mobile phones, cameras, pads, etc. One of Mapillary's value-added features concerns how to automatically recognize more than 40 street-level features using computer vision. After the platform detects each type of feature, feature location is enabled by triangulation on the maps, resulting in a list of road-related objects for everything from road signs to manhole covers.

3.5.3 The Degrees of Confluence Project (DCP)

DCP exemplifies a free, open-access, web-based citizen science initiative (Qian et al. 2020). This project's platform provides geo-tagged images and geospatial details at intersections where integer degrees of latitude and longitude converge worldwide (Iwao et al. 2006). At each visited confluence point, photographs taken in four directions are shared online, along with a description of the observed scenery and the geospatial data. The volunteered data, comprising geo-coordinates, images, and textual descriptions, can function as reference data for land cover analysis. This enables users to acquire essential insights into study areas on a global scale or at specific locations to support mapping and validation efforts (Iwao et al. 2006; Kinoshita et al. 2014). More information about the project platform is available at https://confluence.org/index.php. Figure 3.22 shows several representative land cover and land use categories employed in the DCP classification framework.

3.5.4 Points of Interest (POIs)

POIs refer to specific locations on a map that hold particular significance or interest, such as stores, restaurants, landmarks, hospitals, and parks. These places are typically frequently visited by people

FIGURE 3.22 Common land cover and land use classes used in the DCP classification system.

Reference Data

in their daily lives and offer various services and experiences. POIs data comes from a variety of sources, including map providers, social platforms, location services providers, and user-generated content. Map providers like Google Maps and HERE Technologies[12] collect and curate POIs data, while social platforms like Foursquare and Yelp offer business information and user reviews. User-generated content platforms like OSM allow users to edit map data and add POIs information. Through these channels, users can access a wealth of POIs data for applications such as map display, location search, and navigation.

3.5.5 Virtual Interpretation of Earth Web-Interface Tool (VIEW-IT)

VIEW-IT is a collaborative, LULC reference data collection system that is accessed via a web browser (Clark and Aide 2011). This system has been designed to provide reference data at 250 m resolution (i.e., the nominal scale of pixels in the MOD13Q1 MODIS satellite products with the actual pixel size being 231.7 m). Users visit sample locations centered on MODIS pixels and visually estimate the percentage cover of LULC classes within a 250 m × 250 m grid cell using high-resolution images from Google Earth, which, in many areas of the developing world, consists of scenes from the QuickBird and IKONOS commercial satellites (Clark and Aide 2011). VIEW-IT was created for a Windows operating system, but its components primarily consist of open-source software that could be transferred to a different operating system (see Figure 3.23).

3.5.6 Public Webcams

As outlined by Johansson (2008), webcams seamlessly integrate geographical context with real-time internet dissemination of information, serving as an extension of both camera applications and online information sharing and engagement. Webcams have significantly transformed the practice of repeat photography, which involves comparing images taken at different times but from fixed locations. Traditional repeat photography necessitated revisiting sites and recreating view angles, constraining sampling frequency (Butler, 1994; Butler and DeChano, 2001). With webcams functioning

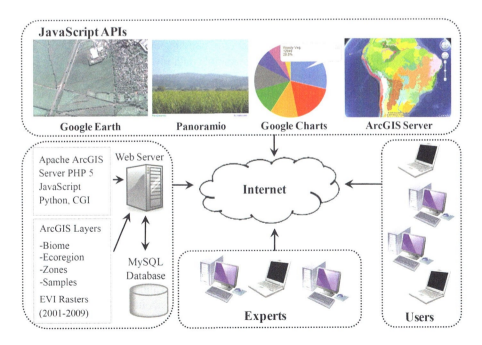

FIGURE 3.23 Technical and user components of the VIEW-IT (Adopted from Clark and Aide 2011).

FIGURE 3.24 Examples of excluded images due to poor illumination, sun over-exposure, and thick fog (Tom et al. 2020).

as internet-connected digital cameras, these images (including video feeds) can be automatically transmitted to databases, offering users real-time access without the delay of manual retrieval (Bradley and Clarke, 2011). Platforms such as the Flickr[13] map interface for user-uploaded and geo-tagged photos enable users to search by content tags (e.g., architecture, urban, forest) and location (e.g., neighborhood, city, state, zip) for selective display.

Recently, there has been a proposal to use ground-based cameras for verifying satellite-based snow information (Portenier et al. 2022; Piazzi et al. 2019; Aalstad et al. 2020), with recognition of their potential as a valuable dataset for accurately monitoring snow in mountainous areas (Liu et al. 2015; Hu et al. 2019; Portenier et al. 2020).

Webcams offer superior detection capabilities with much finer spatial resolution than satellites. However, the locations, orientations, coverage areas, and image qualities (as shown in Figure 3.24) of public webcams are beyond user control. Additionally, there are insufficient webcams for some specific land cover types. Nevertheless, on the upside, webcams are largely unaffected by cloud cover. And for localized studies and as a source of reference data at specific sites, webcams provide a viable alternative (Tom et al. 2020; Hüsler et al. 2012; Helmrich et al. 2021; Piazzi et al. 2019).

3.6 IN SITU DATA

Validating land cover products using ground-truth data is a crucial process in ensuring the accuracy and reliability of the generated information. Ground-truth data, obtained through on-site measurements and observations, are useful as reference data for assessing classification results in land cover products. These ground-truth data include ground sample data revealing actual soil and vegetation types, GPS positioning data for precise location referencing, vegetation photos offering visual validation, and other ground observations for assessing classification accuracy. By systematically comparing and analyzing land cover products with these ground-truth data, researchers and analysts can validate the quality of classifications, identify potential errors, and enhance the overall reliability of the practice of land cover mapping for various applications in environmental monitoring, resource management, and land use planning.

3.6.1 MEASUREMENT DATA

Although less standardized and more labor-intensive, measured data can still serve as valuable auxiliary information for validation, thanks to several advantages of them. First, measured data from peer-reviewed published datasets are representative of major land cover types drawn from groups of scientists or ecosystem observatories. These samples encompass a wide range of indicators for land cover types, making them highly valuable for cover type validation. Second, peer-reviewed measurement samples typically come with detailed information about each sample unit, including geographic location, land use/land cover/ecological environment type, and socioecological information, facilitating data integration with remote sensing imagery. Lastly, peer-reviewed measurement

Reference Data

data tend to be more accurate and reliable due to the wealth of information provided, including sample ID, collector details, geographic location and elevation, land cover type, data collection method, dataset source, citations, and accompanying photos or maps. Some measured sample data are as follows (see Tables 3.9–3.13).

TABLE 3.9
Measurement Data Information Template of Forest

Sample ID	1
Collector	DOKO Tomoko, CHEN Wenbo
Location (Lat./Log.)	18°30′34″ N, 72°24′44″E
Altitude(m)	205
Land Cover Type	Forest
Data Source	http://www.geodoi.ac.cn/WebEn/doi.aspx?Id=167
Description of measurement data	Field measurement
Quote	DOKO Tomoko, CHEN Wenbo. 2014. Presence/Absence point records of Japanese serow (Naemorhedus crispus) with environmental conditions in the Fuji-Tanzawa region, Japan (JpSerow_env_FT), Global Change Research Data Publishing & Repository, DOI:10.3974/geodb.2014.02.06.V1
Other information	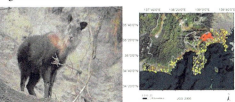

TABLE 3.10
Measurement Data Information Template of Grassland

Sample ID	2
Collector	Shi Kunbo, Wang Wenrui, Yang Yongchun, Shao Rui, Shi Yabo
Location (Lat./Log.)	35°6′10″ N; 102°25′48″ E
Altitude(m)	3070
Land Cover Type	Grassland
Data Source	http://www.geodoi.ac.cn/WebEn/doi.aspx?Id=584
Description of Measurement Data	Field measurement
Quote	Shi Kunbo, Wang Wenrui, Yang Yongchun, Shao Rui, Shi Yabo. 2016. Dataset of tourism activities' impact on the vegetation in the Sangke Prairie (2013) (Tourism ImpactGrassland). Global Change Research Data Publishing & Repository, DOI:10.3974/geodb.2016.07.22.V1
Other information	

TABLE 3.11
Measurement Data Information Template of Barren Land

Sample ID	3
Collector	LV Tingting, LIU Chuang, ZHAO Jinling, DAI Lijun, WANG Jinnian, GU Xingfa
Location (Lat./Log.)	18°30′34″ N; 72°24′44″ W
Altitude(m)	206
Land Cover Type	Barren land
Data Source	http://www.geodoi.ac.cn/WebEn/doi.aspx?Id=151
Description of Measurement Data	High-resolution remote sensing imagery (Google Earth)
Quote	LV Tingting, LIU Chuang, ZHAO Jinling, DAI Lijun, WANG Jinnian, GU Xingfa. 2014. Haiti Earthquake (2010) Landslides Dataset in Riviere_Frorse Basin (HaitiEQ_LS_2010data), Global Change Research Data Publishing & Repository, DOI:10.3974/geodb.2014.02.10. V1
Other information	

TABLE 3.12
Measurement Data Information Template of Water

Sample ID	4
Collector	Lawrence Nderu, LIU Chuang, Mabel Imbaga, Waweru Mwangi, Mika Odido, Willis Owino, Fred Wamunyokoli, ZHOU Xiang, LV Tingting, GU Xingfa, Wenbo Chen, Vladimir TIkunov, ZHU Yunqiang, SHI Ruixiang
Location (Lat./Log.)	0 °21′30″ S; 36°5′30″W
Altitude(m)	1757
Land Cover Type	Water
Data Source	http://www.geodoi.ac.cn/WebEn/doi.aspx?Id=251
Description of Measurement Data	Field measurement
Quote	Lawrence Nderu, LIU Chuang, Mabel Imbaga, Waweru Mwangi, Mika Odido, Willis Owino, Fred Wamunyokoli, ZHOU Xiang, LV Tingting, GU Xingfa, Wenbo Chen, Vladimir TIkunov, ZHU Yunqiang, SHI Ruixiang. 2015. Time-Series Data of the Lake Nakuru (1976–2015) (LakeNakuru1976_2015). Global Change Research Data Publishing & Repository, DOI:10.3974/geodb.2015.02.13.V1
Other information	

Reference Data

TABLE 3.13
Measurement Data Information Template of Wetland

Sample ID	5
Collector	ZHANG Haiying, NIU Zhenguo, GONG Ning, XING Liwei, XU Panpan, CHEN Yanfen, HU Shengjie
Location (Lat./Log.)	61°18′1″N; 67°54′0″E
Altitude(m)	28
Land Cover Type	Wetland
Data Source	http://www.geodoi.ac.cn/WebEn/doi.aspx?Id=243
Description of Measurement Data	High-resolution remote sensing imagery (Google Earth)
Quote	ZHANG Haiying, NIU Zhenguo, GONG Ning, XING Liwei, XU Panpan, CHEN Yanfen, HU Shengjie. 2015.The Wetland Cover Datasets on the Large Wetlands of International Importance in 2001 and 2013 by Remote Sensing Data Integration (RamsarSites_Top100_WetlandCover_2001/2013). Global Change Research Data Publishing & Repository, DOI:10.3974/geodb.2015.02.09. V1
Other information	

3.6.2 THE LAND USE/COVER AREA FRAME SURVEY (LUCAS)

LUCAS is primarily an in situ LULC database for field surveys intended to offer consistent statistics on LULC throughout the European Union. The LUCAS project was implemented following Decision 1445/2000/EC of the European Parliament and of the Council of May 22, 2000, "on the application of area-frame survey and remote sensing techniques to the agricultural statistics for 1999 to 2003" and has continued since then (D'Andrimont et al., 2020).

LUCAS was initially developed to generate early crop estimates for the European Commission, starting with a pilot study involving a limited number of European Union (EU) member states in 2001. Over time, the survey has evolved into a crucial resource for policymakers and statisticians, providing growing volumes of data on various types of LULC in the EU. In 2006, the sampling methodology was modified, and its emphasis expanded from solely an agricultural land survey to encompass a wider scope of land cover, land use, and landscape analysis. Simultaneously, the survey frequency was adjusted to occur every three years.

During the five LUCAS surveys (2006, 2009, 2012, 2015, 2018), a total of 1,351,293 observations from 651,780 unique locations were gathered, encompassing 106 variables. This dataset also includes approximately 5.4 million landscape photos depicting the observer's view in the four cardinal directions, along with point photos illustrating the surveyed point itself (D'Andrimont et al., 2022). The most recent survey, LUCAS 2022, encompasses all EU member states and involves the observation of 400,000 selected points. In addition to capturing data on land cover and land use at each of these points, additional information is collected to evaluate environmental aspects such as grassland and soil quality.

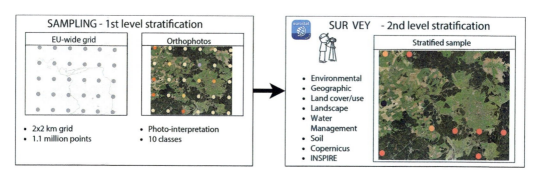

FIGURE 3.25 An illustration of a two-phase sample survey (adopted from D'Andrimont et al. 2022).

LUCAS relies on statistical computations to interpret field observations, employing a standardized survey methodology encompassing a sampling plan, classifications, data collection processes, and statistical estimators aimed at producing consistent and unbiased estimates of LULC. LUCAS operates as a two-phase sample survey (D'Andrimont et al., 2022). The initial phase involves the systematic selection of points on a grid with 2 km spacing in eastings and northings. Each point is then categorized into one of ten land cover classes through visual interpretation of ortho-photos or satellite images. Subsequently, stratified sampling is chosen to achieve the desired statistically representative spatial distribution of sample land cover classes. Figure 3.25 shows an illustration of the two-phase sample.

3.7 SUMMARY

In the process of land cover product validation, reference data play a crucial role. Reference data serve as the benchmark for validating the accuracy and reliability of models or products. This chapter provides a comprehensive overview of various types of reference data sources essential for validating land cover products. It covers VHR imagery, finer-resolution land cover data, large-scale maps, crowdsourcing/VGI data, and in situ data. By considering the authenticity and accuracy, temporal and spatial consistency, temporality, and other characteristics of these reference data sources, this section will further summarize the critical role they play in ensuring the accuracy and reliability of land cover products.

VHR imagery is particularly noted for its exceptional level of detail and spatial resolution. VHR satellite imagery can capture intricate features like individual structures, vehicles, and small vegetation areas with remarkable clarity. Operating at submeter resolution, VHR image interpretation has substantially replaced or at least greatly lessened manual identification of land cover in the field, making it a valuable source for verifying land cover products due to its ability to provide highly detailed and retraceable data for validation purposes.

Finer-resolution land cover data offers a more granular view of Earth's surface, allowing for a more accurate depiction of land cover characteristics. The advantages of using finer-resolution data for land cover product validation are numerous. First, they enable a more comprehensive assessment of land cover changes and trends, capturing smaller-scale variations that may be missed in coarser-resolution data. Second, they enhance the accuracy of land cover classifications and reduce the potential for misinterpretation of land cover types. Additionally, they facilitate more precise spatial analysis and identification of specific land cover features or classes. However, when utilizing finer-resolution land cover data for validation, there are several points for consideration. Third, ensuring data consistency is crucial, meaning that the finer-resolution data used for validation should align spatially and categorically with the product being evaluated. This helps prevent discrepancies and errors during the validation process. Fourth, spatial alignment is vital to ensuring that the spatial reference frameworks of both datasets match each other, enabling accurate comparison and evaluation. Fifth, evaluating the quality and accuracy of the finer-resolution data is essential, as low-quality or inaccurate data can undermine the reliability of the validation results. Sixth, it is crucial to check for classification consistency to ensure that the land cover classification systems used in the validation process are consistent and compatible.

Apart from finer-resolution land cover data, large-scale maps play a significant role in assessing accuracy and validating land cover classifications. Large-scale maps, which have a resolution finer than 10 m, encompass a variety of maps, such as land cover products, thematic maps, and detailed depictions of Earth's surface attributes. These maps are portrayals of land cover types, land use patterns, and terrain characteristics. Through the integration of diverse large-scale map sources, such as topographic maps, soil maps, vegetation maps, land use maps, and hydrological maps, researchers and land managers can conduct thorough validation of land cover classifications. This process enhances the accuracy and accountability of land cover data for a wide range of applications in environmental management, land use planning, and conservation of natural resources.

Crowdsourced data collected from the general public through online platforms, social media, map editing tools, mobile apps, and other channels cover various topics, such as geographic locations, environmental features, and landscape changes. When it comes to validating land cover products, crowdsourced data can provide reference information, especially in obtaining real-time and wide-area coverage data. However, ensuring the reliability and quality of these data is crucial, as inconsistencies or errors may be present. Attention to spatial and temporal resolution, along with effective processing and integration, is necessary to align crowdsourced data with other validation sources. Additionally, safeguarding individuals' privacy and data security is a key concern when utilizing crowdsourced data, ensuring compliance with relevant laws and ethical standards.

In situ data, also known as field-collected data, refer to information gathered through direct measurements or observations at the location of interest. This type of data provides firsthand ground-truth information about environmental conditions, land cover characteristics, and other relevant factors. In the context of validating land cover products, in situ data serves as a crucial reference for assessing the accuracy and reliability of remotely sensed data or modeled results. When utilizing in situ data as reference data for land cover product validation, several matters are worth considering. First, the representativeness of the in situ data is essential. Care must be taken to ensure that the collected field measurements or observations are indicative of the areas being studied and are not biased toward local conditions. Additionally, the spatial distribution of in situ data points should reflect the variability of the landscape under investigation. Furthermore, the quality and consistency of in situ data are paramount. Proper calibration and validation procedures should be followed to minimize errors and biases in the collected measurements. Standardized protocols for data collection, including metadata documentation, are essential for ensuring the reliability and reproducibility of in situ data. Another important point for consideration is the temporal aspect of in situ data. Temporal consistency and currency of the collected field data are crucial, especially with respect to historical land cover products. Changes in land cover over time can impact the relevance of in situ data for validation purposes.

In the future, advancements in reference data sources for validating land cover products are likely to focus on integrating multiple data types and leveraging cutting-edge technologies. This may involve the development of advanced fusion techniques that combine VHR imagery, finer-resolution land cover data, large-scale maps, crowdsourcing/VGI data, and in situ data to create comprehensive and accurate reference datasets. Additionally, the emergence of new remote sensing platforms and sensors, such as hyperspectral and LiDAR technologies, could enhance the richness and diversity of reference data, enabling more detailed and precise validation of land cover products. Furthermore, the integration of artificial intelligence and machine learning algorithms holds promise for computerizing the validation process, thus increasing efficiency and expanding the scope of validation for large-area applications over longer time periods.

NOTES

1 https://www.google.cn/
2 https://worldwind.arc.nasa.gov/
3 https://www.microsoft.com/en-us/maps/bing-maps
4 https://www.tianditu.com/
5 https://dynamicworld.app/
6 https://viewer.esa-worldcover.org/worldcover/
7 https://livingatlas.arcgis.com/landcover/
8 https://www.openstreetmap.org/
9 https://ideditor.com/
10 https://josm.openstreetmap.de/
11 https://www.mapillary.com/
12 https://www.here.com/
13 https://www.flickr.com/map/

REFERENCES

Aalstad, K., Westermann, S. and Bertino, L. 2020. Evaluating satellite retrieved fractional snow-covered area at a high-Arctic site using terrestrial photography. *Remote Sensing Environment*, 239: 111618.

Agapiou, A. 2017. Remote sensing heritage in a petabyte-scale: Satellite data and heritage Earth Engine applications. *International Journal of Digital Earth*, 10: 85–102.

Alsamhi, S.H., Shvetsov, A.V., Kumar, S., et al. 2022. UAV computing-assisted search and rescue mission framework for disaster and harsh environment mitigation. *Drones-Basel*, 6:154.

Battersby, S. E., Finn, M. P., Usery, E. L. and Yamamoto, K.H. 2014. Implications of web Mercator and its use in online mapping. *Cartographica: The International Journal for Geographic Information and Geovisualization*, 49: 85–101.

Bey, A., Sánchez-Paus Díaz, A., Maniatis, D., Marchi, G., Mollicone, D., Ricci, S., Bastin, J.F., Moore, R., Federici, S., Rezende, M. and Patriarca, C. 2016. Collect earth: Land use and land cover assessment through augmented visual interpretation. *Remote Sensing*, 8: 807.

Biradar, C. M., Thenkabail, P. S., Noojipady, P., Li, Y., Dheeravath, V., Turral, H., Velpuri, M., Gumma, M.K., Gangalakunta, O.R.P., Cai, X.L. and Xiao, X. 2009. A global map of rainfed cropland areas (GMRCA) at the end of last millennium using remote sensing. *International Journal of Applied Earth Observation and Geoinformation*, 11: 114–129

Bradley, E. S. and Clarke, K. C. 2011. Outdoor webcams as geospatial sensor networks: Challenges, issues and opportunities. *Cartography and Geographic Information Science*, 38: 3–19.

Brown, C. F., Brumby, S. P., Guzder-Williams, B., Birch, T., Hyde, S.B., Mazzariello, J., Czerwinski, W., Pasquarella, V.J., Haertel, R., Ilyushchenko, S. and Schwehr, K.. 2022. Dynamic world, near real-time global 10 m land use land cover mapping. *Scientific Data*, 9(1): 251.

Butler, D. R. 1994. Repeat photography as a tool for emphasizing movement in physicalgeography. *Journal of Geography*, 93: 141–151.

Butler, D. R. and L. M. DeChano. 2001. Environmental change in Glacier National Park, Montana: An assessment through repeat photography from fire lookouts. *Physical Geography*, 22: 291–304.

Carroll, M.L., Townshend, J. R., Dimiceli, C.M., Noojipady, P. and Sohlberg, R.A. 2009. A new global raster water mask at 250 m resolution. *International Journal of Digital Earth*, 2: 291–308

Chen, Y. W., Yap, K. H. and Lee, J. Y. 2013. Tianditu: China's first official online mapping service. *Media Culture & Society*, 35: 234–249.

Clark, M. L., and Aide, T. M. 2011. Virtual interpretation of Earth Web-Interface Tool (VIEW-IT) for collecting land-use/land-cover reference data. *Remote Sensing*, 3(3): 601–620.

Conroy, G. C., Anemone, R. L., Regenmorter, J. V. and Addison, A. 2008. Google Earth, GIS and the Great Divide: A new and simple method for sharing paleontological data. *Journal of human evolution*, 55: 751–755.

Costa, M.P.F., Niemann, O., Novo, E. and Ahern, F. 2002. Biophysical properties and mapping of aquatic vegetation during the hydrological cycle of the Amazon floodplain using JERS-1 and Radarsat. *International Journal of Remote Sensing*, 23: 1401–1426.

Craglia, M., Goodchild, M. F., Annoni, A., Camara, G., Gould, M.F., Kuhn, W., Mark, D., Masser, I., Maguire, D., Liang, S. and Parsons, E. 2008. Next-generation digital Earth—A position paper from the Vespucci initiative for the advancement of geographic information science. *International Journal of Spatial Data Infrastructures Research*, 3: 146–167.

Reference Data

D'Andrimont, R., Yordanov, M., Martinez-Sanchez, L., Eiselt, B., Palmieri, A., Dominici, P., Gallego, J., Reuter, H.I., Joebges, C., Lemoine, G. and van der Velde, M. 2020. Harmonised LUCAS in-situ land cover and use database for field surveys from 2006 to 2018 in the European Union. *Science Data*, 7: 352.

D'Andrimont, R., Yordanov, M., Martinez-Sanchez, L., Haub, P., Buck, O., Haub, C., Eiselt, B. and van der Velde, M. 2022. LUCAS cover photos 2006–2018 over the EU: 874 646 spatially distributed geo-tagged close-up photos with land cover and plant species label. *Earth System Science Data*, 14: 4463–4472.

Defourny, P., Lamarche, C., Bontemps, S., et al. 2017. Land Cover CCI Product User Guide –Version 2.0. http://maps.elie.ucl.ac.be/CCI/viewer/

Dhonju, H.K., Walsh, K. B. and Bhattarai, T. 2023. Web mapping for farm management information systems: A review and Australian Orchard case study. *Agronomy-Basel*, 13: 2563.

Dronova, I. 2015. Object-based image analysis in wetland research: A review. *Remote Sensing*, 7: 6380–6413.

Esch, T., Bachofer, F., Heldens, W., Hirner, A., Marconcini, M., Palacios-Lopez, D., Roth, A., Üreyen, S., Zeidler, J., Dech, S. and Gorelick, N. 2018. Where we live—A summary of the achievements and planned evolution of the global urban footprint. *Remote Sensing*, 10: 895.

Fonte, Cidália C., Bastin, L., Foody, G. and Lupia, F. 2015. Usability of VGI for validation of land cover maps. *International Journal of Geographical Information as Science*, 29(7): 1269–1291.

Frazier, A. E. and Hemingway, B.L. 2021. A technical review of planet smallsat data: Practical considerations for processing and using PlanetScope imagery. *Remote Sensing*, 13: 3930.

Fritz, S., McCallum, I., Schill, C., Perger, C., See, L., Schepaschenko, D., Van der Velde, M., Kraxner, F. and Obersteiner, M. 2012. Geo-Wiki: An online platform for improving global land cover. *Environmental Modelling & Software*, 31: 110–123.

Fritz, S., See, L., Perger, C., McCallum, I., Schill, C., Schepaschenko, D., Duerauer, M., Karner, M., Dresel, C., Laso-Bayas, J.C. and Lesiv, M. 2017. A global dataset of crowdsourced land cover and land use reference data. *Scientific Data*, 4: 170075.

Gong, Peng, Liu, H., Li, C., Wang, J., Huang, H., Clinton, N., Ji, L., Li, W., Bai, Y. and Chen, B., 2019. Stable classification with limited sample: Transferring a 30-m resolution sample set collected in 2015 to mapping 10-m resolution global land cover in 2017. *Science Bulletin*, 64(6): 370–373.

Goodchild, M. F. 2008. The use cases of digital earth. *International Journal of Digital Earth*, 1: 31–42.

Gorelick, N., Hancher, M., Dixon, M., Ilyushchenko, S., Thau, D., and Moore, R. 2017. Google Earth Engine: Planetary-scale geospatial analysis for everyone. *Remote sensing of Environment*, 202: 18–27.

Goudarzi, M.A. and Landry, R. J. 2017. Assessing horizontal positional accuracy of Google Earth imagery in the city of Montreal, Canada. *Geodesy and Cartography*, 43: 56–65.

Grenier, M., Demers, A.-M., Labrecque, S., Benoit, M., Fournier, R.A. and Drolet, B. 2007. An object-based method to map wetland using RADARSAT-1 and Landsat ETM images: Test case on two sites in Quebec, Canada. *Can. J. Remote Sensing*, 33: 28–45.

Grenier, M., Labrecque, S., Garneau, M. and Tremblay, A. 2008. Object-based classification of a SPOT-4 image for mapping wetlands in the context of greenhouse gases emissions: The case of the Eastmain region, Quebec, Canada. *Canadian Journal of Remote Sensing*, 34: 398–413.

Hansen, M. C., Potapov, P. V., Moore, R., Hancher, M., Turubanova, S.A., Tyukavina, A., Thau, D., Stehman, S.V., Goetz, S.J., Loveland, T.R. and Kommareddy, A. 2014. High-resolution global maps of 21st-century forest cover change. *Science*, 344: 850–853.

Helmrich, A. M., Ruddell, B. L., Bessem, K. and Albrecht, S. 2021. Opportunities for crowdsourcing in urban flood monitoring. *Environmental Modelling & Software*, 143: 105124.

Hu, Z., Dietz, A. and Kuenzer, C. 2019. The potential of retrieving snow line dynamics from Landsat during the end of the ablation seasons between 1982 and 2017 in European mountains. *International Journal of Applied Earth Observation and Geoinformation*, 78: 138–148.

Hüsler, F., Jonas, T., Wunderle, S., et al. 2012. Validation of a modified snow cover retrieval algorithm from historical 1-km AVHRR data over the European Alps. *Remote Sensing of Environment*, 121: 497–515.

Iqbal, U., Riaz, M. Z. B., Zhao, J. H., Barthelemy, J. and Perez, P. 2023. Drones for flood monitoring, mapping and detection: A bibliometric review. *Drones-Basel*, 7: 32.

ISO 19157. 2013. *Geographic information — Data quality*. Geneva: International Organization for Standardization.

Iwao, K., Nishida, K., Kinoshita, T. and Yamagata, Y. 2006. Validating land cover maps with degree confluence project information. *Geophysical Research Letters*, 33(23).

Johansen, K., Phinn, S. and Witte, C. 2010. Mapping of riparian zone attributes using discrete return LiDAR, QuickBird and SPOT-5 imagery: Assessing accuracy and costs. *Remote Sensing Environment*, 114: 2679–2691.

Johansson, T. 2008. The live outdoor webcams and the construction of virtual geography. *Knowledge, Technology & Policy*, 21: 181–189.

Karra, K., Kontgis, C., Statman-Weil, Z., Mazzariello, J.C., Mathis, M. and Brumby, S.P. 2021. global land use/land cover with sentinel 2 and deep learning. In *2021 IEEE International Geoscience and Remote Sensing Symposium IGARSS*: 4704–4707.

Kinoshita, T., Iwao, K. and Yamagata, Y. 2014. Creation of a global land cover and a probability map through a new map integration method. *International Journal of Applied Earth Observation and Geoinformation*, 28: 70–77.

Kobayashi, T., Tsend-Ayush, J. and Tateishi, R. 2014. A new tree cover percentage map in Eurasia at 500 m resolution using MODIS data. *Remote Sensing*, 6: 209–232.

La Rosa, D. and Privitera, R. 2013. Characterization of non-urbanized areas for land-use planning of agricultural and green infrastructure in urban contexts. *Landscape and Green Infrastructure in Urban Contexts*, 109(1): 94–106.

Lesiv, M., See, L ., Bayas, J. C. L., Sturn, T., Sturn, T., Schepaschenko, D., Karner, M., Moorthy, I., McCallum, I. and Fritz, S. 2018. Characterizing the spatial and temporal availability of very high resolution satellite imagery in Google Earth and Microsoft Bing maps as a source of reference data. *Land*, 7: 118.

Li, Y. and Niu, Z.G. 2022. Systematic method for mapping fine-resolution water cover types in China based on time series Sentinel-1 and 2 images. *International Journal of Applied Earth Observation and Geoinformation*, 106: 102656.

Li, Zhuohong., He, W., Cheng, M. F., Yang, G. and Zhang, H., 2023. SinoLC-1: the first 1 m resolution national-scale land-cover map of China created with a deep learning framework and open-access data. *Earth System Science Data Discussions*, 15(11): 1–38.

Liu, J. F., Chen, R. S. and Wang, G. 2015. Snowline and snow cover monitoring at high spatial resolution in a mountainous river basin based on a time-lapse camera at a daily scale. *Journal of Mountain Science*, 12: 60–69.

Luo, L., Wang, X. Y., Guo, H. D., Lasaponara, R., Shi, P., Bachagha, N., Li, L., Yao, Y., Masini, N., Chen, F. and Ji, W. 2018. Google Earth as a powerful tool for archaeological and cultural heritage applications: A review. *Remote Sensing*, 10: 1558.

Luo, L., Wang, X., Guo, H., Liu, C., Liu, J., Li, L., Du, X. and Qian, G. 2014. Automated extraction of the archaeological tops of Qanat shafts from VHR imagery in Google Earth. *Remote Sensing*, 6: 11956–11976.

Mccoy, M.D. 2017. Geospatial big data and archaeology: Prospects and problems too great to ignore. *Journal of Archaeological Science*, 84: 74–94.

Melchiorri, M., Florczyk, A.J., Freire, S., Schiavina, M., Pesaresi, M. and Kemper, T. 2018. Unveiling 25 years of planetary urbanization with remote sensing: Perspectives from the global human settlement layer. *Remote Sensing*, 10: 768.

Methakullachat, D. and Witchayangkoon, B. 2019. Coordinates comparison of Goolge® Maps and orthophoto maps in Thailand. *International Transaction Journal of Engineering, Management, & Applied Sciences & Technologies*, 10: 1–8.

Mohammed, N. Z., Ghazi, A. and Mustafa, H. E. 2013. Positional accuracy testing of Google Earth. *International Journal of Multidisciplinary Sciences and Engineering*, 4: 6–9.

Mohsan, S. A. H., Khan, M. A., Noor, F., Ullah, I. and Alsharif, M.H. 2022. Towards the unmanned aerial vehicles (UAVs): A comprehensive review. *Drones-Basel*, 6: 147.

Mulu, Y.A. and Derib, S.D. 2019 Positional accuracy evaluation of Google Earth in Addis Ababa, Ethiopia. *Artificial Satellites: Journal of Planetary Geodesy*, 54: 43–56.

Neuhold, Gerhard, Ollmann, Tobias, Haub, P., Buck, O., Haub, C., Eiselt, B. and van der Velde, M. 2017. *In Proceedings of the IEEE international conference on computer vision*: 4990–4999.

Pengra, B., Jordan L., Devendra D., Stephen V. S. and Thomas R. L., 2015. A global reference database from very high resolution commercial satellite data and methodology for application to landsat derived 30 m continuous field tree cover data. *Remote Sensing of Environment*, 165(August): 234–48.

Pekel, J. F., Cottam, A., Gorelick, N. and Belward, A.S. 2016. High-resolution mapping of global surface water and its long-term changes. *Nature*, 540: 418–422.

Piazzi, G., Tanis, C. M., Kuter, S., Simsek, B., Puca, S., Toniazzo, A., Takala, M., Akyürek, Z., Gabellani, S. and Arslan, A.N. 2019. Cross-country assessment of H-SAF snow products by sentinel-2 imagery validated against in situ observations and webcam photography. *Geosciences*, 9: 129.

Portenier, C., Hüsler, F., Härer, S. and Wunderle, S. 2020. Towards a webcam-based snow cover monitoring network: Methodology and evaluation. *Cryosphere*, 14: 1409–1423.

Portenier, C., Hasler, M., and Wunderle, S. 2022. Estimating Regional Snow Line Elevation Using Public Webcam Images. *Remote Sensing*, 14(19): 4730

Reference Data

Pulighe, G., Baiocchi, V. and Lupia, F. 2016. Horizontal accuracy assessment of very high resolution Google Earth images in the city of Rome, Italy. *International Journal of Digital Earth*, 9: 342–362.

Qian, T. N., Kinoshita, T., Fujii, M. and Bao, Y. 2020. Analyzing the uncertainty of degree confluence project for validating global land-cover maps using reference data-based classification schemes. *Remote Sens*, 12: 2589.

Renner, R. D., Hemani, Z. Z. and Tjouman, G. C. 2009. Extending advanced geospatial analysis capabilities to popular visualization tools. *Technology Review Journal Spring/Summer*, 17: 89–106.

Schepaschenko, D. G., Shvidenko, A. Z., Lesiv, M. Y., Ontikov, P.V., Shchepashchenko, M.V. and Kraxner, F. 2015. Estimation of forest area and its dynamics in Russia based on synthesis of remote sensing products. *Contemporary Problems of Ecology*, 8: 811–817.

Schneider, A., Friedl, M. A. and Potere, D. 2010. Mapping global urban areas using MODIS 500-m data: New methods and datasets based on 'urban ecoregions'. *Remote Sensing of Environment*, 114: 1733–1746.

Schramm, M., Pebesma, E., Milenković, M., Foresta, L., Dries, J., Jacob, A., Wagner, W., Mohr, M., Neteler, M., Kadunc, M., Miksa, T., Kempeneers, P., Verbesselt, J., Gößwein, B., Navacchi, C., Lippens, S., and Reiche, J. 2021. The openeo api–harmonising the use of earth observation cloud services using virtual data cube functionalities. *Remote Sensing*, 13(6), 1125.

See, L., Fritz, S., Perger, C., et al, Schepaschenko, D., Schill, C., McCallum, I., Schepaschenko, D., Duerauer, M., Sturn, T., Karner, M., Kraxner, F. and Obersteiner, M. 2015. Harnessing the power of volunteers, the internet and Google Earth to collect and validate global spatial information using Geo-Wiki. *Technological Forecasting and Social Change*, 98: 324–335.

See, L., Georgieva, I., Duerauer, M., Kemper, T., Corbane, C., Maffenini, L., Gallego, J., Pesaresi, M., Sirbu, F., Ahmed, R. and Blyshchyk, K. 2022. A crowdsourced global data set for validating built-up surface layers. *Scientific Data*, 9(1): 13.

Shin, N., Katsumata, C., Miura, T., Tsutsumida, N., Ichie, T., Kotani, A., Nakagawa, M., Khoon, K.L., Kobayashi, H., Kumagai, T.O. and Tei, S. 2023. Perspective: Improving the accuracy of plant phenology observations and land-cover and land-use detection by optical satellite remote-sensing in the Asian tropics. *Frontiers in Forests and Global Change*, 6: 1106723.

Tom, M., Prabha, R., Wu. T., Baltsavias, E., Leal-Taixé, L. and Schindler, K.. 2020. Ice monitoring in swiss lakes from optical satellites and webcams using machine learning. *Remote Sensing*, 12: 3555.

Tsendbazar, N., Herold, M., Lesiv, M., et al. 2017. *Validation report of moderate dynamic land cover, collection 100M, version 1*. Wageningen: Wageningen University and Research.

Venter, Z. S., Barton, D. N., Chakraborty, T., Simensen, T., and Singh, G. 2022. Global 10 m Land Use Land Cover Datasets: A Comparison of Dynamic World, World Cover and Esri Land Cover, *Remote Sensing*, 14, 4101.

Wang, B., Choi, J., Choi, S., Lee, S., Wu, P. and Gao, Y. 2017. Image fusion-based land cover change detection using multi-temporal high-resolution satellite images. *Remote Sensing*, 9: 804.

Wang, D. L., Shao, Q. Q. and Yue, H. Y. 2019. Surveying wild animals from satellites, manned aircraft and unmanned aerial systems (UASs): A review. *Remote Sensing*, 11: 1308.

Wood, J., Dykes, J., Slingsby, A. and Clarke, K. 2007. Interactive visual exploration of a large spatio-temporal dataset: Reflections on a geovisualization mashup. *IEEE Transactions on Visualization and Computer Graphics*, 13: 1176–1183.

Yang, B., Hawthorne, T., Torres, H. and Feinman, M. 2019. Using object-oriented classification for coastal management in the East Central Coast of Florida: A quantitative comparison between UAV, satellite, and aerial data. *Drones-Basel*, 3: 60.

Yang, B.Y., Kim, M. and Madden, M. 2012. Assessing optimal image fusion methods for very high spatial resolution satellite images to support coastal monitoring. *GIScience & Remote Sensing*, 49: 687–710.

Yu, L. and Gong, P. 2012. Google Earth as a virtual globe tool for Earth science applications at the global scale: Progress and perspectives. *International Journal of Remote Sensing*, 33: 3966–3986.

Zhang, Z. X. and Zhu, L. X. 2023. A review on unmanned aerial vehicle remote sensing: Platforms, sensors, data processing methods, and applications. *Drones-Basel*, 7: 398.

4 Sample Judgment

Lijun Chen
National Geomatics Center of China, Beijing, China

Zhengxing Wang
Chinese Academy of Sciences, Beijing, China

Maria Antonia Brovelli
Politecnico di Milano, Milan, Italy

Gorica Bratic
Politecnico Milano, Milan, Italy

Hanfa Xing
South China Normal University, Foshan, China

Wen Meng and Chao Wei
Tongji University, Shanghai, China

4.1 THE FRAMEWORK FOR OBTAINING REFERENCE CLASSIFICATIONS

4.1.1 MODES OF INTERPRETATION

There are three modes of interpretation: blind, plausibility, and enhanced plausibility (See et al. 2015). In blind interpretation, interpreters are not provided with any information about sample units, avoiding interference of prior knowledge. Plausibility interpretation means that interpreters consider the knowledge of sample units' class labels and can then make better use of prior knowledge. Enhanced plausibility interpretation goes further so that interpreters not only judge whether the classification is correct but also determine the true class labels of sample units.

Each of these three modes has its advantages and disadvantages. Interpreters should choose an appropriate one according to the situation. Due to the difference between these modes, inconsistency caused by interpreters' knowledge or ability can be observed.

4.1.2 PROCEDURES AND GUIDELINES

The basic procedure of sample judgment starts with the collection of high-resolution remote sensing images, finer-resolution land cover maps, field sampling data, and other data sources. These data sources are then interpreted, with the validity of resulting classifications quantified. In general, interpretation for determining land cover types at sample units in a study area goes through five steps, as shown in Figure 4.1.

As shown in Figure 4.1, after the selection of reference data sources, such as high-resolution images, large-scale land cover maps, geotagged photos, and in situ data, the focus is to label sample units through interpretation. The purpose of reference class interpretation is to rely on the experience

Sample Judgment

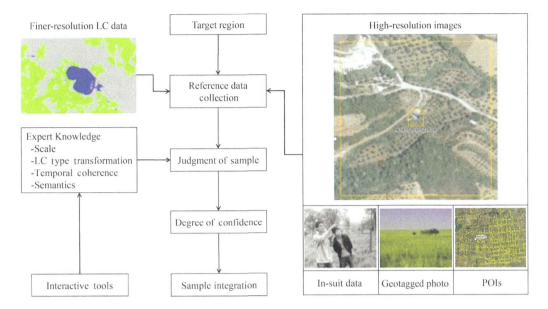

FIGURE 4.1 The procedure for reference sample judgment.

and knowledge of experts to label sample units. When interpreting sample units, interpreters need to determine reference classes from various aspects, such as scale uncertainty, classification system differences, temporal coherence, and semantic ambiguity. Useful interactive tools are often used to assist in the interpretation process.

The credibility of the resulting reference sample data needs to be quantified. The credibility of reference data furnished is divided into several levels to support the evaluation of reference data accuracy. Finally, a database of detailed and standardized reference sample data in images and photos is also built to assist in the accuracy assessment process.

Due to inconsistency in classification systems, spatial resolution, and map classifications, interpretations often suffer from biases and misclassification errors. It is thus necessary to establish a set of guidelines to standardize the interpretation process:

(1) Area priority

Area priority is to set a threshold to determine the reference class of a sample unit with mixed land cover. For example, within a sample pixel (e.g., 30 m × 30 m), there are four types, say bare land 85%, grass 5%, forest 5%, and shrubs 5%. Assuming a threshold of 70%, the reference class of this sample pixel is bare land since a proportion of 85% of the pixel is bare land.

(2) Majority rule

For mixed pixels in which there is more than one land cover type, a majority decision rule can be applied, as used for the validation of DISCover products (Scepan 1999). Based on this rule, three experts should interpret each sample unit independently, and the class label for the sample unit is verified as the mode label, on which at least two of them agree. However, it remains uncertain if their interpretations are inconsistent. This method can reduce the error caused by subjectivity, though it incurs increased costs.

(3) Scale

Geographic phenomena and spatial features are usually in multiscale existence. Resolution is thus an important concept for land cover classification and mapping. Generally, the land cover map to be validated needs to be compared and analyzed with reference data of finer resolution. It is also necessary to consider whether land cover types under study are consistent at different resolutions (scales). Sometimes, classification is based on the principles of area priority or class rarity priority.

(4) Classification system

When using an existing land cover map for reference class interpretation and verification, it is necessary to convert the classification systems to a desired one. This gets complicated if the classification systems of the maps involved are inconsistent with that of the reference sample data to compile.

(5) Temporal coherency

Land cover is dynamic, especially for classes known to have experienced changes. Land cover types may be different over time. This requires that reference sample data be temporally aligned with land cover maps if they are to be compared and synthesized for improved land cover mapping.

(6) Semantic ambiguity

The conceptual fuzziness and cognitive uncertainty of land cover type are important factors for consideration, as they affect reference classifications to be furnished from reference data sources. They are caused by the expert's individual understanding and knowledge of land cover patterns in the study area. The results of interpretation and classification usually reflect the expert's subjective judgment. The challenges are also due to ambiguity due to spectral confusion, immeasurability of land cover existing inherently as continua, conceptual semantic ambiguity, and other kinds of complexity inherent in land cover classification. In general, it is necessary to adopt a majority decision rule to address these issues.

4.1.3 Sample Assessment Units (SAUs)

The sample units obtained through sampling in Chapter 2 actually refer to sample pixels on maps or remote sensing images. They are called SAUs. In general, the suitable size of SAUs depends on the resolution of the reference data sources being analyzed and verified. If the resolution of the land cover map to be verified is 30 m, then the size of SAUs is 30 m × 30m.

The interpretation results of SAUs of different sizes may vary. For example, two SAUs are shown in Figure 4.2; one of them is 30 m × 30 m, and the other is 300 m × 300 m. Two land cover type labels were assigned to each sample unit. According to the area priority rule, the first SAU is assigned based on the cover type that covers the largest areal extent of the 30 m × 30 m square, and the same rule for

FIGURE 4.2 Example SAUs of 30 m × 30 m and 300 m × 300 m.

Sample Judgment

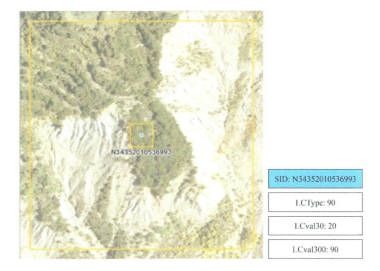

FIGURE 4.3 Example 1 for SAUs interpretation.

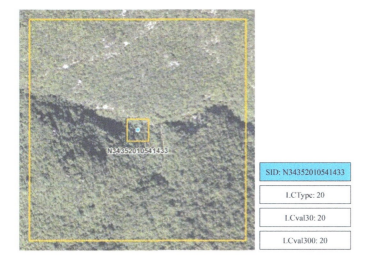

FIGURE 4.4 Example 2 for SAUs interpretation.

the second SAU, though its areal extent is 300 m × 300 m. Information about area priority is saved in new columns in the attribute table of the provided shape file, named LCval30 and LCval300.

Figures 4.3 and 4.4 show typical Mediterranean landscapes where land cover mosaics are complex, even at fine resolution, making land cover classification a demanding exercise. Three land cover class labels are shown for both larger and smaller SAUs in Figures 4.3 and 4.4. In Figure 4.3, the cover type is labeled 90 for the map being verified. The cover type label is 20 for the SAU of 30 m resolution, while the class label is 90 for the SAU of 300 m × 300 m. In the same way, as is shown in Figure 4.4, the class label is 20 for the original sample unit and for SAUs of 30 m and 300 m.

4.1.4 INTERACTIVE TOOLS

A number of web-based tools have been developed to provide interactive functions for land cover map validation. These tools integrate images, land cover maps, digital elevation models (DEMs), and other auxiliary data into a system through web service technology architecture. Users do not

FIGURE 4.5 Globeland30 service platform.

need to transfer and process a large amount of data and can easily use them online. Further detail of these interactive tools is given in Chapter 6.

For example, the Globeland30 service platform provides a tool, GLCVal, as shown in Figure 4.5, which has spatial labeling functions using split-screen displays. It is used to support online aggregation and orderly transmission of information for multiuser collaborative interpretation. By using data browsing and split-screen comparison, interpreters can visually interpret and compare the land cover map with online high-resolution images such as MapWorld, Google Earth, or other reference data. It also supports interactive comparison, spatial analysis, and statistical analysis based on relevant knowledge rules, assisting experts in making more accurate interpretations (type, location, contextual relationship) of samples. It also supports online publication and aggregation of sample information, making it convenient to integrate and process the interpreted sample results for subsequent accuracy validation of land cover maps.

4.2 INTERPRETING HIGH-RESOLUTION IMAGES

4.2.1 Image Interpretation Elements

While a universal interpretation key might be desired to ensure consistent labeling, it is very hard to design and implement such a key due to the variation of the same classes in different regions. Nevertheless, early experiences gained from aerial photo interpretation can be helpful to SAU interpretation based on high-resolution images. For instance, Lillesand et al. (2015) summarized seven elements for aerial photo interpretation, whose applications with adaptation in Roof of the World Region (RoTW) are listed in Table 4.1.

In remote sensing images, different ground objects possess different image features, as shown in Figure 4.6. These features somehow reflect the nature and shapes of the objects on the earth's surface and are the basis for interpreting cover types. To assist interpretation of sample units (i.e, SAUs), a square of side length 100 m is made to enclose a 30 m × 30 m square SAU. These SAUs (30 m and 100 m) can be stored in ArcMap, identified by SID, and exported to Google Earth's kmz format, allowing for interpretation on Google Earth. For GlobeLand30, ten cover types are

TABLE 4.1
Eleven Interpretation Elements in RoTW

Elements	Application in RoTW
1. Shape	Shape refers to the general form, configuration, or outline of individual objects. Some land cover types (e.g., cropland, water bodies, and artificial surfaces) may be identified by their shapes solely.
2. Size	The size of objects within the boundary of an area covered by 30 m pixel in high spatial resolution imagery can be estimated by reference to this 30 m by 30 m square. Therefore, the percentage of various objects identified by their size can be calculated. This information, together with information from other elements, helps to label some land cover types.
3. Pattern	Spatial arrangement of objects, when combined with geographic knowledge, can be helpful to interpretation. Some examples are permanent snow and ice: this pattern usually has snow on top of high mountains, with very steep slopes below snow-capped peaks. Sometimes glaciers develop in lower valleys. Barren lands often coexist with permanent snow and ice.Human activities:some small objects can aid in interpretation. Road systems are direct indicators of human impacts. A forest divided by roads may be degraded into shrubs and grassland; an irrigation system can differentiate cropland in fallow from barren land.
4. Shadow	Shadows block objects and make interpretation impossible, such as shadows from huge clouds and reliefs.Shadows present illusive information, as some shadows from sparse vegetation may be misinterpreted as vegetation and lead to overestimation of vegetation coverage. Shadows provide useful information since the displacement of shadows and objects can be used to estimate the relative height of an object.
5. Tone	Tone refers to the color or relative brightness of objects in photographs or images. By intuition, an interpreter may expect a green color for green vegetation from true-color Google Earth images. This may hold true for some sample units below 4,000 m altitude. The reason is that the growing season in this high-altitude region is relatively short and frequently covered by clouds. Therefore, the majority of the Google Earth images are from a dormant season when vegetation is less or not greenish. However, still other tones can be applied to distinguish vegetation from non-vegetation, and to differentiate various land covers.
6. Texture	Unlike pattern, texture is a term used to describe the internal structure of an object. An object is usually characterized by one or more elements (primitives) and their spatial arrangement. For instance, the crown of a tree or the crown of a bush can be such elements. Given a basic interpretation unit, the amount, size, and shape, as well as the spatial arrangement of various elements, can be good indicators of land cover structures.
7. Site	The location of an object is important for its interpretation, at least in two aspects. On the one hand, the same land cover types may vary greatly in different places; this explains why it is hard to develop a universal interpretation key for large regions. On the other hand, the location of an object does provide some information for interpretation. For instance, an object above 5,000 m altitude excludes the possibility of forest and cropland, and an object below 3,500 m altitude has little chance of remaining as permanent snow and ice in RoTW. In addition, the site can provide information at a microscale. In arid regions mainly covered with grasslands, bush may grow along a small stream.
8. Temporal	Multitemple images are helpful for interpretation and are often prerequisites for discriminating some classes. For instance, snow and ice can only be interpreted on the images flown in summer.
9. Spatial	A sample unit can be better interpreted in a spatial context.
10. Altitude	Altitude is a key indicator in two ways. On one hand, some classes only appear in some altitude ranges. On the other hand, some altitude ranges exclude other classes.
11. Stereoscopy	Stereoscopy helps to identify three-dimensional structures.

Source: Adapted from Lillesand T. M. and Kiefer R. W. *Remote sensing and image interpretation*. John Wiley & Sons, New York, 1979.

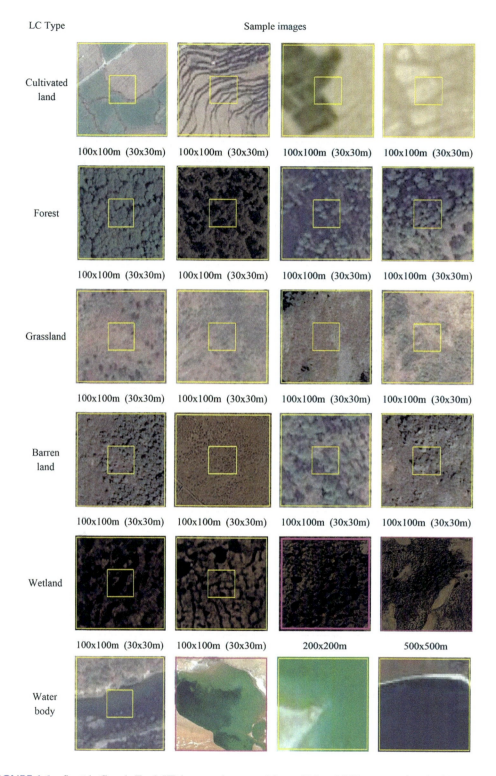

FIGURE 4.6 Sample Google Earth HR image subsets used for verifying SAU interpretations in GlobeLand30 product.

Sample Judgment 111

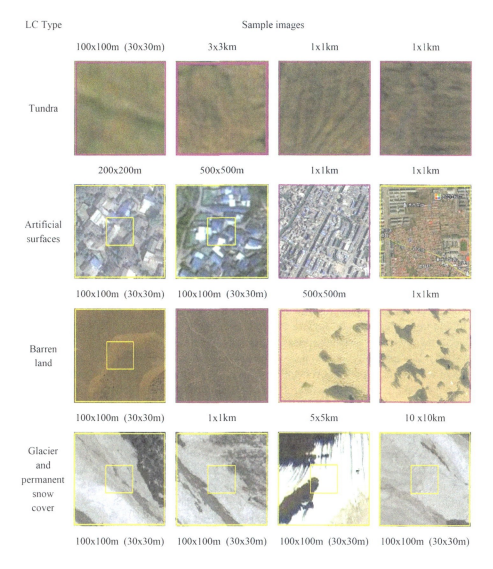

FIGURE 4.6 (CONTINUED) Sample Google Earth HR image subsets used for verifying SAU interpretations in GlobeLand30 product.

interpreted using ten interpretation elements based on Google Earth high-resolution images, as described next.

(1) Cultivated land: It mainly depends on shape. Generally, patches of cultivated land have distinctive geometric shapes. However, there exist difficult cases. One is about cultivated land that is reclaimed in forest areas, where cultivated land patches have no ridges, the ridges are small, or the ridges are submerged during the crop growth period. The other difficulty occurs when cultivated land appears spectrally similar to bare land, especially in arid areas. In addition, there are often human settlements near agricultural areas, making it difficult to discern cultivated land from mixed agricultural-residential settings.
(2) Forest: One of the major challenges in interpreting forested land is due to the presence of shadows, which tend to exaggerate forest cover, especially in forests with a low proportion

of detected tree crowns (say below 10%). Trees at the edge of the forest areas and residual forests in agricultural areas are often mixtures of forest, shrubs, and grass. It is necessary to carefully observe whether the tree covers exceed 10% of forested land in arid areas. When tree canopies are relatively small, the difference between them and shrubs tends to decrease. Given the definition of forest as tree forest canopy density >0.1, there may be forests that appear sparse with a bright background, which differs significantly in appearance from fully covered forests in forest zones.

(3) Grassland: The difficulty in interpreting grassland is that the texture between grassland and sloping farmland, grassland, and shrubs is hard to distinguish. The shrubs' crowns appear as cluster-shaped features, while patches of grassland usually appear as smooth in images.

(4) Shrubs: In forest zones below an elevation of 4,000 m, shrubs are usually distributed at the edges of forests with different degrees of degradation or remain at the edges of agricultural areas. The are usually mixed with forests and grasslands but rarely with pure forests. It needs to be confirmed that the forest cover threshold is not met, but the canopy density of shrubs is greater than 0.3. In grasslands above 4,000 m, shrubs are usually distributed on shady slopes and around water bodies. In the dormant season, the residual snow in shrubs lasts longer but still shows a rough texture. In forest areas below 4,000 m, in growing seasons, shrub forests appear slightly lighter than forest areas. In nongrowing seasons (when Google images are usually flown), in grasslands above 4,000 m, shrubs generally appear dark brown, while grasslands appear smooth in a yellowish tone.

(5) Wetland: It is generally located near water bodies, and in areas with seasonal changes, it is mixed with water bodies and vegetation. This ecosystem is best extracted using the object-oriented method, allowing the overall characteristics of a wetland ecosystem to be ascribed to the sample units concerned. Interpretation should consider all sample points of wetlands to avoid being misclassified as water bodies or grasslands.

(6) Water bodies: Rivers, lakes, reservoirs, and ponds are all instances of water bodies. Generally, water bodies can be interpreted through shape and color. It should be noted that, due to differences in the compositions of water bodies and differences in water pollution, they may appear in different colors.

(7) Tundra: Tundra refers to an alpine and humid grassland ecosystem outside the world's polar regions. In general, the interpretation of tundra cannot rely solely on images but also on its geographic distribution characteristics.

(8) Artificial surfaces: Generally, artificial surfaces can be interpreted through shape. The difficulty of interpretation lies in the small difference between residential areas and the background in arid and humid areas.

(9) Bare land: For bare land, the difficulty of interpretation lies in the transition zones, both vertical and horizontal. Between the tree line and the snow line (vertical), there is a gradual transition from shrubs at the bottom, through grass and bare land in the middle, to ice and snow on the top. Horizontally, there is only a slight difference between sparse grasslands (especially in the dormant season, for low and withered grasslands) and bare land. Therefore, it is necessary to employ multitemporal images for more accurate interpretation.

(10) Glacier and permanent snow cover: It would be easy to discriminate glacier and permanent snow cover from other cover types using "hottest month" images. When such discriminatorily powerful images are not available, other interpretative information and reasoning need to be applied. Below 4,000 m above sea level, glaciers and permanent ice and snow are unlikely to occur. In nonsummer months, there may only be temporary snow cover. Another useful interpretation rule is that no matter what season of the year, as long as snow cover is determined to be seasonal for a location once, that location will not be labeled permanent ice and snow.

Sample Judgment 113

4.2.2 CONTEXT-BASED INTERPRETATION

(1) Correct formation of stereo models

Google Earth images come with various dates and hence different sun-surface–sensor observation geometry. Some of the observation geometry causes certain images to form false stereo (negative stereo or pseudostereo) pairs, whereby protruding objects in the real world look concave in negative stereo models. In this case, vegetation cover will be underestimated.

The solution to the false stereo phenomenon is simple. If an image pair is suspected to be a false stereo pair, it can be rotated 180 degrees to see whether a positive stereo model can be obtained. For example, in Figure 4.7 (sample unit "N462520101431586"), the SAU dated January 6, 2010, looks like bare land in the negative stereoscopic model, though being forest in the positive stereo model. The SAU dated November 5, 2014, looks like shrub

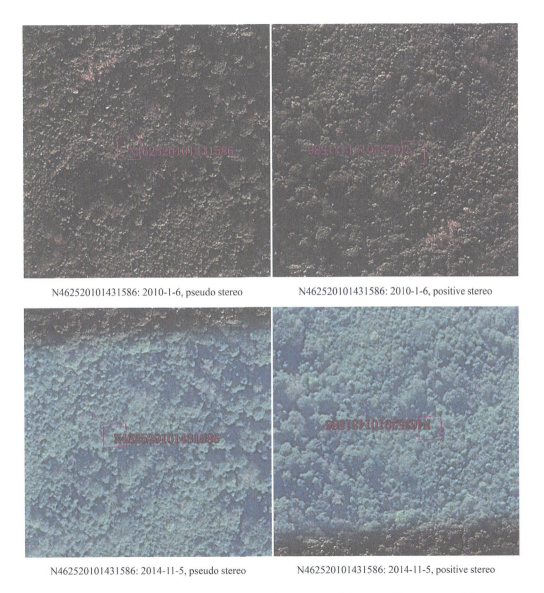

N462520101431586: 2010-1-6, pseudo stereo N462520101431586: 2010-1-6, positive stereo

N462520101431586: 2014-11-5, pseudo stereo N462520101431586: 2014-11-5, positive stereo

FIGURE 4.7 Image pairs of forest landscape as negative stereo models vs. positive stereo models.

N452520101500925: 2013-3-21, Bare land   N452520101500925: 2014-10-26, Water

FIGURE 4.8 Seasonal changes in wetlands.

or grassland in the negative stereo model. But, in the positive stereo model, it is labeled as forest since the tree crowns appear very obvious.

(2) Temporal series information interpretation

While reference images need to be temporally aligned with the products to be validated, time-series reference images are better for interpreting the reference class labels. Temporal information is mainly manifested as seasonal changes and inter-annual changes or trends, as demonstrated in the following examples.

- Seasonal permanent snow and ice: Permanent ice and snow can only be determined by using reference images of the hottest season. The ice and snow seen in other seasons are probably seasonal ice and snow, especially when the sample unit is located at a low altitude (<4,000 m).
- Seasonal wetland: Wetland is the seasonally inundated land. Thus, reference images through seasons are usually needed for the correct interpretation of SAUs of wetland class based on the wetland's defining characteristic of periodic inundation. Otherwise, SAUs of wetlands may be wrongly identified as water, vegetation, or bare land. Figure 4.8 shows that the reference image subset on the left may be labeled as bare land and that on the right water when they are interpreted separately. But when interpreted together, the wetland is ascribed correctly to this SAU, as it is shown to be seasonally inundated.
- Seasonal vegetation: The majority of vegetation is seasonal, through spring, summer, autumn, and winter, and over dry and wet periods. The patterns of seasonal vegetation change can vary greatly. For instance, for sample unit N442520102084560 in Figure 4.9, reference image subset May 28, 2012, may be classified as grassland, subset February 25, 2014, as bush, and subset November 25, 2016, as forest. Therefore, unless the interpreter is very familiar with the seasonality of the imaged area, time-series reference images should be used for accurate interpretation.
- Inter-annual change and trend: According to the land cover classification system (Table 4.1), the forest logging area is still defined as forest, with the underlying rationale being that it will be reforested. Single-date reference images cannot help to label it correctly. Time-series reference images are needed to determine whether the forest logging site is to be reforested or just logging without reforestation.

Sample Judgment 115

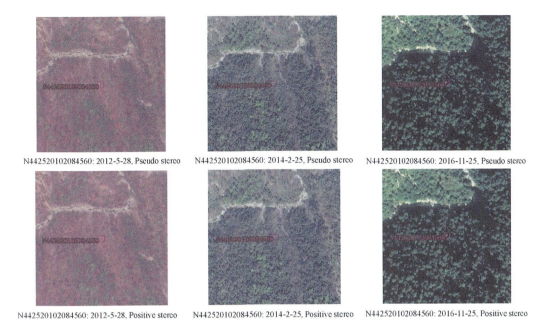

FIGURE 4.9 Seasonality of vegetation seen in time-series images.

In Figure 4.10, for sample unit N472020102638903, reference image subset February 1, 2013, can be interpreted as a logging site. After five years, reforestation is clearly shown in image subset November 13, 2017. Thus, this unit is reasonably classified as forest. However, not all forest logging areas are reforested. Figure 4.11 (sample unit: N472525102677953) shows that roads appear in an imaged logging site dated December 24, 2015.

In addition to the aforementioned applications, time-series reference images can also be used for retrospective reasoning of some land cover types. For example, under normal circumstances, forests cannot form in a short period. In high-altitude mountains or arid areas, shrubs and even grassland cannot be formed in a short time. If there are no high-quality reference images circa 2010 but at a later year, interpretation based on later date images may still be used as a better substitute for low-quality reference images circa 2010.

(3) Spatial information for enhancing interpretation

The question is how a specific land cover class can be interpreted relatively reliably in a changing environment. In other words, this concerns whether there are any relatively stable and unchanging features that can assist the interpretation of reference data. Here, this kind of features refer those of a class relative to the typical features on a specific date.

Under this premise, the sample unit can be interpreted in three steps with the help of spatial information:

- find the best quality reference image;
- interpret the most typical land cover classes in the best reference image; and
- interpret the sample unit based on its relative position at these typical land cover classes.

For instance, as shown in Figure 4.12, based on reference image subset September 28, 2012, sample unit N433020103648681 may be interpreted as forest if one only uses information within and near the sample unit. However, there are more typical arbor forests in the same scene. Therefore, a more accurate interpretation should be shrub for the primary land cover class (Class_1) and forest for the secondary land cover class (Class_2).

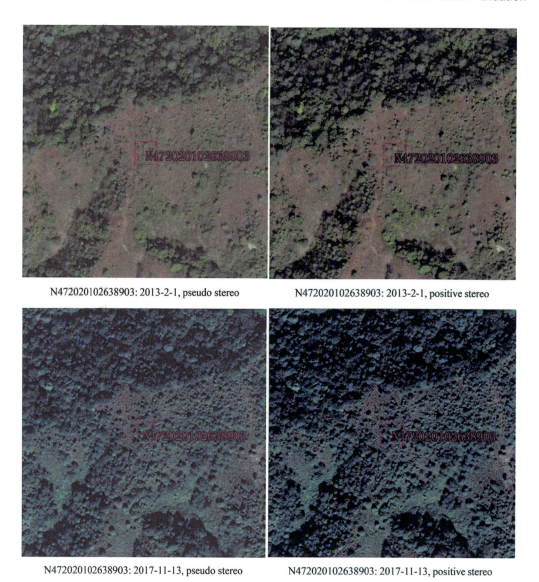

FIGURE 4.10 Reforestation after logging, from 2013 to 2017.

(4) Interpretation based on local geospatial knowledge

Geospatial knowledge is useful for increasing the accuracy of land cover classification. Similarly, for product validation, geospatial knowledge can also help to improve the classification accuracy of reference images. For example, in the RoTW (the region including the Pamirs, the Himalayas, the Tibetan Plateau, the Hindu Kush, the Tian Shan, and the Altai Mountains), common geospatial knowledge includes the following:

- Vertical zonalization of land cover: From top to bottom, the general order of cover types is permanent ice and snow, bare land, grassland, shrub, and forest. However, in high-altitude areas, a site with steep slope and high terrain shadow ratio is likely misclassified as a water body.
- Permanent snow and ice: Temperature on the plateau is generally low, so the snow cover time is long. Only ice and snow that still exists in the hottest season (June–August) can be labeled as permanent ice and snow.

Sample Judgment 117

N472525102677953: 2010-11-1, Logging area   N472525102677953: 2015-12-24, Bare land

FIGURE 4.11 Forests degenerated into bare land or grassland after logging.

N433020103648681: 2010-4-6   N433020103648681:2012-9-28

FIGURE 4.12 Interpretation based on spatial information of forests vs. shrubs.

- Vegetation: the growing season of vegetation in RoTW areas is short, and there are more clouds and fog in the growing season, leading to fewer high-quality remote sensing images available. Therefore, vegetation in Google Earth images often does not appear green. Moreover, interpretation should not be based on color (tone) but on size, texture, and other elements (see Table 4.2).
- Shrubs in high-altitude areas: Vegetation on the plateau is generally small, with shrubs on the plateau being shorter than those at low altitudes. The effect of this differentiation has been accommodated in the classification system (see Table 1.3). Compared with grassland, shrubs are generally distributed near the catchment of small watersheds, where the shady slope is flat, and the image grains are coarser; shrub is often accompanied by greater soil humidity (appearing darker in images) and longer snow cover time.

TABLE 4.2

Globeland30 Land Cover Classification System (Excerpt) and Samples in the RoTW Region

LCType	Land Cover	Land Cover Descriptions	Valid Sample	Invalid Sample
10	Cropland	This includes regularly harvested artificial grassland. The cash forest dominated by trees in cultivated land belongs to forest, and the cash forest dominated by bush in cultivated land belongs to cropland.	24	1
20	Forest	Tree crown >10%: logging site(in the process of natural expansion of forests).	127	4
30	Grassland	Dominated by herbaceous plants with total vegetation coverage >10%.	331	31
40	Bush	Dominated by bush, with shrub crown >10% in desert areas, >30% in all others.	116	11
50	Wetland	Seasonally inundated land, excluding paddy fields and persistent dry beaches.	2	
60	Water bodies	Exclude paddy fields and permanent snow and ice.	10	1
80	Artificial surface	Exclude large water bodies and green land.	1	
90	Bare land	Total vegetation coverage <10%, excluding seasonally inundated land (tidal flat).	171	28
100	Permanent Snow/ice	Snow and ice that exist throughout the year.	19	
Total			801	76

4.2.3 FURTHER RULES FOR INTERPRETING COVER TYPES

Taking the Globeland30 validation in the RoTW as an example, we used the sampling method in Chapter 2 to obtain 877 sampling points distributed in the region. Next, we need to judge these samples. There are some interpretation techniques when relying on high-resolution images for sample judgment. The following six rules are set forth to address the possible ambiguities and confusion in reference classification at SAUs:

(1) The definition of reference classification is consistent with that of map classification.
(2) The class of dominant species has priority if there is more than one cover type within an SAU.
(3) The class with the largest percentage of the areal extent has priority if there are more than one cover type in a SAU.
(4) Two class labels are allowed on the ecological gradient and near a boundary.
(5) The confidence level for each sample unit's interpretation and resultant reference classification should be recorded.
(6) A blind mode of interpretation should be adopted for an SAU whose land cover the interpreter has no knowledge of.

Rule 1 seems to be self-evident in principle, but it is often ignored in practice. In particular, when validating global datasets such as Globeland30, interpretations may be inconsistent due to the differences in the understanding of the classification system used.

Rule 2 mainly applies to the situation where two or more land cover types belong to the forest-shrub-grassland-bare land mixture. Based on Table 4.1, cover types should be interpreted in the

Sample Judgment 119

N433020103634042: 2013-11-25, 1194m

FIGURE 4.13 The high-level land cover type should be interpreted first (i.e., in the order of forest, shrub, grassland, and bare land).

order of forest-shrub-grassland-bare land. This rule requires that the class of dominant species be ascribed to a sample unit should it belong to a mixed class.

In Figure 4.13, sample unit N4330201036334042 is a sparse woodland located at the edge of a forest. Although grassland and bare land dominate the sample unit, priority should be given to arbor forest being the label. This is because the crown coverage is more than 10% (thus the sample unit is classified as forest).

Rule 3 applies to a mosaic class, with the sample labeled as the class of the largest areal extent. If no classes occupy the largest areal proportion within an SAU, Rule 4 is referred to.

Rule 4 says that two class labels may be assigned to a unit if neither one alone but two together assume the largest areal proportion. Rule 4 applies to two situations: the sample unit is on a natural ecological gradient, or the sample unit is near a natural boundary.

If a sample unit is on a natural ecological gradient, it is quite common in nature that both typical and atypical classes coexist for a given classification system. For instance, a sample unit located in a forest-bush continuum is not easily classified as forest or bush. The omission of either forest or bush is clearly a loss of information, while the inclusion of both forest and bush seems to be a reasonable compromise.

If a sample unit is near a boundary, the unit likely resembles two cover types therein. If this occurs in plains, Rule 3 may be invoked. If this happens in mountainous areas, it is safer to retain both types of land covers for this sample unit since geolocation accuracy may be lower than that in plains, and there is uncertainty in the area percentage of these two cover types. If neither of the two types can be determined, the site belongs to a low-quality verification point and will be eliminated from the precision evaluation (Table 4.2, invalid sample).

As shown in Figure 4.14(a), sample unit N462520101435 is located in the forest-shrub transition zone, as shown in the image subset October 21, 2009, on the left. In Figure 4.14(b), the sample unit is near the boundary of farmland and building (image subset November 28, 2015 on the right). Considering the possible position errors, it is difficult to label the unit by using area proportion alone. The robust interpretation should include the types on both sides of the boundary.

Rule 5 concerns the confidence level of interpretation. Due to the complexity of land cover, the quality of reference data, and the difference between interpreters concerning the problem domain,

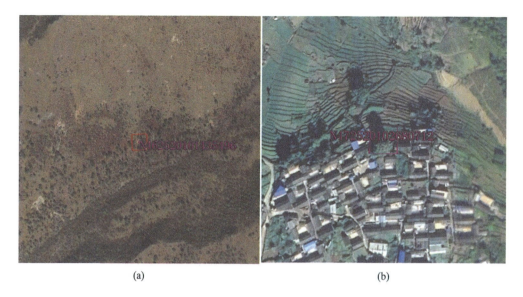

FIGURE 4.14 Possible scenarios for assigning two LC classes at a SAU. (a) The transition zone of "forest – shrub." (b) The transition zone of "farmland – artificial building."

the confidence level of interpretation will be different and limited. This determines which interpretation can finally be taken as the reference classifications. Specifically, the confidence levels are defined as follows:

- High (Conf. value = 1): The interpreter is SURE that there is only one dominant class in the sample unit, based on Table 4.1 and Rule 3 1~3.
- Medium (Conf. Value = 2): The interpreter is sure that there are only two dominant classes (Class_1, Class_2) in the SAU, based on Table 4.1 and Rule 4.
- Low(Conf. Value = 3: The interpreter is sure that there are neither one nor two dominant classes in the SAU, based on Table 4.1 and Rule 4.

Rule 6 refers to the assertion that blind interpretation should be the default process for reference class interpretation. Thus, the interpreter is not informed of map classification for a specific sample unit but relies solely on reference images and other auxiliary data to determine reference classification at the unit, as it is believed that blind interpretation can prevent interpreters from biases of preconception. Finally, some of the key points are summarized in Table 4.2.

4.3 REMAPPING LAND COVER USING FINER-RESOLUTION MAPS

4.3.1 KNOWLEDGE-BASED METHODS

The knowledge here mainly refers to common sense, evolution laws, and processes related to the distributions and changes of land cover. Existing reference data, such as a finer-resolution land cover map and their derivatives, can be used to augment information about land cover and form prior knowledge about distributions, patterns, and changes in land cover. The knowledge includes physical geographic knowledge of land cover, cultural knowledge of land cover, and knowledge of land cover changes, etc. After summarizing and refining these kinds of knowledge, rules for reference sample classification are formed to support the remapping of existing mapped land cover.

Natural conditions, human activities, and ecological environments affect the geospatial distribution and temporal transformation of land cover (Lu and Weng 2007). These factors are related to the

Sample Judgment

121

TABLE 4.3

Examples of Geospatial Knowledge and Classification Rules

Knowledge Types	Knowledge Statements	Classification Rules
Nature-based knowledge	The elevation of cultivated land is usually distributed below 4,700 m in Tibet.	Elevation cultivated land < 4,700 m
	Water bodies generally lie in relatively flat areas.	Slope water < 5°
	Ice and snow exist in the mountains where the elevation is higher than the local snow line.	Elevation ice and snow < Elevation snowline
	......	......
Culture-based knowledge	Cultivated land and artificial surfaces (AS) usually accompany each other in a certain region.	Cultivated land and AS
	There is consistency between ice and snow in GlobeLand30 and World Glacier Inventory.	Ice (WGI)
	The difference between GlobeLand30 and statistical data (SD) of land cover should be at a certain threshold.	$|SD_{GlobeLand30} - SD_{other}| <$ Threshold
	......	......
Temporal-constraint knowledge	Most AS continue to exist.	$Area_{AS-2010} - Area_{AS-2000} > 0$
	Permanent snow and ice (PSI) have a trend of shrinking with global warming.	$Area_{PSI-2010} - Area_{PSI-2000} < 0$
	......	......

aforementioned three kinds of knowledge, which can be represented as rules. Some examples of knowledge-based rules are listed in Table 4.3.

(1) Physical geographic knowledge-based rules

The geographic distribution of vegetation on the earth's surface is controlled by climate that is affected by latitude, longitude, and topography. The vegetation distribution can thus be characterized by the latitudinal, longitudinal, and vertical zonalization. From the west to the east of China, vegetation type changes from desert through steppe and forest-steppe to forest. In fact, land cover types in a region are often bound to environmental variables, such as precipitation, elevation, aspect, and slope (Judex et al. 2006). One example of geospatial knowledge is that water bodies and wetlands are usually distributed in relatively flat areas where the slope is less than 5. A classification rule can thus be formulated as slopewater < 5°, i.e., slopes of the locations where water bodies are located should be less than 5°. Otherwise, these locations might represent potential misclassification errors. This rule can be realized by calculating slopes from a DEM and conducting a rule-based comparison. Cultivated land usually lies below 4,700 m in Tibet because the climate above that altitude is not suitable for cultivating. This is represented as a rule: elevation cultivated land <4,700 m. Ice and snow exist only in the mountains where the elevation exceeds the snow line. A careful verification is needed if the elevation of ice and snow is less than the snow line.

(2) Cultural knowledge-based rules

Certain relations exist between land cover and cultural activities, as well as other social, economic, and political events. Such geospatial knowledge may help in the verification of GLC datasets and thus improve the classification of land cover (Chen et al. 2015). For instance, cultivated land and AS usually accompany one another. If only cultivated land is present in a large area and AS are missing, a misclassification might be indicated. In addition, verification can be conducted with the help of land cover–related census data or inventory statistics, such as the global SD for forests, cultivated land, and wetlands. Some gross classification errors may be revealed by comparing statistics from GlobeLand30 and land cover–related census data. Moreover, a number of national, regional, and global land cover/use databases have been developed in the past two decades, such as the national land use/cover database of China at 1:100,000 scale (Zhang et al. 2014), GlobCover2009 (Bontemps et al. 2011), and the World Glacier Directory (Cogley and Graham 2010). These databases provide valuable knowledge with which to verify GlobeLand30 and other GLC products.

(3) Temporal-constraint-based rules

Temporal consistency needs to be checked in post-classification verification of GLC products, especially those with multiple base-year maps. This is approached by summarizing the temporal constraints between successive land cover data sets. For instance, most AS will continue to exist for quite a long time and will not suddenly disappear once in confirmed presence. This can be formulated as a verification rule for the temporal consistency of AS: AreaAS-2010-AreaAS-2000 > 0. Using this verification rule, the spatial difference between GlobeLand30-2000, GlobeLand30-2010, and GlobeLand30-2020 can be checked to find potentially misclassified regions where AS decreased significantly from 2000 to 2010. Another example is that glaciers and permanent snow cover show a trend of shrinking with global warming. The resulting verification rule is that the area of ice and snow in 2010 should be less than that in 2000. This helps to reduce misdetected changes caused by misclassification or inconsistency in the original GLC products.

4.3.2 Using Maps of Different Spatial Resolutions

In general, the resolution of reference source maps of land cover needs to be finer than that of the land cover products to be verified. This section describes methods of classification when using maps of different resolutions.

Resampling (i.e., changing map resolution) is a common preprocessing operation before assessing the accuracy of a land cover map using another map as reference. Unquestionably, when resolution is made finer, some information will be lost in the presence of mixed pixels or low-resolution (LR) bias. Boschetti et al. (2004) introduced a measure of LR bias based on omission error (OE) and commission errors (CE) together with Pareto boundary that can be used for determining optimal trade-off between OE and CE.

Figure 4.15 helps explain the concept of LR bias and how to graph the Pareto boundary using a binary map with only one class of interest denoted ω_1 (against the background, denoted ω_0).

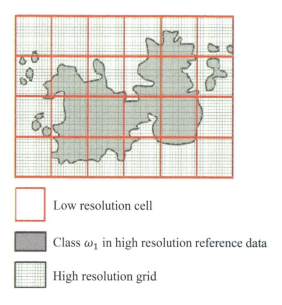

Low resolution cell

Class ω_1 in high resolution reference data

High resolution grid

FIGURE 4.15 A HR map overlaid with LR map and HR grid cells.

Specifically, Figure 4.15 shows a high-resolution (HR) binary map with finer grid cells, with coarser grid cells (in red) representing the resampled (upscaled) LR map.

During the resampling process, the HR pixels of the HR map falling in the cell of the LR grid cell of the LR map will merge in one LR pixel. The newly assigned values of the LR pixel depend on the prevalent class among the HR pixels. However, the prevalence threshold t must be determined.

Figure 4.16 shows that the LR grid cells will be populated by ω_1 labels, which are proportional to prevalence thresholds chosen. The thresholds are the percentage of an LR cell occupied by ω_1. It also shows the number of HR cells containing ω_1 in the total number of HR cells within an LR cell.

Different thresholds will give rise to a different hypothetical LR map after resampling. This is illustrated in Figure 4.17.

Furthermore, computing CE and OE for different hypothetical LR maps using the HR map as a reference map results in the Pareto boundary. This is shown in Figure 4.18.

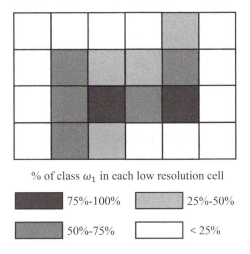

FIGURE 4.16 Resampling from HR to LR concerning different prevalence thresholds.

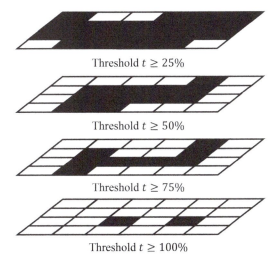

FIGURE 4.17 Different prevalence thresholds lead to different LR maps.

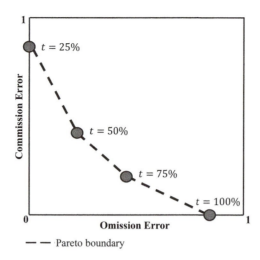

FIGURE 4.18 An example Pareto boundary derived based on omission (OE) and CE pairs of four different prevalence thresholds.

TABLE 4.4
Equations for Computing OE and CE in Continuous and Discrete Cases

Continuous Case	Discrete Case	
$A_L(t) = \int_t^1 N_{Li} d_i$	$A_L(t) = \sum_{i=t}^{1} N_{Li}$	(4.1)
$A_H = \int_{0+}^1 i N_{Li} d_i$	$A_H = \sum_{i=t}^{1} i N_{Li}$	(4.2)
$O_L(t) = \int_{0+}^1 i N_{Li} d_i$	$O_L(t) = \sum_{i>0}^{t} i N_{Li}$	(4.3)
$C_L(t) = \int_t^1 (1-i) N_{Li} d_i$	$C_L(t) = \sum_{i=t}^{1} (1-i) N_{Li}$	(4.4)

Every time an LR cell is partially covered by ω_1, its accuracy decreases. If such a cell/pixel is classified as ω_1, CE increases; CE decreases if classified as ω_0. Therefore, in the presence of mixed pixels, errors are inevitable. In addition, the magnitude of OE and CE depends on the prevalence threshold.

In terms of total error, all solutions on the Pareto boundary are considered optimal. The user's cost function can be used to estimate what is more efficient from the user's perspective. (For more information on user's cost function, please refer to Boschetti et al. 2004).

Since a decrease in OE leads to an increase in CE and vice versa, the Pareto boundary is useful for understanding which threshold of prevalence keeps both of the errors optimally balanced. OE and CE can be computed as shown by Eqs (4.1)–(4.4) (Table 4.4), taking advantage of the area covered by the class of interest in the two maps (HR and LR) to overcome the problem due to different resolutions.

Sample Judgment **125**

In Eqs (4.1)–(4.4), i is the fraction of LR cell coverage by ω_1, N_{Li} is the number of LR cells having coverage of i exactly, $A_L(t)$ is the area of LR map occupied by ω_1 at threshold t, $A_H(t)$ is the area of HR map occupied by ω_1, OL is the area omitted by applying threshold t, and A_L is the area committed by applying threshold t.

Finally, omission and CE are computed as follows:

$$C_e(t) = C_L(t) / A_L(t) \tag{4.5}$$

$$O_e(t) = O_L(t) / A_H(t) \tag{4.6}$$

The application of the equations to each prevalence threshold leads to the derivation of the Pareto boundary based on the OE/CE pair for each threshold t. The optimal solution is indicated by the Pareto boundary, meaning that the area below the Pareto boundary (closer to 0.0) is not reachable. In general, the closer the Pareto boundary is to 0.0 the smaller the errors due to the mixed pixels in general.

At this point, it is possible to calculate the OE and CE of the resampled classified map using the error matrix and compare it with the Pareto. If the OE/CE pair lies on the Pareto frontier, it means that all the errors in the class are due to the mixed pixels. On the contrary, if the OE/CE pair is located away from the Pareto frontier, only part of the error will be incurred because of mixed pixels, while the rest is due to misclassification or other reasons. Thus, the Pareto boundary is not only useful for giving information about the type of errors present but also for quantifying it.

Although the procedure is described as a binary map, it is also applicable to multiclass maps by reclassifying them to 1's for a particular class of interest and 0's for the rest. This procedure of reclassification is run iteratively for each of the classes present in the map.

4.3.3 USING MAPS OF DIFFERENT CLASSIFICATION SYSTEMS

When the classification system of a reference land cover map is different from that of the land cover product to be verified, conversion of classification systems is required. Consider using ChinaCover30 to verify GlobeLand30. ChinaCover30, circa 2010, is a 30 m resolution product developed by the Institute of Geographic Sciences and Resources, Chinese Academy of Sciences. It includes 6 first-level categories and 37 second-level categories of land cover, with a reported accuracy of 90%.

Table 4.5 shows the ChinaCover30 land cover classification system. A comparison between the classification systems of ChinaCover30 and GlobeLand30 indicates that most of the land cover classes of ChinaCover30 (second level) can be directly converted to GlobeLand30. However, there is some uncertainty in the class of sparse grassland (code 63). In ChinaCover30, sparse grassland belongs to bare land with a vegetation cover of 4%–15%; In GlobeLand30, bare land refers to land with vegetation cover below 10%. Clearly, there is a slight discrepancy in the definition of these two classes. This may be ignored in other regions, except RoTW. The main reason is that there is an ecological gradient in low vegetation cover in RoTW. If we simply convert sparse grassland in ChineCover30 to bare land or grassland in GlobeLand30, misclassification errors may occur. A safer approach is to use visual interpretation to further determine the correct classes.

Therefore, according to the classification definition of Globeland30 (Table 1.3), the Chinacover30 classification system can be converted to the Globeland30 classification system (see Table 4.6). It should be noted that not all classes of land cover maps could be converted into those of reference source maps. For example, vegetation cover of the three first-level classes in Chinacover30 (code 1, 2, 3) cannot be directly translated into Globeland30 classes.

TABLE 4.5
The Classification System of ChinaCover30

Code	First-Level Class	Code	Second-Level Class	Description
1	Woodland	101	Evergreen broad-leaved forest	Natural or seminatural, $H = 3–30$ m, $C > 15\%$
		102	Deciduous broad-leaved forest	Natural or seminatural, $H = 3–30$ m, $C > 15\%$
		103	Evergreen coniferous forest	Natural or seminatural, $H = 3–30$ m, $C > 15\%$
		104	Deciduous coniferous forest	Natural or seminatural, $H = 3–30$ m, $C > 15\%$
		105	Mixed forest	Natural or seminatural, $H = 3–30$ m, $C > 15\%$, $25\% < F < 75\%$
		106	Evergreen broad-leaved shrub forest	Natural or seminatural, $H = 0.3–5$ m, $C > 15\%$
		107	Deciduous broad-leaved shrub forest	Natural or seminatural, $H = 0.3–5$ m, $C > 15\%$
		108	Evergreen needle prickly shrub forest	Natural or seminatural, $H = 0.3–5$ m, $C > 15\%$
		109	Orchard (tree)	Artificial vegetation, $H = 3–30$ m, $C > 15\%$
		110	Orchard (shrubs)	Artificial vegetation, $H = 0.3–5$ m, $C > 15\%$
		111	Urban green space (trees)	Artificial vegetation, $H = 3–30$ m, $C > 15\%$
		112	Urban green space (shrubs)	Artificial vegetation, $H = 0.3–5$ m, $C > 15\%$
2	Grassland	21	Pastoral grassland	Natural or seminatural, $K > 1.5$, $H = 0.03–3$ m, $C > 15\%$
		22	Grassland	Natural or seminatural, $K = 0.9–1.5$, $H = 0.03–3$ m, $C > 15\%$
		23	Grassland	Natural or seminatural, $K > 1.5$, $H = 0.03–3$ m, $C > 15\%$
		24	Grassland (city)	Artificial vegetation, $H = 0.03–3$ m, $C > 15\%$
3	Wetland	31	Wetland (trees)	Natural or seminatural, $T > 2$, $H = 3–30$ m, $C > 15\%$
		32	Wetland (shrubs)	Natural or seminatural, $T > 2$, $H = 0.3–5$ m, $C > 15\%$
		33	Wetland (grassland)	Natural or seminatural, $T > 2$, $H = 0.03–3$ m, $C > 15\%$
		34	Lake	Nature, still water
		35	Reservoir/pond	Artificial, still water
		36	River	Nature, still water
		37	Canal/waterway	Artificial, still water
4	Cropland	41	Cropland (rice field)	Artificial, aquatic crops
		42	Cropland (dryland)	Artificial, dry crops
5	Artificial	51	Residence	Built-up area, residential
		52	Industrial land	Built-up area, industrial land
		53	Transport land	Built-up area, transport land
6	Bare land	61	Sparse wood	Natural or seminatural, $H = 3–30$ m, $C = 4\%–15\%$
		62	Sparse shrubs	Natural or seminatural, $H = 0.3–5$ m, $C = 4\%–15\%$
		63	Sparse grassland	Natural or seminatural, $H = 0.03–3$ m, $C = 4\%–15\%$
		64	Moss/green moss	Nature, moss/lichen
		65	Exposed rocks	Nature, exposed rocks
		66	Bare soil	Nature, bare soil
		67	Desert/sandy land	Natural loose sand
		68	Saline alkali land	Nature, saline alkali, and alkaline soil
		69	Permanent snow and ice	Nature, permanent snow, and ice (in the hottest month)

Note: H; Height (meters); *C*: Coverage rate (%); *F*: Percentage of evergreen coniferous forest (%); *T*: Flood month; *K*: Humidity index

TABLE 4.6
Classification System Conversion ChinaCover30 to GlobeLand30 in RoTW

ChinaCover30			GlobeLand30
	First Level	**Second Level**	**LCType: Code**
1	Woodland	101 ：102 ：103 ：104 ：105	20
		106 ：107 ：108	40
		109 ：110	10
		111	20
		112	40
2	Grassland	21 ：22 ：23 ：24	30
3	Wetland	31 ：32 ：33	50
		34 ：35 ：36 ：37	60
4	Cropland	41 ：42	10
5	Artificial land	51 ：52 ：53	80
6	Bare land	61 ：62 ：63	Visual interpretation is required to determine vegetation coverage below 10%
		64 ：65 ：66 ：67 ：68	90
		69	100

4.4 LABELING LAND COVER CLASSES USING POI DATA

With the application and development of geospatial big data, many scholars have attempted to make use of heterogeneous geospatial data, such as geotagged photos, trajectories, and points of interest (POIs), for land cover mapping (Sitthi et al. 2016; See et al., 2015; Lu et al., 2015; Hu et al. 2016; Meng et al. 2017; Xing et al. 2017a; Xing et al. 2017b). Existing methods primarily focus on location, image features, human activities, and socioeconomic characteristics, which are part of or derived from geospatial data, while overlooking the semantic information carried by the texts themselves. For instance, the semantics of POI texts, such as restaurant, golf course, and seaside resort, correspond to AS, grassland, and water bodies, respectively. Clearly, the semantic information in POI texts can be utilized to obtain reference sample data for land cover verification. Research efforts are exemplified by studies on using textual and location information in POI datasets as the basis for classification, POI documents, and latent Dirichlet allocation (LDA) topic modeling (Blei et al. 2003), as illustrated in Figure 4.19. These methods are described below.

4.4.1 LABELING BASED ON POI CATEGORIES AND LDA MODEL

LDA modeling can be applied for categorizing POI texts into topics based on their semantic similarities (Adams and Janowicz 2015; Lansley and Longley 2016). As Figure 4.20 shows, the LDA model defines documents as input data. It assumes that documents are determined by a certain distribution of topics Z and words W, where words W are constructed from topics Z. The parameters θ and ϕ represent the distribution probabilities of topics Z in documents and words W in topics Z, respectively. Additionally, hyperparameters α and β, as model input parameters, influence the distribution probabilities of topics $Z(\theta)$ and words $W(\phi)$. In Figure 4.20, M represents the number of documents, N the number of words W in each document, and K the number of topics Z.

The LDA model is applied to classify the semantics of POI texts and compute their topic distributions. POI texts are treated as words W, whereas POI texts with similar semantics are considered as a single topic Z. First, the model needs to construct document d for model input. Considering that nearby POIs may contain similar or identical text semantics, the study units are often represented as convex hull polygons. Thus, all POI text words W falling within the same convex hull polygon are

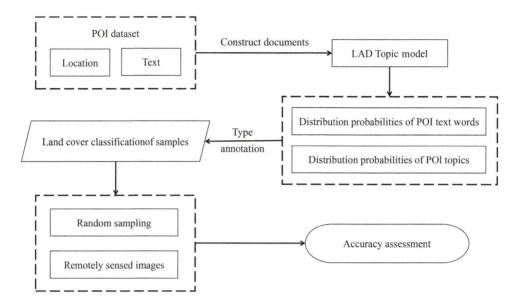

FIGURE 4.19 Land cover labeling using textual analysis of POI data.

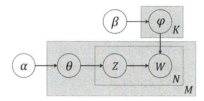

FIGURE 4.20 An LDA topic model.

treated as a single document. Then, the documents are applied to the LDA topic model to compute the distribution probabilities θ of topics Z in documents and the distribution probabilities ϕ of POI text words W in each topic. The formulas concerned are as follows:

$$\theta = \frac{n_d^{(K)} + \alpha_Z}{\sum_{Z=1}^{K} n_d^{(K)} + \alpha_Z}, \tag{4.7}$$

$$\phi = \frac{n_K^{(W)} + \beta_W}{\sum_{W=1}^{V} n_K^{(W)} + \beta_W}, \tag{4.8}$$

where $n_d^{(K)}$ represents the number of occurrences of topic Z in document d, $n_K^{(W)}$ the number of occurrences of word W in topic Z, and V the total number of words W.

Parameters should be set properly for LDA modeling. In the LDA topic model, hyperparameters such as α, β, and the number of topics K determine the distribution probabilities θ of topics Z in documents and the distribution probabilities ϕ of the POI text words W in each topic. The values of α and β can be optimized based on the existing trained models, and in this case, we adopt the optimized parameters from literature (Griffiths et al. 2004), setting $\alpha = 50/K$ and $\beta = 0.1$. To determine the appropriate number of topics K, which depends on the semantic content of the POI texts, we use the perplexity algorithm to evaluate the classified topics with specific K values. The perplexity value indicates the uncertainty of a document belonging to a particular topic, with a lower perplexity value indicating that topic number K is more suitable to the LDA topic model. Perplexity is defined as:

$$\text{perplexity}\,(d) = \exp\left\{-\frac{\sum_{d=1}^{M}\log p\,(W_d)}{\sum_{d=1}^{M} N_d}\right\}, \tag{4.9}$$

where $p(W_d)$ denotes the probability that words W is within the document d, N_d the total count of words W in document d, and M the total number of documents in the corpus.

Finally, consider topic calculation based on POI texts. Once model parameters α, β, and K are determined, it is necessary to compute the probability matrix P for the distribution of topics Z in documents, as well as the probability matrix R for the distribution of POI text words W in each topic. The construction of these matrices is as follows:

$$P = \begin{bmatrix} \theta_{ii} & \theta_{ij} & \theta_{ik} & \theta_{il} \\ \theta_{ji} & \theta_{jj} & \theta_{jk} & \theta_{jl} \\ \theta_{ki} & \theta_{kj} & \theta_{kk} & \theta_{kl} \\ \theta_{li} & \theta_{lj} & \theta_{lk} & \theta_{ll} \end{bmatrix} \tag{4.10}$$

$$R = \begin{bmatrix} \phi_{ii} & \phi_{ij} & \phi_{ik} & \phi_{il} \\ \phi_{ji} & \phi_{jj} & \phi_{jk} & \phi_{jl} \\ \phi_{ki} & \phi_{kj} & \phi_{kk} & \phi_{kl} \\ \phi_{li} & \phi_{lj} & \phi_{lk} & \phi_{ll} \end{bmatrix} \tag{4.11}$$

where the row indices i, j, k, l of matrix P represent the ith, jth, kth, and lth topics, respectively, while the column indices represent the ith, jth, kth, and lth documents. For matrix R, the row indices represent POI types, and the column indices represent topics. For instance, θ_{ij} represents the distribution probability of the ith topic in the jth document, while ϕ_{jk} represents the distribution probability of the jth POI type under the kth topic. The probability matrix R for POI text words W reflects the credibility of topic classification, while the probability matrix P for topic distribution in documents reflects the distribution of sample types.

4.4.2 EXPERIMENTS

The study area is in Shandong Province, China. The experimental dataset consists of POI data from Sina Weibo and Baidu Maps, with the former being a popular social network and the latter a navigational map. POI data were obtained through the Weibo API and Baidu Map API, respectively, which are open to the public. We used POIs with their categories as textual information for topic analysis. Before putting them into the topic model, POIs with unknown classifications (hence unable to have land cover types determined) were removed. The left and right subfigures of Figure 4.21 show POIs distribution within urban and forested areas, respectively.

Before extracting sample tags from the POI data, sample units were identified through spatial clustering methods, such as the DBSCAN algorithm. This approach is believed to be instrumental in identifying areas with a high density of POIs, followed by using the convex hull technique to delineate the boundaries of these areas. Figure 4.22 presents some sample units identified by the methods, showing 26,185 of them for both urban and forest settings extracted within Shandong Province.

In this study, we adopted the value for parameter α as proposed by Griffiths et al. (2004) and set $K = 30$ based on a sensitive analysis using a range of K values from 10 to 100 to compute perplexity. As shown in Table 4.7, $K = 30$ has a relatively low value of perplexity.

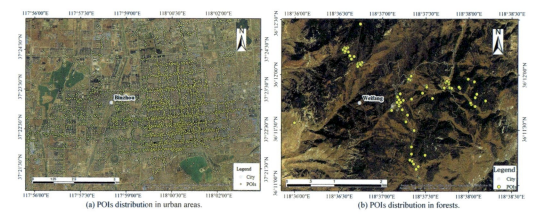

FIGURE 4.21 POI distributions within urban (a) and forest (b) cover types in the study area.

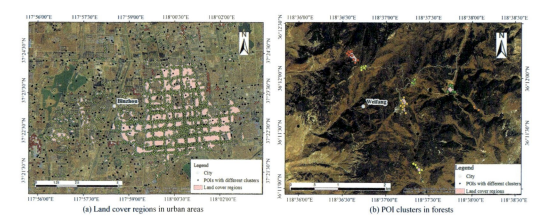

FIGURE 4.22 POI clusters in urban (a) and forest (b) cover types in the study area.

TABLE 4.7
Perplexity Values With Different Numbers of Topics K

K	10	20	30	40	50
Perplexity	370.2218	323.1828	283.7695	267.0446	255.2151
K	60	70	80	90	100
Perplexity	243.9839	238.6872	231.415	227.0478	222.6675

The POI topics with spatial distributions were calculated by proposing an LDA topic model. Three land cover topics with high weights of distributions in Figure 4.21 are shown in Table 4.8. Topic 6 shows POI categories that are related to city entertainment, and topic 13 indicates the topic corresponding to food. Both topics can refer to AS, as they are related to human activities. These two topics occupy a large proportion of land cover in Figure 4.22(a). On the other hand, topic 23 represents outdoor travel-related categories, which can give a clue to non-AS. This topic is dominant in areas in Figure 4.22(b).

While POIs are utilized to classify land cover types, an accuracy assessment should be done to evaluate land cover classification. Random sampling with 500 sample units was carried out three

Sample Judgment

131

TABLE 4.8
Distributions of Topics in Several Sample Locations

Topic 6	Topic 13	Topic 23
Shopping Malls	Chinese Restaurant	Tourist Attractions
Supermarket	Shandong Restaurant	Scenic Area
Beauty Salon	Winehouse	Ticketing
Dessert Shop	Cinema	Ancient Artifacts
Photo Studio	Café	Parking Areas

TABLE 4.9
Land Cover Classification Accuracy (NS: Nonartificial Surfaces, AS: Artificial Surfaces)

Samples	500 (%)		500 (%)		500 (%)	
Land Cover Type	NS	AS	NS	AS	NS	AS
PA	28.89	85.93	25.53	85.21	20.20	86.44
UA	16.88	92.43	15.19	91.69	14.08	90.68
OA	80.81		79.60		79.84	

times in the study area and was interpreted using remotely sensed images. The accuracy assessment results are shown in Table 4.9, where producer's accuracy (PA) of AS is around 85%, and user's accuracy (UA) of this cover type is even greater, with an accuracy of about 91%. On the other hand, PA and UA of non-AS are much lower, both below 30%. The unsatisfying accuracy of non-AS is due to the low proportion of nonartificial regions in the areas analyzed. OA obtained with the three samples in Table 4.9 is 80.8%, 79.6%, and 79.8%.

4.5 CONFIDENCE IN INTERPRETATION FOR REFERENCE CLASSIFICATIONS

4.5.1 DEGREE OF TRUST

Validation for GLC products is influenced by many factors, such as global ecological distribution, global phonological conditions, and time differences, which can lead to uncertainty. The blind, plausibility, and enhanced plausibility modes of interpretation by experts or volunteers have been studied. The confidence level of interpretation can be established under uncertain conditions. These contribute to the credible evaluation of global land cover map accuracy.

The sample credibility is divided into four levels: Level A, absolutely correct; Level B, absolutely incomplete; Level C, uncertain (due to reference data source quality); and Level D, uncertain (due to personal knowledge and interpretation ability).

Therefore, the four levels of the degree of trust in sample interpretation can be determined to show the accuracy of the land cover map. Potential erroneous sample data that will influence the final accuracy evaluation can be identified, helping to ensure the credibility of sample data quality and subsequent product validation.

4.5.2 QUALITY ISSUES

Reference sample image subsets interpretation involves interpreters' logical and subjective judgment based on the given source data. Two possible factors can affect the results: inadequate sources and insufficient knowledge. For interpretation based on HR images, in particular, there are issues with low-quality reference data sources, hard-to-interpret mixed cover types, and differences in personal skills and/or experience of image interpreters.

(1) Issues with reference data sources

Some samples are difficult to interpret due to limitations of reference data sources, such as no data, poor quality, and cloud contamination (Figure 4.23). Despite the increasing maturity of remote sensing technology, it remains a challenge to acquire quality HR images over any place at any time. Without quality HR images, it is hard to determine reference classes through visual interpretation. Poor quality of image mosaicking and chromatic aberration also affect reliability of image interpretation. Finally, cloud contamination makes sample image subsets unusable for visual interpretation.

(2) Sample image subsets unusable for visual interpretation

The collection of reference sample data relies mainly on visual interpretation. Sample image subsets (e.g., SAUs) often fall on transition zones of multiple land types and locations of mixed cover types, which make unambiguous labeling extremely difficult, if not impossible. Such sample image subsets are therefore considered invalid or unusable (Figure 4.24). Boundary areas between two land types are prone to misclassification errors. An example is shown in the first row of Figure 4.24, illustrating boundaries between wetlands and water bodies, between grassland and snow/ice, and between grassland and forest. Some boundaries change with climate, phenology, and seasons. For example, winter ice and snow have more obvious boundaries with grasslands. However, when ice and snow melt and disappear while grass grows green, the aforementioned boundaries will change, leading to ambiguity in the class labels obtained.

Sample units over locations of highly mixed cover types are also difficult to interpret for reference class labels. Examples include mixtures of cultivated land, shrubs, grasslands, vegetation, crops, and forests in the second row of Figure 4.24. There are also differences

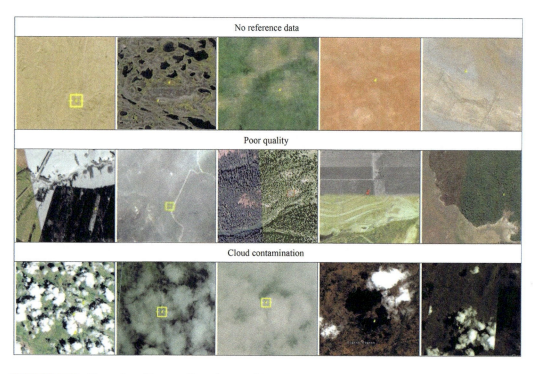

FIGURE 4.23 Examples of low-quality reference data sources.

Sample Judgment 133

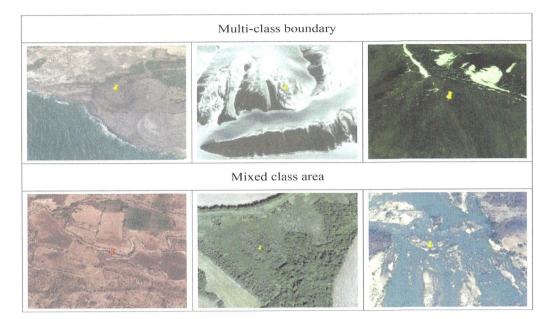

FIGURE 4.24 Examples of problematic sample image subsets.

in cover types in different biomes. Over a diverse and complex African landscape, various cover types, such as bare land, grassland, shrubs, and cultivated land, are often intertwined. Imaged scenes over land cover mixture are bound to be hard to interpret visually alone.

(3) Differences among individual interpreters

For the same cover types, different interpreters often come up with different class labels. Figure 4.25 shows images of the same object (a tree) from different sources, where the left one is an HR image subset from Google Earth, while the right one is sourced from photo archives in the Degree Confluence Project (DCP), with photos provided by volunteers worldwide.

For the left image in Figure 4.25, due to the lack of three-dimensional information or contextual information about the surroundings, the location may be interpreted as grassland or shrub. On the other hand, it is straightforward to label the object as a tree based on what is shown in the right image. This clearly pinpoints the inconsistency in class labeling due to different interpretations using different source images.

FIGURE 4.25 Different class labeling using different source images.

4.6 SUMMARY

Sample judgment through interpretation and verification is vital to the collaborative validation of land cover maps. Usually, validation experts or volunteers make subjective judgments on reference samples' most probable classes through the integrated use of reference source data, geospatial knowledge, and enhanced protocols and tools.

This chapter mainly discusses how to interpret or obtain reference sample classifications based on HR images, finer-resolution land cover maps, and crowdsourcing information. Generally, there are three interpretation modes: blind, plausibility-based, and enhanced plausibility-based. There are several aspects of interpretation for consideration, serving as guidelines: area priority, majority decision, temporal coherence, and classification system transformation and hominization. The whole procedure of interpretation also includes the selection of SAUs and the use of advanced online verification tools. Interpretation, judgment, reasoning, and analysis are all involved in getting reliable reference class labels for sample units, although this demanding process is not elaborated on in much detail in this chapter. Then, based on the availability of reference data and the trustworthiness of interpreters, the reliability of resultant reference sample interpretation and classification is graded, helping to facilitate subsequent validation of the land cover map.

With the popularization of internet technology and the development of geographic big data, crowdsourcing information, such as POI, social media information, and geotagged photos, can be used to extract usable reference sample data through geographic information modeling. This method does not benefit from statistical sampling and verification due to limited samples. Nevertheless, it is cost-effective to obtain proxy reference land cover maps through internet mining and geographic information modeling, making it one of the promising research directions in the field.

REFERENCES

Adams, B. and Janowicz, K. 2015. Thematic signatures for cleansing and enriching place-related linked data. *International Journal of Geographical Information Science*, 29(4): 556–579.

Blei, D.M., Ng, A.Y. and Jordan, M.I. 2003. Latent dirichlet Al-location. *Journal of Machine Learning Research,* 3: 993–1022.

Bontemps, J. D., Hervé, J. C., Leban, J. M. and Dhôte, J. F. 2011. Nitrogen footprint in a long-term observation of forest growth over the twentieth century. *Trees*, 25: 237– 251

Boschetti, L., Flasse, S.P. and Brivio, P.A. 2004. Analysis of the conflict between omission and commission in low spatial resolution dichotomic thematic products: The Pareto boundary. *Remote Sensing of Environment*, 91(3–4): 280–292.

Chen, J., Chen, J., Liao, A., et al. 2015. Global land cover mapping at 30 m resolution: A POK-based operational approach. *ISPRS Journal of Photogrammetry and Remote Sensing*, 103: 7–27.

Cogley, J. G. 2010. A more complete version of the World Glacier Inventory. *Annals of Glaciology*, 50(53): 32–38.

Griffiths, T.L. and Steyvers, M. 2004. Finding scientific topics. *Proceedings of the National Academy of Sciences of the United States of America*, 101(S1): 5228–5235.

Hu, T., Yang, J., Li, X., et al. 2016. Mapping urban land use by using landsat images and open social data. *Remote Sensing*, 8(2): 151–168.

Judex, M., Thamm, H. P. and Menz, G. (2006, September). Improving land cover classification with a knowledge based approach and ancillary data. In Proceeding of the workshop of the EARSeL sig on Land Use and Land Cover (pp. 28–30).

Lansley, G. and Longley, P.A. 2016. The geography of twitter topics in London. *Computers Environment and Urban Systems*, 58: 85–96.

Lillesand, T., Kiefer, R. W. and Chipman, J. (2015). *Remote sensing and image interpretation*. John Wiley & Sons.

Lu Guozhen, Chang Xiaomeng, Li Qingquan, et al. 2015. Land use classification based on massive human-activity spatio-temporal data. *Journal of Geo-information Science*, 17(12): 1.497–1.505.

Lu, D. and Weng, Q. (2007). A survey of image classification methods and techniques for improving classification performance. *International Journal of Remote Sensing*, 28(5): 823–870.

Meng, Y., Hou, D. and Xing, H. 2017. Rapid detection of land cover changes using crowdsourced geographic information: A case study of Beijing, China. *Sustainability*, 9(9): 1547–1562.

Scepan, J., 1999. Thematic validation of high resolution global land-cover data sets. *Photogrammetric Engineering & Remote Sensing*, 65(9): 1050–1060.

See L, Schepaschenko D, Lesiv M, et al. 2015. Building a hybrid land cover map with crowdsourcing and geographically weighted regression. *ISPRS Journal of Photogrammetry and Remote Sensing*, 103: 48–56.

Sitthi A, Nagai M, Dailey M, et al. 2016. Exploring land use and land cover of geotagged social-sensing images using Naive Bayes classifier. *Sustainability*, 8(9): 921–942.

Xing, H., Meng, Y., Hou, D., et al. 2017a. Exploring point-of-interest data from social media for artificial surface validation with decision trees. *International Journal of Remote Sensing*, 38(23): 6945–6969.

Xing, H., Meng, Y., Hou, D., et al. 2017b Employing crowd-sourced geographic information to classify land cover with spatial clustering and topic model. *Remote Sensing*, 9(6): 602–621.

Zhang Z , Wang X , Zhao X, et al. 2014. A 2010 update of National Land Use/Cover Database of China at 1:100000 scale using medium spatial resolution satellite images. *Remote Sensing of Environment*, 149: 142–154.

5 Accuracy Assessment

Xiaohua Tong and Chao Wei
Tongji University, Shanghai, China

Jingxiong Zhang
Wuhan University, Wuhan, China

Maria Antonia Brovelli
Politecnico di Milano, Milan, Italy

Zhenhua Wang
Shanghai Ocean University, Shanghai, China

Yanmin Jin
Tongji University, Shanghai, China

Gorica Bratic
Politecnico Milano, Milan, Italy

5.1 INTRODUCTION

Accuracy assessment is a crucial component of product validation, referring to the independent evaluation of the quality of data products under study. Assessing the accuracy of global land cover (GLC) products is vital for understanding their potential utility and the possible impact of errors on their intended applications. For GLC products, accuracy assessment aims to provide information using indicators about how accurately map classifications reflect true cover types as indicated in reference classifications (Strahler et al. 2006).

Accuracy assessment is based on comparing land cover products to reference data (Stehman and Foody 2009). The accuracy of a product is often quantified as the percentage of the map areas that have been correctly classified in comparison with reference data. This accuracy measure is typically obtained by evaluating the correctness of map classifications against reference classifications at a set of sample units (pixels or polygons; Story and Congalton 1986).

Historically, the developments of methodology and practice of accuracy assessment have gone through four stages (Congalton 1994). The first is the initial stage, which consists of a visual inspection of maps without conducting any accuracy measurements. The second is non-site-specific assessment. This involves comparing the areal extents of land cover classes on the map to reference data. While informative, the focus here is not on pinpointing specific locations. The third is site-specific assessment. This means comparing actual ground locations to corresponding locations on the map to determine overall accuracy, typically expressed as a percentage of correctly classified pixels. This technique was dominant until the late 1980s. The fourth is the so-called current approach, which works on the basis of confusion matrices estimated by comparing reference classifications at sample units to corresponding map classifications. This method offers a more comprehensive evaluation of map accuracy.

Accuracy Assessment

Acknowledging the assumptions underlying accuracy assessment in land cover maps is crucial for a comprehensive understanding and meaningful interpretation of validation results. Fulfillment of the requirements related to these assumptions, such as accuracy of reference data, accurate co-registration, purity of map pixels, and temporal consistency, helps to ensure the reliability and validity of accuracy assessments.

Many widely promoted and used accuracy indicators are primarily derived from confusion matrices. Based on these accuracy indicators, several accuracy assessment methods have been proposed. Accuracy indicators are to be described in Section 5.2. These include individual-class accuracy indicators and all-class accuracy indicators, which are derived from confusion matrices and other statistical constructs concerning classification. In Section 5.3, we delve into accuracy assessment methods. Figure 5.1 shows the major accuracy indicators and accuracy assessment methods.

5.2 ACCURACY INDICATORS

5.2.1 Confusion Matrices, Overall Accuracy, and Kappa Coefficients of Agreement

Confusion matrices, also known as error matrices or contingency tables, provide the counts or proportions for assessing classification accuracy and have been used in remote sensing for a long time (Congalton 1991; Congalton, Oderwald, and Mead 1983; Czaplewski and Catts 1992, Rosenfield and Fitzpatrick-Lins 1986). A confusion matrix is derived by comparing reference classifications and map classifications (or image classifications) and counting the number of spatial units for each class combination (Hay 1988).

Table 5.1 shows an example confusion matrix, where columns represent reference classifications, and rows represent map classifications. For example, cell n_{21} in the confusion matrix represents the count of spatial units that are classified in class 2 but belong to class 1 per the reference labeling. Cells on the diagonal represent counts of correctly classified spatial units, while cells off the diagonal represent counts of incorrectly classified spatial units. In the case of an ideally classified map, the confusion matrix would have the values of all off-diagonal cells equal to zero.

When a confusion matrix includes all spatial units in a study area, it becomes the population confusion matrix (Pontius Jr. and Millones 2011) since it includes the whole population of spatial units in the study area when tallying the confusion matrix. This kind of confusion matrices is tallied when another land cover map is used as a reference. However, it is more often the case that reference data are collected through field surveys or photo interpretations since it would be time-consuming and costly to collect information for each and every spatial unit. This is especially true for large study areas (e.g., global or continental land cover maps). Therefore, a sampling strategy is usually employed to determine the number of locations where representative reference sample data are to be collected (Stehman 2009, Stehman and Czaplewski 1998). To account for the area omitted by sampling, the confusion matrix is estimated by computing the proportions (probabilities) of spatial units of all combinations of reference classes and map classes (Congalton and Green 2019). To do so, it is necessary to apply Equation (5.1) on every cell of the sample confusion matrix tallied using the sample data.

$$p_{ij} = \pi_i \left(\frac{n_{ij}}{n_{i+}} \right), \tag{5.1}$$

where π_i is the proportion of class i in the study area; π_i is also called marginal proportion for class i, and it is not necessarily derived from the confusion matrix but is preferably obtained from more reliable sources. Based on Equation (5.1), the confusion matrix is derived with individual cell probabilities estimated, as shown in Table 5.2.

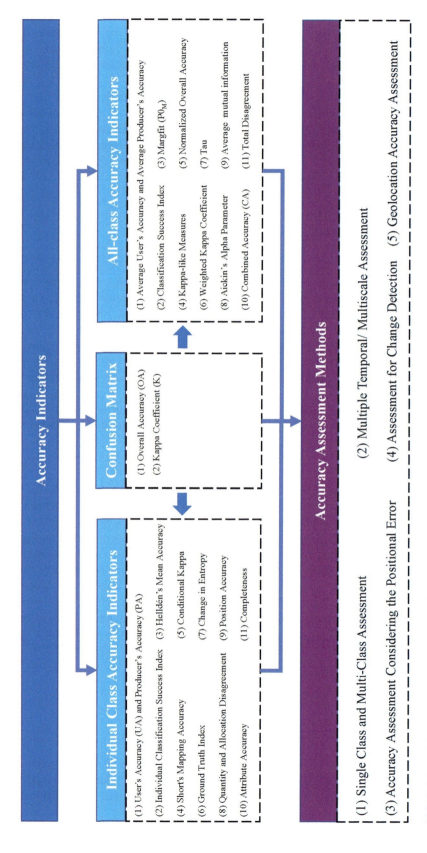

FIGURE 5.1 Major accuracy indicators and accuracy assessment methods.

TABLE 5.1
An Example Confusion Matrix

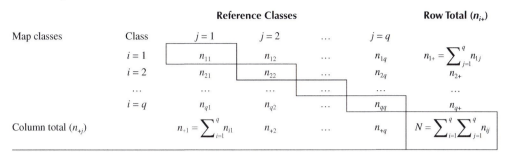

TABLE 5.2
Confusion Matrix of Class Probabilities

	Class	Reference Classes $j=1$	$j=2$	...	$j=q$	Marginal Proportion (p_{i+})
Map classes	$i=1$	p_{11}	p_{12}	...	p_{1q}	$p_{1+} = \sum_{j=1}^{q} p_{1j}$
	$i=2$	p_{21}	p_{22}	...	p_{2q}	p_{2+}
	...	...	...	...	...	...
	$i=q$	p_{q1}	p_{q2}	...	p_{qq}	p_{q+}
Marginal proportion (p_{+j})		$p_{+1} = \sum_{i=1}^{q} p_{i1}$	p_{+2}	...	p_{+q}	

The confusion matrix is the basis for quantifying various accuracy indicators. Overall accuracy (OA) and Kappa coefficient (K) are probably the most well-known, as described next.

(1) Overall Accuracy (OA)

OA is the most commonly used index of accuracy for land cover maps (Myint et al. 2011; Rodriguez-Galiano et al. 2012; Wickham et al. 2013). It is computed by simply dividing the number of spatial units that are correctly classified by the total number of spatial units in the study area. Mathematically, it equals the sum of the diagonal elements in the confusion matrix of class probabilities (Table 5.2), as stated in Equation (5.2).

$$P_0 = \frac{1}{N} \sum_{i}^{q} n_{ii} = \sum_{i}^{q} p_{ii} \qquad (5.2)$$

The maximum value of the P_0 is 1, indicating a perfect classification. However, since perfect classification does not exist, several researchers have proposed thresholds for acceptable accuracy. For example, Anderson (1971) advised that minimum accuracy should be at least 85% for the land cover maps at an intermediate scale from 1:250,000 to 1:1,000,000. The same threshold is recommended by Thomlinson, Bolstad, and Cohen (1999) when revising the validation procedure for MODIS-based land cover products at 1 km spatial resolution. However, in practice, it is shown that these thresholds are hard to reach, especially for high-resolution global-scale land cover maps (Chen et al. 2015; Yu et al. 2013).

(2) Kappa Coefficient of Agreement (K)

The Kappa coefficient of agreement was defined by Cohen (1960) for psychological research and introduced by Congalton, Oderwald, and Mead (1983) to remote sensing communities. It is estimated by adjusting actual agreement with chance agreement, where actual agreement is nothing else but OA,

$$p_0 = \frac{1}{N} \sum_{i=1}^{q} n_{ii} = \sum_{i=1}^{q} p_{ii} \tag{5.3}$$

and chance agreement is computed based on row and column totals,

$$p_c = \frac{1}{N^2} \sum_{i=1}^{q} n_{i+} * n_{+i} = \sum_{i=1}^{q} p_{i+} * p_{+i}. \tag{5.4}$$

Kappa estimation relies on a multinomial sampling model (when reference data are samples), and accordingly, the Kappa coefficient is estimated as

$$\hat{K} = \frac{p_0 - p_c}{1 - p_c}. \tag{5.5}$$

One of the advantages of the Kappa coefficient is the possibility to calculate its variance (Bishop, Fienberg, and Holland 2007; Czaplewski 1994; Everitt 1968; Hudson 1987) using the Delta method,

$$\hat{\text{var}}\left(\hat{K}\right) = \frac{1}{N} \left\{ \frac{\theta_1(1-\theta_1)}{(1-\theta_2)^2} + \frac{2(1-\theta_1)(2\theta_1\theta_2 - \theta_3)}{(1-\theta_2)^3} + \frac{(1-\theta_1)^2(\theta_4 - 4\theta_2^2)}{(1-\theta_2)^4} \right\}, \tag{5.6}$$

where

$$\theta_1 = \frac{1}{N} \sum_{i=1}^{q} n_{ii}, \tag{5.7}$$

$$\theta_2 = \frac{1}{N} \sum_{i=1}^{q} n_{i+} * n_{+i}, \tag{5.8}$$

$$\theta_3 = \frac{1}{N^2} \sum_{i=1}^{q} n_{ii} \left(n_{i+} + n_{+i}\right), \tag{5.9}$$

and

$$\theta_4 = \frac{1}{N^3} \sum_{i=1}^{q} \sum_{j=1}^{q} n_{ij} \left(n_{j+} + n_{+i}\right)^2 \tag{5.10}$$

The Kappa variance can be used to compute confidence intervals since Kappa is asymptotically normally distributed per the law of large numbers. Based on Kappa variance, it is possible to infer if a confusion matrix derived starting from remote sensing data is similar to or different from one resulting from a mere random class assignment. If it is similar to the result of a random class assignment,

Accuracy Assessment

map classifications are too erroneous to be of any use. The test of significance of the Kappa coefficient is formulated as

$$Z = \frac{\hat{K}}{\sqrt{v\hat{a}r\left(\hat{K}\right)}}, \tag{5.11}$$

where Z represents standard normal deviate. The null hypothesis of Kappa being equal to 0 ($H_0 : K_1 = 0$) is rejected if $Z \geq Z\alpha_{/2}$, where $\alpha/2$ is the confidence level of the two-tailed Z test, and the degrees of freedom are assumed to be infinity. The alternative hypothesis of the test is that Kappa is different from ($H_1 : K_1 \neq 0$) (Congalton and Green 2019).

Standard normal deviations for Kappa coefficients can also be exploited to compare if one of the two independent classification maps is significantly more accurate than the other. The two classifications do not necessarily have anything in common (e.g., different imagery classified, different dates of imagery acquisition). If $\widehat{K_1}$ is Kappa coefficient for the first classification map with variance $var\left(\widehat{K_1}\right)$ and $\widehat{K_2}$ is Kappa coefficient for the second classification map with variance $var\left(\widehat{K_2}\right)$, the test of significance of difference between Kappa coefficients can be calculated as

$$Z = \frac{\left|\widehat{K_1} - \widehat{K_2}\right|}{\sqrt{v\hat{a}r\left(\widehat{K_1}\right) + v\hat{a}r\left(\widehat{K_2}\right)}}, \tag{5.12}$$

where Z is standardized and normally distributed. The null hypothesis that the difference between $\widehat{K_1}$ and $\widehat{K_2}$ is equal to 0 ($H_0 : (K_1 - K_2) = 0$) is rejected if $Z \geq Z\alpha_{/2}$. The alternative to the null hypothesis is that the difference between $\widehat{K_1}$ and $\widehat{K_2}$ is different from 0 ($H_0 : (K_1 - K_2) \neq 0$) if $Z < Z\alpha_{/2}$ (Congalton and Green 2019).

Kappa coefficients range from -1 to 1, but only positive values indicate a positive correlation between reference classifications and map classifications. In their study, Landis and Koch (1977) split the possible Kappa values and assigned them labels as a kind of agreement characterization to aid interpretation concerning Kappa coefficients. According to them, "almost perfect" agreements correspond to Kappa coefficients above 0.8, "substantial" to Kappa coefficients between 0.61 and 0.8, "moderate" to Kappa coefficients between 0.41 and 0.6, "fair" to Kappa coefficients between 0.21 and 0.4, "slight" to Kappa coefficients between 0.0 and 0.2, and "poor" for Kappa coefficients below 0.0.

Kappa coefficients have been widely used as an accuracy indicator (Tung and LeDrew 1988; Smits, Dellepiane, and Schowengerdt 1999; Klemenjak et al. 2012; Vorovencii 2014; Grzegozewski et al. 2016; Yu and Shang 2017; Congalton and Green 2019), despite criticisms. One of the major problems with Kappa coefficients is the inclusion of the actual agreement into chance agreement since this falsely overestimates the latter (Aickin 1990; Brennan and Prediger 1981; Foody 1992). Some other issues concern Kappa coefficients' calculation and interpretation (Pontius Jr. and Millones 2011), the influence of unequal class probabilities (Ma and Redmond 1995), etc.

Apart from OA and the Kappa coefficient, several other accuracy indicators are described in the following sections using the notations in Table 5.1 or Table 5.2. Some of these accuracy indicators focus on class-specific accuracy, while others assess the accuracy of the map being assessed as a whole.

5.2.2 Individual-Class Accuracy Indicators

(1) User's Accuracy (UA) and Producer's Accuracy (PA)

Regarding individual-class accuracies, user's accuracy (UA) and producer's accuracy (PA) are the most commonly used indicators (Comber et al. 2012).

UA is the ratio of the number of correctly classified spatial units in a class over the total number of spatial units in that class as per the classified map (Equation 5.13) (Aronoff 1982; Story and Congalton 1986). UA for class i, UA_i is estimated as

$$UA_i = \frac{n_{ii}}{n_{i+}},\tag{5.13}$$

where n_{i+} and n_{ii} are numbers of spatial units belonging to map class i and those of spatial units belonging to map class i, which also belong to reference class i, as shown in Table 5.1. Equation (5.13) is applicable for sample data collected with simple random sampling and random sampling stratified with map classes.

From the practical point of view, UA of a class expresses the probability that one will find a pixel on a map classified as class i belonging actually to reference class i. It is related to a measure of commission error (i.e., percent of spatial units labeled as a certain class by mistake), which equals 1 minus UA. The minimum acceptable value of a per-class agreement such as UA is 0.7, according to Anderson (1976).

Similarly, PA is the ratio between the number of correctly classified spatial units in a class over the total number of spatial units in that class as per reference classifications (Equation 5.14; Aronoff 1982; Story and Congalton 1986). PA for class i, PA_i, is estimated as

$$PA_i = \frac{n_{ii}}{n_{+i}},\tag{5.14}$$

where n_{+i} and n_{ii} are numbers of spatial units belonging to reference class i and those of spatial units belonging to reference class i, which also belong to map class i, as shown in Table 5.1. However, Equation (5.14) is only applicable for simple random sampling design. To extend its applicability, it is necessary to properly estimate confusion matrices populated with cell probabilities, as shown in Table 5.2. Then, PA' can be computed using Equation (5.14) but with p's (in Table 5.2) substituting n's.

In relation to PA, there is a measure of omission error (i.e., proportion of spatial units excluded from a certain reference class by mistake). Both indicators PA and UA have values between 0 and 1, where 1 represents maximum accuracy and 0 represents completely inaccurate labeling of a class from a user's or a producer's perspective. PA for a class i should be at least 0.7 so that the classification of a class is considered satisfactory (Anderson 1976).

(2) Individual Classification Success Index

The individual classification success index (ICSI) was proposed by Koukoulas and Blackburn (2001). They identified the need for an index that takes into account both omission and commission errors when studying the best band combinations for classifying certain types of vegetation. The ICSI is computed as follows:

$$ICSI_i = 1 - UA_i + PA_i.\tag{5.15}$$

The ICSI value ranges from -1 to 1, where a value of 1 indicates that a class is correctly classified, and a value of -1 indicates that a class is completely misclassified from both the user's and producer's perspectives.

The same index with minor modification was introduced by (Liu, Frazier, and Kumar 2007). They called the index the average of user's and producer's accuracy (AUP), which in essence is the ICSI when the range of values is rescaled to range from 0 to 1.

Accuracy Assessment

143

(3) Helldén's Mean Accuracy

Another indicator that considers both the user's and producer's perspectives is Helldén's mean accuracy (MAH). It is the harmonic mean of UA and PA, which indicates to which extent a certain class overlaps in both reference classifications and map classifications (Helldén 1980). It is given by

$$\mathrm{MAH}_i = 2\left(\frac{1}{UA_i} + \frac{1}{PA_i}\right) = \frac{2p_{ii}}{p_{i+} + p_{+j}}. \tag{5.16}$$

Values of MAH_i range from 0 (no overlap) to 1 (complete overlap).

(4) Short's Mapping Accuracy

Short's mapping accuracy (MAS_i) is the ratio between the intersection of the map classes and reference classes and their union in terms of cardinality (Equation 5.17; Short 1982).

$$\mathrm{MAS}_i = \frac{p_{ii}}{p_{i+} + p_{+j} - p_{ii}} \tag{5.17}$$

There is a linear relation between MAH and MAS expressed with the following equation:

$$\mathrm{MAS}_i = \frac{\mathrm{MAH}_i}{2 - \mathrm{MAH}_i}. \tag{5.18}$$

(5) Conditional Kappa Coefficients

Conditional Kappa ($\hat{K}_i$) coefficient for class i is one of the chance-adjusted measures of classification accuracy measured by OA. Chance-adjusted measures aim at excluding the agreements between reference classifications and map classifications that happen by chance.

Conditional Kappa coefficient $\hat{K}_i$ is closely related to the Kappa coefficient. It can be computed relative to map classes (Equation 5.19) or to reference classes (Equation 5.20).

$$\hat{K}_i = \frac{n_{ii} - n_{i+}n_{+j}}{n_{i+} - n_{i+}n_{+j}}, i = j \tag{5.19}$$

$$\hat{K}_i = \frac{n_{ii} - n_{i+}n_{+j}}{n_{+j} - n_{i+}n_{+j}}, i = j \tag{5.20}$$

(6) Ground Truth Index

Ground truth (GT) index was formulated by Turk (1979) as a proportion of individuals in a given GT category that will be correctly and surely identified by the classifier under study. The GT index is chance corrected like conditional Kappa coefficients. However, it is not an index of accuracy but a measure of the diagnostic capability of a classification algorithm employed for land cover mapping (Türk 2002). GT index cannot be used in the case of binary classification (it works with at least three classes) or in the case of perfect classification (no errors at all). Another limitation regarding GT is the statistical requirement of quasi-independence, which is seldom met in practice (Liu, Frazier, and Kumar 2007). The index is computed as

$$\mathrm{GT}_i = \frac{\left(UA_{i\_}R_{+i}\right)}{\left(1 - R_{+i}\right)} * 100, \tag{5.21}$$

where R_{+i} is chance agreement, calculated using the formula in Türk (1979). In principle, R_{+i} represents the column total in class i of the matrix of expected frequencies F, which can be derived using the algorithm described next.

The algorithm begins by modifying Table 5.1 such that all diagonal elements of the matrix are set to zero ($n_{ii} = 0$, $\forall\, i = 1, 2, \ldots, q$), resulting in a new matrix D. Using Equations (5.22) and (5.23), the marginal totals of the rows (n_{i+D}) and columns (n_{+iD}) of matrix D are calculated.

Next, two sets of intermediate quantities, U_i and V_i, can be determined. These will be used later to calculate the row and column parameters, R_{i+} and R_{+i}. U_i is an intermediate quantity related to map classes (rows in the confusion matrix), while V_i is an intermediate quantity related to the reference classes (columns in the confusion matrix).

The calculation of U_i and V_i by iterative fitting is as follows: The first step is to set the initial values of U_i and V_i equal to the corresponding marginal totals of matrix D—that is,

$$U_i^{(0)} = n_{i+}^{D} = n_{i+} - n_{ii}, \tag{5.22}$$

$$V_i^{(0)} = n_{+i}^{D} = n_{+i} - n_{ii}. \tag{5.23}$$

Then, iterations begin from $m = 1$ and continue until convergence.

V_i will always be updated in an odd-numbered step of the iteration

$$V_i^{(2m-1)} = n_{+i}^{D} \big/ \left[U_{i+}^{(2m-2)} - U_i^{(2m-2)} \right], \tag{5.24}$$

where $U_{i+}^{(2m-2)} = \sum_{i=1}^{q} U_i^{(2m)}$, and superscripts indicate the iteration number, with q representing the number of rows/columns.

For other, U_i always represents an even-numbered step within iteration:

$$U_i^{(2m)} = n_{i+}^{D} \big/ \left[V_{+i}^{(2m-1)} - V_i^{(2m-1)} \right] \tag{5.25}$$

and

$$V_{+i}^{(2m-1)} = \sum_{j=1}^{q} V_j^{(2m-1)}.$$

To clarify, an example of the equations for computing U_i and V_i for $m = 1$ and $m = 2$ is given below, with initial values calculated using Equations (5.22) and (5.23):

$m = 1$

$$V_i^{(1)} = n_{+j}^{D} \big/ \left[U_{i+}^{(0)} - U_i^{(0)} \right] \tag{5.26}$$

$$V_{+i}^{(1)} = \sum_{j}^{q} V_i^{(1)} \tag{5.27}$$

$$U_i^{(2)} = n_{i+}^{D} \big/ \left[V_{+i}^{(1)} - V_i^{(1)} \right] \tag{5.28}$$

$$U_{i+}^{(2)} = \sum_{i}^{q} U_i^{(2)} \tag{5.29}$$

Accuracy Assessment

$m = 2$

$$V_i^{(3)} = n_{+j}^D / \left[U_{i+}^{(2)} - U_i^{(2)} \right] \tag{5.30}$$

$$V_{+i}^{(3)} = \sum_j^q V_i^{(3)} \tag{5.31}$$

$$U_i^{(4)} = n_{i+}^D / \left[V_{+i}^{(3)} - V_i^{(3)} \right] \tag{5.32}$$

$$U_{i+}^{(4)} = \sum_i^q U_i^{(4)} \tag{5.33}$$

In practice, it is necessary to follow the sequence indicated by the superscript numbers to maintain iteration until the desired accuracy is achieved.

R_{i+} and R_{+i} are calculated from the values of U_i and V_i using the following equations:

$$R_{i+} = \frac{U_i}{\sum_i^q U_i} \tag{5.34}$$

$$R_{+i} = \frac{V_i}{\sum_i^q V_i} \tag{5.35}$$

Then, R_{+i} can be inserted into Equation (5.21); the value of the GT index for class i is obtained.

(1) Change in Entropy

The indicator change in entropy (EC) originates from the information theory, where information contained in the map is considered uncertainty about the possible classes at each spatial unit of a map (Finn 1993). Information about class frequencies is quantified as the entropy of the map, with the spatial juxtaposition of the classes not taken into account. The entropy of a map with q classes and the proportion of the area in the i_{th} class being p_{+i} can be calculated as follows:

$$H(\text{GT}) = -\sum_{i=1}^q p_{+i} * \log(p_{+i}) \tag{5.36}$$

According to thermodynamics, entropy is at the maximum when energy is distributed evenly in the system (every class has equal p_{+i}). On the opposite, entropy is minimum when it is contained in a single point. Analogously, entropy is at the maximum when each class occupies an equal area, and it is at the minimum when a map consists of a single class. Unlike the indices described previously, this indicator has units that differ depending on the logarithm base in Equation (5.36). Units are bits for base 2, nats for base e, and hartleys for base 10.

To exemplify the computation of change in entropy, let us assume that we are comparing a map (M) with reference ground truth (GT). Class j has probability p_{+j} in GT, and class i has probability p_{i+} in M (Table 5.2). If map M is known, the uncertainty of the GT data is reduced as the amount of information that the data GT contains in M is identified. In other words, knowing that a spatial unit belongs to class *i* in map *M* gives an indication that it belongs to class *j* of the GT data. This is related to the concept of conditional probability, formulated as

$$p_{ij}(GT_j|M_i) = \frac{p_{ij}}{p_{i+}}. \tag{5.37}$$

Thus, the posterior entropy value of *GT* when the matching class in *M* is known to be *i* (M_i) can be calculated as follows:

$$H(GT|M_i) = -\sum_{j=1}^{q} p_{ij}(GT_j|M_i) * \log\left[p_{ij}(GT_j|M_i)\right] \tag{5.38}$$

Likewise, the posterior entropy of map *M* can be calculated when GT_i is known by swapping their places in the previous equation.

Finally, the change in entropy based on the entropy and posterior entropy values is computed as follows:

$$EC_i = \frac{(H(GT) - H(GT|M_i))}{H(GT)} \times 100 \tag{5.39}$$

The difference between *H*(GT) and *H*(GT|M_i) indicates a change in the uncertainty about the class in GT after knowing map class M_i. If the information about M_i is added to the information about GT, it will reflect in the decrease of the uncertainty. The value of EC_i is the proportion in the change in the entropy for data GT given M_i. It can take negative values if the entropy increases and positive values if entropy decreases given M_i.

(2) Quantity and Allocation Disagreement

Quantity disagreement (QD) and allocation disagreement (AD) are defined by Pontius Jr. and Millones (2011). QD is the number of differences between reference classifications and map classifications data that occurs because of a variation in the proportions of the categories. AD is the number of differences between reference classifications and map classifications regarding the variation in the spatial allocation of the categories. The difference between the two indicators is described in Figure 5.2.

The mapped domain in the example consists of nine pixels (spatial units), which belong to either class A (white) or class B (black). *QD* occurs because one black pixel is classified as white, resulting in a missing black pixel on the map compared with the reference. Meanwhile, a black pixel in the third row and the first column of the reference is allocated to the third row and the third column of the map. It is an omission error for class A and a commission error for class B. Conversely, the white pixel from the third row and the third column of the reference falls on the third row and the first

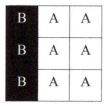

FIGURE 5.2 An example of reference classification (left) and map classification (right).

Accuracy Assessment

147

column of the map. This is an omission error for class B and a commission error for class A. Ultimately, AD occurs for two pixels due to commission and omission errors.

Calculation of the aforementioned disagreement indicators is based on the estimated population confusion matrix, with cell values obtained through

$$p_{ij}^s = \left(\frac{n_{ij}}{n_{i+}}\right)\left(\frac{N_{i+}^p}{N^p}\right), \tag{5.40}$$

where $N_{i+}^p \neq n_{i+}$, which represents all population units (instead of sample units) belonging to class i. In rare cases where the confusion matrix is built using the entire population rather than sample units, $p_{ij}^s = p_{ij}$.

The quantity disagreement (QD) for an arbitrary category, which is the difference between the proportion of category i (p_{+i}^s) in the reference and that of category i in the map (p_{i+}^s). It is calculated as

$$QD_i = \left|p_{i+}^s - p_{+i}^s\right|. \tag{5.41}$$

The allocation disagreement (AD) for an arbitrary category i is computed by multiplying smaller values among commission error and omission error by a factor of 2, as commission and omission errors come in pairs (Pontius Jr., Shusas, and McEachern 2004). This is formulated as

$$AD_i = 2 * \min\left[\left(p_{i+}^s - p_{ii}^s\right), \left(p_{+i}^s - p_{ii}^s\right)\right], \tag{5.42}$$

where p_{i+}^s and p_{+i}^s are the same as in Equation (5.41), while p_{ii}^s refers to the proportion of units that are correctly classified for class i.

(3) Positional Accuracy

Positional accuracy refers to the planimetric accuracy of point features, line feature segments, polygon feature centroid distances, and polygon feature surface distances. Positional accuracy for these feature types is quantified as follows.

The positional accuracy of point features is computed as the average of Euclidean distance between coordinates of point features in the map and the reference (Fox and Sabbagh 2002),

$$\text{Position}_{\text{point}} = \frac{1}{p}\sum_{i=1}^{p}\sqrt{\left(x_i - x_j\right)^2 + \left(y_i - y_j\right)^2}, \tag{5.43}$$

where p is the total number of point features considered, i the point feature in the map, j the corresponding point feature in the reference, (x_i, y_i) the position coordinate of point feature i on the map, (x_j, y_j) the position coordinate of point feature j on the reference.

The positional accuracy of line features is commonly assessed using buffer analysis (Hughes, McDowell, and Marcus 2006; Tveite 1999; Zhang, Leung, and Ma 2019). Specifically, this is done by calculating the offset distance between the geometric center of the assessed features and their reference counterparts (Alganci, Besol, and Sertel 2018).

$$\text{Position}_{\text{line}} = \frac{1}{q}\sum_{n=1}^{q}\sqrt{\left(x_n - x_m\right)^2 + \left(y_n - y_m\right)^2}, \tag{5.44}$$

where q is the total number of line features, n the line feature n being assessed, m the reference line feature m matching line feature n, (x_n, y_n) the coordinates of the geometric center of reference line feature n, (x_m, y_m) the coordinates of the geometric center of map line feature m.

The positional accuracy of polygon features is assessed based on centroid distance or surface distance (Ali and Vauglin 2000; Eon 2006; Ye, Pontius Jr., and Rakshit 2018). Centroid distance between polygon features is calculated by the offset distance between the centroid of the polygon being assessed data and that of the reference, reflecting the degree of spatial offset of the polygons being assessed relative to their reference (Ye, Pontius Jr., and Rakshit 2018):

$$\text{Position}_{\text{centroid}} = \frac{1}{r} \sum_{k=1}^{r} \sqrt{\left(x_k - x_l\right)^2 + \left(y_k - y_l\right)^2}, \tag{5.45}$$

where r is the total number of polygon features, k the polygon feature being assessed, l the corresponding reference, (x_k, y_k) the coordinates of the centroid of the map polygon feature, (x_l, y_l) the coordinates of the centroid of reference polygon l.

The surface distance of polygon features is calculated using the area overlap method to determine the ratio of the intersection area over the union area of map polygon features and their references (Eon 2006):

$$\text{Position}_{\text{surface}} = \frac{1}{r} \sum_{k=1}^{r} \text{Polygon}_k \times 100\% \tag{5.46}$$

$$\text{Position}_l = \frac{\text{Inter}\left(\text{Area}\left(A, B\right)\right)}{\text{Union}\left(\text{Area}\left(A, B\right)\right)}, \tag{5.47}$$

where Poloygon_l is surface distance of the polygon feature l, Inter(Area(A, B)) the intersection area of map polygon feature A and its reference B, Union(Area(A, B)) the union area of map polygon feature A and its reference B.

Note that positional accuracy is only meaningful for well-defined features in the map being assessed and does not convey any information about classification accuracy. Nevertheless, positional accuracy does have an impact on land cover mapping and change monitoring. And validation of land cover products is affected by the level of positional accuracy in image registration and image-map co-registration, and so is uncertainty in subsequent spatial queries and applications.

(4) Attribute Accuracy

Attribute accuracy refers to the degree of correctness of the attribute values (Guptill and Morrison 2013). Clearly, attribute accuracy includes classification accuracy. Feature classification accuracy means the proportion of correctly classified features

$$\text{Attribute}_{\text{accuracy}} = \frac{n}{N} \times 100\%, \tag{5.48}$$

where n is the number of correctly classified features; N the total number of sample size. Feature classification accuracy can be divided into point feature classification accuracy, line feature classification accuracy, and polygon feature classification accuracy, whose calculation formula is the same.

Accuracy Assessment

(5) Completeness

Completeness refers to redundancy or omission of data elements (Mao et al. 2023). Attribute completeness measures the proportion of attribute values that are effectively realized in the assessed data (Kounadi 2009, Wang et al. 2013):

$$\text{Completeness}_{\text{attribute}} = \frac{\sum_{i=0}^{N} x_i}{P \times Q} \times 100\%,$$ (5.49)

$$x_i = \begin{cases} 1, x_i \text{ is effective} \\ 0, x_i \text{ is ineffective} \end{cases},$$ (5.50)

where N is the total number of attributes, i the attribute value i, and x_i the two situations of attribute values.

5.2.3 ALL-CLASS ACCURACY INDICATORS

(1) Average User's Accuracy and Average Producer's Accuracy

Many class-level accuracy indicators can be summarized at a map level by taking the average of the values of the indicator for each class. An example of such a summary is the average user's accuracy (AUA) (Tung and LeDrew 1988), which can be computed as

$$\text{AUA} = \frac{1}{q} \sum_{i=1}^{q} \text{UA}_i.$$ (5.51)

Analogously, the average producer's accuracy (APA) can be computed as (Tung and LeDrew 1988)

$$\text{APA} = \frac{1}{q} \sum_{i=1}^{q} \text{PA}_i.$$ (5.52)

(2) Classification Success Index

Classification success index (CSI) is the average of the ICSI:

$$\text{CSI} = 1 - \left(\frac{\sum_{i=1}^{q} (1 - \text{UA}_i) + (1 - \text{PA}_i)}{q} \right) = \frac{1}{q} \sum_{i=1}^{q} \text{ICSI}_i$$ (5.53)

CSI can be computed for particular classes of interest. For some applications, certain classes and their accuracy indicators might be more important than others (Koukoulas and Blackburn 2001). Thus, the group classification success index (GCSI) becomes useful and is defined as

$$\text{GCSI} = 1 - \left(\frac{\sum_{i=1}^{h} (1 - \text{UA}_i) + (1 - \text{PA}_i)}{h} \right) = \frac{1}{h} \sum_{i=1}^{h} \text{ICSI}_i,$$ (5.54)

where $i = 1, 2, \ldots, h$ denotes the ordinal number of classes of interest selected. According to the authors, a CSI index value of at least 0.8 is needed to claim an effective classification (Koukoulas and Blackburn 2001).

The value of CSI can range from -1 to 1. If this range of values is converted to scale from 0 to 1, CSI would be equal to the index known as the double average of user's and producer's accuracy (DAUP; Liu, Frazier, and Kumar 2007).

(3) Margfit (P0$_M$)

Margfit is often applied to normalize confusion matrices for comparison purposes. It uses an iterative proportional fitting procedure that forces each row and each column of the confusion matrix to sum to a predetermined value. When the predetermined value is 1, each cell value represents a proportion of 1 and can easily be multiplied by 100 to represent percentages or accuracies. Alternatively, the predetermined value could be set to 100 to directly refer to percentages or to any other value the analyst chooses.

In this normalization process, differences in sample sizes used to create the matrices are reconciled, making individual cell values within the matrix directly comparable. Additionally, since the rows and columns are summed to marginals as part of the iterative process, the resulting normalized matrix better reflects the off-diagonal cell values. Essentially, all the values in the matrix are iteratively balanced by rows and columns, incorporating information from each row and column into the resulting cell values. This adjustment changes the cell values along the main diagonal of the matrix, allowing for the computation of normalized OA by summing the main diagonal and dividing by the total sum of the matrix's cell values. In the case of zero-valued cells, it is recommended to add to them some negligibly small positive values to avoid potential problems in computing.

To compute normalized accuracy (P0$_M$), Equation (5.55) is applied to each cell to normalize the matrix with respect to the row totals. Next, Equation (5.56) is applied to every cell of the previously normalized matrix. The procedures are repeated using Equations (5.56) and (5.57) until the column and row totals converge to the predetermined value within an acceptable error margin, typically 0.001. The computing is described next.

The initial step of the iterative procedure involves computing the following using Equation (5.55):

$$\hat{p}_{ij}\left(0\right) = \frac{n_{ij}}{n_{i+}}, \tag{5.55}$$

where n_{ij} and n_{i+} are defined as previously, and $\hat{p}_{ij}$ is the estimated cell value. As a result of the initial step, the resultant matrix is the same as in Table 5.2.

Then, Equations (5.56) and (5.57) are applied iteratively, where number of iteration is $m = 1, 2, \ldots$ until convergence. The superscript in the following equations denotes step serial number, with Equation (5.56) applied to odd-numbered steps and Equation (5.57) applied to even-numbered steps:

$$\hat{p}_{ij}^{(2m-1)} = \frac{\hat{p}_{ij}^{(2m-2)}}{\hat{p}_{+j}^{(2m-2)}}, \tag{5.56}$$

where $\hat{p}_{+j}$ is the estimated values of column totals, and

$$\hat{p}_{ij}^{(2m)} = \frac{\hat{p}_{ij}^{(2m-1)}}{\hat{p}_{i+}^{(2m-1)}}, \tag{5.57}$$

where $\hat{p}_{+j}$ is the estimated value of row totals.

Accuracy Assessment **151**

An example of two iterations, where $m = 1, 2$, is presented below. The initial step (step 0) is the same as shown in Equation (5.55). When

$$m = 1,$$

Step 1.

$$\hat{p}_{ij}^{(1)} = \frac{\hat{p}_{ij}^{(0)}}{\hat{p}_{+j}^{(0)}} \tag{5.58}$$

Step 2.

$$\hat{p}_{ij}^{(2)} = \frac{\hat{p}_{ij}^{(1)}}{\hat{p}_{i+}^{(1)}} \tag{5.59}$$

When

$$m = 2,$$

Step 3.

$$\hat{p}_{ij}^{(3)} = \frac{\hat{p}_{ij}^{(2)}}{\hat{p}_{+j}^{(2)}} \tag{5.60}$$

Step 4.

$$\hat{p}_{ij}^{(4)} = \frac{\hat{p}_{ij}^{(3)}}{\hat{p}_{i+}^{(3)}} \tag{5.61}$$

Therefore, one may assert that the normalized accuracy is a better representation of accuracy than OA computed from the original matrix, as it incorporates information about the off-diagonal cell values. In addition to computing a normalized accuracy indicator, the normalized matrix can also be used to directly compare cell values between matrices.

Using iterative proportional fitting (IPF) on the rows of a confusion matrix is similar to post-stratified estimation wherein the marginal proportions employed for adjustment are the known map population marginal proportions (i.e., the area proportion of each map class within the study area). Applying the IPF algorithm to both rows and columns amounts to using a raking estimator (Stehman 2004). However, the benefits of raking in terms of precision gains are achieved only when relevant population information is available for both rows and columns in the matrix.

Within a confusion matrix, the true reference column marginals are not known, rendering the construction of raking estimators unfeasible. Adjusting the estimated cell proportions to match uniform homogeneous marginal proportions, a typical practice for normalized confusion matrices, is hardly helpful because these margins do not contain relevant information about the map being evaluated.

(4) Kappa-like Measures

Measures like the Kappa coefficient, adjusted for chance agreements, are suggested to address the comparability issue in raw measures of accuracy. Chance-adjusted agreements are calculated by excluding the chance agreements, supposedly leading to a more accurate indicator for accuracy assessment. In

(Stehman 1997), chance-adjusted was considered a better terminology than chance corrected. Türk's GT index and the conditional Kappa mentioned earlier are actually chance-corrected measures.

(5) Normalized Overall Accuracy

Normalized overall accuracy (PO_M) is calculated based on a confusion matrix normalized by Margfit described previously. Once the matrix is normalized with Margfit, it is possible to apply Equation (5.2) to compute PO_M, which is considered a better accuracy indicator than OA (Congalton, Oderwald, and Mead 1983).

Indeed, confusion matrix normalization is found useful when two confusion matrices are to be compared (e.g., estimating the best classifier among several candidates; Smits, Dellepiane, and Schowengerdt 1999; Zhuang et al. 1995). It was also helpful for studying the effect of different marginal proportions on the Kappa statistics (Agresti, Ghosh, and Bini 1995) or simply measuring map accuracy (Fahsi et al. 2000; Ustin et al. 1996).

However, Stehman and Czaplewski (1998), Foody (2002), and Stehman (2004) raised issues regarding confusion matrix normalization. Stehman and Czaplewski (1998) pointed out that normalization could cause inconsistent estimates of accuracy and suggested using conditional probabilities (using either row or column marginal proportions) for better standardization. (Foody 2002) noticed that normalization of the confusion matrix could lead to bias, as well as to unrealistic equalization of UA and PA. Stehman (2004) cautioned that the bias of the normalized matrix can be considerable. He also remarked that the bias of the normalized matrix is larger for UA and PA than the bias of PO_M.

(6) Weighted Kappa Coefficients

Weighted Kappa coefficients (Kw) were proposed several years after Kappa coefficients (Cohen 1960, Fleiss, Cohen, and Everitt 1969). Unlike the ordinary Kappa coefficients, weighted Kappa coefficients are based on the reasoning that not all confusion among the candidate classes has the same importance. The weight of each diagonal element of the confusion matrix of class probabilities is set to 1, while the weight of off-diagonal elements can range from 0 to 1, with 1 indicating the maximum importance (Fleiss, Cohen, and Everitt 1969).

If the weight of the matric element ij belonging to class i in the map and class j in the reference is w_{ij}, the actual agreement can be calculated by using the following equation:

$$p_0^* = \sum_{i=1}^{q} \sum_{j=1}^{q} w_{ij} * p_{ij} \tag{5.62}$$

Weighted chance agreement is computed as

$$p_c^* = \sum_{i=1}^{q} \sum_{j=1}^{q} w_{ij} * p_{i+} * p_{+j}. \tag{5.63}$$

Finally, the weighted Kappa coefficient is defined as

$$\widehat{K_w} = \frac{p_0^* - p_c^*}{1 - p_c^*}. \tag{5.64}$$

As for unweighted Kappa coefficients, large sample variance for a weighted Kappa coefficient can be computed by proper weighting:

$$\hat{\text{var}}\left(\hat{K}_w\right) = \frac{1}{N\left(1 - p_c^*\right)^4} * \left\{ \sum_{i=1}^{q} \sum_{i=1}^{q} p_{ij} \left[w_{ij} \left(1 - p_c^*\right) - \left(\bar{w}_{i+} - \bar{w}_{+j}\right) \right.\right.$$

$$\left.\left. \left(1 - p_0^*\right) \right]^2 - \left(p_0^* p_c^* - 2p_c^* + p_0^*\right)^2 \right\} \tag{5.65}$$

Accuracy Assessment   **153**

Tests of significance in differences between weighted Kappa coefficients can be performed as for unweighted Kappa coefficients.

However, weighted Kappa coefficients are not widely accepted indicators of accuracy due to the fact that it is hard to select appropriate weights that are universally justified. Moreover, every change of weights can significantly change the results, only adding to confusion if one weighting scheme is better than the other one for the same confusion matrix (Congalton and Green 2019).

(7) Tau

Tau (τ) is another Kappa-like indicator that relies on prior class probabilities to compute chance agreements rather than on posterior class probabilities derived after the construction of confusion matrices (as the Kappa coefficient does; Ma and Redmond 1995). Initially, for computing this indicator, it is assumed that marginal probabilities of reference classes (columns in confusion matrices) are known *a priori* since reference data are collected prior to classifications. On the contrary, prior probabilities of map classes are not known. For computing Tau i, it is necessary to obtain information about prior probabilities, which may be based on previous surveys of class probability, estimation of class probability from aerial photos, or assignments of prior probability according to class importance. Tau is defined as

$$\tau = \frac{Q_1 - Q_2}{1 - Q_2}, \tag{5.66}$$

where

$$Q_1 = \sum_{i=1}^{q} p_{ii}, \tag{5.67}$$

$$Q_2 = \sum_{i=1}^{q} p_i * p_{+i}, \tag{5.68}$$

$$Q_3 = \sum_{i=1}^{q} p_{ii} \left(p_i * p_{+i} \right), \tag{5.69}$$

$$Q_4 = \sum_{i=1}^{q} \sum_{j=1}^{q} p_{ij} \left(p_{+i} * p_j \right)^2. \tag{5.70}$$

These equations are the same as those of Kappa coefficients, except that row marginal p_{i+} are replaced by prior class probability p_i. In the absence of reliable information about prior class probabilities, they may be set to equal probability for all classes (i.e., $p_i = 1/q$, q being the number of classes considered). Clearly, assigning suitable prior probabilities is critical because the resulting value of Tau is largely dependent on them.

Based on the Q coefficients (Equations 5.67–5.70), the variance of Tau can be computed as

$$\hat{\text{var}}(\tau) = \frac{1}{N} \left[\frac{Q_1(1-Q_1)}{(1-Q_2)^2} + \frac{2(1-Q_1)(2Q_1Q_2 - Q_3)}{(1-Q_2)^3} + \frac{(1-Q_1)^2 (Q_4 - 4Q_2^2)}{(1-Q_2)^4} \right]. \tag{5.71}$$

We can also assess if one classification is significantly more accurate than another by substituting τ and var(τ) for $\hat{K}$ and var$\left(\widehat{K}\right)$, respectively, in Equation (5.12).

Næsset (1996) introduced an extension of the Tau index by treating row marginals as row prior probabilities and introducing prior probabilities to column marginals to assess the producer's reliability of each class. The computation related to Tau can also be applied to the Næsset's version of Tau index with modification.

(8) Aickin's Alpha Parameter

Aickin's alpha (α) parameter is similar to Kappa coefficients, except for the way chance agreements are estimated. When formulating the α parameter, Aickin (1990) assumed that only a small portion of observed class probabilities result in agreement by chance. This means that all spatial units in the study area can be categorized as being either easy to classify or hard to classify. Map classifications and reference classifications will always agree for the units of the first type, while for the second type, they will agree only sometimes (i.e., randomly).

Since it is not possible to estimate hard-to-classify units, Aickin (1990) proposed an iterative algorithm that estimates conditional probabilities of hard-to-classify units in the map (M) and the reference (GT) that are later used for estimation of chance agreement. The iterative algorithm for the estimation of $p_{i|H}^{(M)}$, $p_{i|H}^{(GT)}$, and, eventually, the value of α is based on a set of three equations listed below. m in the equations denotes numbers of iterations, which starts at 0 and increases by 1 until convergence.

$$\hat{\alpha}_A^{(m+1)} = \frac{p_a - p_e^{(m)}}{1 - p_e^{(m)}},$$ (5.72)

where

$$p_e^{(m)} = \sum_{i=1}^{q} p_{i|H}^{M(m)} * p_{i|H}^{GT(m)},$$ (5.73)

$$p_{i|H}^{M(m+1)} = \frac{p_{i+}}{\left(1 - \hat{\alpha}_A^{(t)}\right) + \hat{\alpha}_A^{(m)} * p_{i|H}^{GT(m)} / p_e^{(m)}},$$ (5.74)

$$p_{i|H}^{GT(m+1)} = \frac{p_{+j}}{\left(1 - \hat{\alpha}_A^{(m)}\right) + \hat{\alpha}_A^{(m)} * p_{i|H}^{M(m)} / p_e^{(m)}}.$$ (5.75)

In the first iteration, the value for $p_{i|H}^{M(m)}$ is p_{i+} and the starting value for $p_{i|H}^{GT(m)}$ is p_{+i}. Thus, the alpha value $\hat{\alpha}_A^{(0)}$, when $m = 0$ (in the first iteration), is equal to the Kappa coefficient. In the next iteration, the alpha value $\hat{\alpha}_A^{(1)}$, when $m = 1$, is calculated using $\hat{\alpha}_A^{(0)}$ and other probability values according to the previous equations. The iterative process is repeated until the difference between two successive alpha values $\hat{\alpha}_A^{(m+1)}$ and $\hat{\alpha}_A^{(m)}$ is below a predetermined value (e.g., 0.001), below which the difference between the two indicators is considered practically negligible (Gwet 2014).

The presence of 0's in confusion matrix elements can prevent algorithms from convergence. Thus, Aickin (1990) recommended adding a pseudocount of 1 to all counts of spatial units when estimating the confusion matrix.

(9) Average Mutual Information

The accuracy of a land cover map can be assessed by assessing the amount of information shared between a map and its assumed reference. Average mutual information (AMI) originates from

Accuracy Assessment

information theory (Finn 1993, Code 1963). When reference data are denoted with GT and map as M, the shared information is computed as

$$\text{AMI} = I\left(\text{GT};M\right) = H\left(\text{GT}\right) - H(\text{GT} \mid M) = \sum_{i=1}^{q}\sum_{j=1}^{q} p_{ij} * \log\left[\frac{p_{ij}(\text{M}_i \mid \text{GT}_j)}{p_{i+}}\right], \tag{5.76}$$

where p_{ij} and p_{i+} are defined in the Table 5.2, while $p_{ij}(\, M_i|\text{GT}_j)$ can be computed by modifying Equation (5.38). *AMI* is symmetrical due to the fact that the amount of information that *GT* contains about *M* is equal to the amount of information that *M* contains about *GT*. AMI can be expressed as the proportion of uncertainty or entropy in the reference, $H(GT)$ (Finn 1993)—i.e.,

$$\text{NMI} = \frac{\text{AMI}}{H\left(\text{GT}\right)}. \tag{5.77}$$

NMI can also be computed by substituting the entropy of the map (i.e., $H(M)$) for the entropy of the reference (i.e., $H(GT)$) in Equation (5.77).

(10) Combined Accuracy

The combined accuracy (CA) is the mean of OA and either AUA as per Equation (5.51) or APA as per Equation (5.52). It has been employed to reduce the intrinsic biases of OA and average accuracy (Nelson 1983).

$$\text{CA}_U = \frac{\text{OA} + \text{AUA}}{2} \tag{5.78}$$

$$\text{CA}_P = \frac{\text{OA} + \text{APA}}{2} \tag{5.79}$$

(11) Total Disagreement

Total disagreement takes into account all quantity disagreement (QD_i) and allocation disagreement (AD_i) of individual classes. Total QD is computed as

$$\text{QD}_{\text{tot}} = \left(\frac{\sum_{i=1}^{q} \text{QD}_i}{2}\right), \tag{5.80}$$

where class-level QD is divided by 2 to account for the double count due to the fact that the commission error of one class induces omission error in another and vice versa.

Likewise, the sum of the AD of all individual classes, divided by 2, is total AD:

$$\text{AD}_{\text{tot}} = \left(\frac{\sum_{i=1}^{q} AD_i}{2}\right), \tag{5.81}$$

where, again, double count is accounted for in computing AD.

Total disagreement (D) is the sum of both total quantity and total allocation disagreements:

$$D = \text{qd}_{\text{tot}} + \text{ad}_{\text{tot}} \tag{5.82}$$

It is interesting to notice that total disagreement is the difference between OA and 1, with 1 being the maximum of OA. So,

$$D = 1 - C. \tag{5.83}$$

Nevertheless, total disagreement can be decomposed to allocation disagreements and QDs, providing additional information about map accuracy.

5.3 EXTENDED METHODS FOR ACCURACY ASSESSMENT

In addition to the basic accuracy assessment indicators reviewed in 5.2, some of the extended methods for accuracy assessment have been developed, taking into account spatial and temporal characteristics of GLC maps and errors in them. This section gives an overview of them.

5.3.1 SPATIAL CORRESPONDENCE OF MAP CLASSIFICATIONS AND THEIR PROBABILISTIC SYNTHESIS

To evaluate spatial accuracy and variation in map accuracy, the spatial correspondences between a GLC map and the reference data are encoded using indicator variables. Sample units are assigned an indicator code 1 when map classes and reference classes agree on them; indicator code 0 is assigned otherwise.

Spatial analysis may then be performed on the aforementioned indicators of spatial correspondence. For this, variogram modeling is usefully explored. Nested variogram models are fitted to experimental variograms using the method of moment approach with binning of 3–5, 10–15, and intervals of 25 km (Pebesma and Wesseling 1998). The variograms are fitted by weighted least squares, with weights defined as N_j/h^2 j, where N_j represents the number of point pairs in the j_{th} lag and h_j is the corresponding lag distance.

Weighted voting (WeVo) was also proposed to assess GLC maps, given reference data (Tsendbazar et al. 2015). Let $sc_i(x)$ denote the spatial correspondence of the i_{th} GLC map ($i = 1, \cdots, 4$) at location x. $W_i(x)$, the weight given to map i at location x, is calculated as

$$W_i(x) = \frac{sc_i(x)}{\sum_{i=1}^{4} sc_i(x)}. \tag{5.84}$$

Then, land cover classes are encoded as multiple 1 or 0 indicator variables, with 1 indicating that a land cover class k ($k = 1, \cdots, K$) is present and 0 when class k is absent. For each k, a total weight of the land cover class at x location is created by summing the class weights of the four GLC maps considered.

$$W_{i,k}(x) = W_i(x) \times k_i(x), \tag{5.85}$$

$$W_k(x) = \sum_{i=1}^{4} W_{i,k}(x), \tag{5.86}$$

where k is the land cover class, $W_k(x)$ is the total weight of class k at location x, $W_{i,k}(x)$ is the class weight of i_{th} map. This method also serves to generate classifications by labeling land cover classes, which registers the greatest total weight ($W_k(x)$), at individual locations.

It is constructive to pursue the integration of existing GLC maps. For example, there exist multiple medium resolution (300–1000 m) GLC maps, such as GLC2000, MODIS, Glob cover, and LC-CCI. Integration of these maps can also take advantage of reference datasets used to calibrate and validate them. Some of these reference datasets are available through the Global Observation of Forest and Land Cover Dynamics (GOFCGOLD) Reference Data portal, the Geo-Wiki platform (an online

Accuracy Assessment

platform for improving GLC), and the International Steering Committee for Global Mapping. The integration of GLC maps and reference datasets was performed using regression kriging, which predicts local land cover class presence probabilities, as described in Tsendbazar et al. (2015).

In another research endeavor, the overall trend in the presence probabilities of land cover classes was estimated using a multinomial logistic (MNL) regression (Kempen et al. 2009). The predicted probabilities were locally refined by interpolating regression residuals through simple kriging.

$$p_k(x) = \pi_k(x) + \varepsilon_k(x),\tag{5.87}$$

where $p_k(x)$ is the presence probability of land cover class k at location x, $\pi_k(x)$ is predicted probability of land cover class k, and $\varepsilon_k(x)$ is the regression residual for that class.

The MNL regression extends logistic regression by simultaneously predicting probabilities for more than two categories (Kempen et al. 2009). In this approach, with different land cover classes transformed into separate indicator variables, binary logistic regression models were fitted using GLC map classifications at sample locations as the explanatory variables.

5.3.2 Assessment with Sampling Accommodating Boundary/Interior Pixels

Images often contain mixed pixels around patch boundaries, leading to reduced classification accuracy. To address this, a stratified sampling method that accommodates spatial heterogeneity due to the presence of boundary pixels is useful (Tran, Julian, and De Beurs 2014). It includes the following steps: pixels divided into boundary pixels and interior pixels, stratification based on the combination of pixel type and whether it is a boundary pixel, optimal allocation of sample pixels to individual strata, and random sampling within strata. These are described in a bit more detail next.

According to the classification accuracy obtained by presampling and the number of sample points, it is first possible to calculate stratum variance (hence standard deviation). Then, the number of sampling points needed for each stratum is calculated. When the classification accuracy difference between boundary pixels and interior pixels is large, the weighted square sum of the classification accuracy difference between boundary pixels and interior pixels can be calculated, leading to a quantity abbreviated as SWDA. Where SWDA values are higher, the classification accuracy difference between the boundary region and the interior region of a class tends to become more balanced as a larger proportion of equilibrium is observed. This method can be used to evaluate the classification accuracy of land cover data with fewer samples while achieving higher accuracy and stability.

5.3.3 Geolocation Accuracy Assessment

There are a variety of errors concerning patch boundaries and patch classification in GLC products. Misclassification and/or boundary positional errors of patches can lead to low map accuracies. Efficient quality inspection is one of the main challenges concerning GLC accuracy assessment. Therefore, developing robust, efficient, accurate, automatic, and cost-effective quality inspection methods is needed. Boundary errors can be detected in the vicinity of boundary pixels and their neighbors. Relationships between neighboring pixels can be analyzed through buffer zones (usually a width of one pixel), as pixels on both sides of boundaries are checked for boundary positional accuracy. This is described next.

First, all boundary lines are converted to raster lines. If the slope of the line is greater than $\pi/4$, the X-axis coordinates of the line are discretized into integers with an increment of one (raster cell side length) using the line equation, with the rasterized Y-axis coordinates left unchanged. When the slope of the line is less than $\pi/4$, Y coordinates are treated using the line equation, while rasterized X coordinates are kept. The resulting raster coordinates are the location of the pixels on the raster line. The pixels on this raster line are called middle pixels.

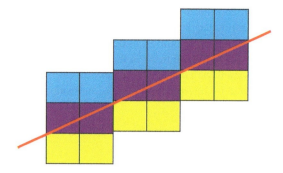

FIGURE 5.3 Pixels within the buffer zone (the diagnal line is the vector line, the boxes overlapped with the diagnal line is the middle pixels, the boxes above the diagnal line represents the out pixels, and the boxes below the diagnal line is the interior pixels).

TABLE 5.3
Decision Rules for Checking Classification Correctness in Boundary Pixels

Cases	Out Pixel	Middle Pixel	Interior Pixel	Decision
1	A	A	A	False
2	A	A	$\bar{A}$	True
3	A	$\bar{A}$	A	True
4	$\bar{A}$	A	A	True
5	$\bar{A}$	$\bar{A}$	$\bar{A}$	False
6	$\bar{A}$	$\bar{A}$	A	True
7	$\bar{A}$	A	$\bar{A}$	True
8	A	$\bar{A}$	$\bar{A}$	True

For each middle pixel, two pixels across the direction of the boundary are selected as the interior pixel and the out pixel. Figure 5.3 shows an example of a three-pixel buffer zone. As one can see from Figure 5.3, the the diagnal line indicates the vector line of the boundary. The boxes along the diagonal line are the middle pixels; the boxes above and below the diagonal line are the out pixels and interior pixels, respectively. The correctness of the boundary may be analyzed by checking the pattern of these three pixels.

It is necessary to refine pixel class labels within the buffer zone. However, if the number of classes is large, this process is time-consuming. Thus, a patch of pixels with class labels is extracted. A wider buffer zone—e.g., a width of five pixels—is rendered, and class labels within this buffer zone are chosen as reference class candidates. Spectral angles of the test pixels are estimated using the reference spectrum for this candidate class. Assume that a patch enclosed in a boundary is A, $\bar{A}$ otherwise. Considering all the eight cases that may occur. The decision rules are listed in Table 5.3. For example, classification for pixels is determined to be false only when the attributes of the three pixels are the same (see Cases 1 and 5).

5.4 SUMMARY

One of the primary challenges in accuracy assessment lies in selecting appropriate accuracy indicators. Multiple yet complementary indicators provide valuable insights into the accuracy of GLC products. It is essential to understand their application scenarios as well as their underlying

Accuracy Assessment

principles. Moreover, there exist various extended methods for accuracy assessment, making the processes of product validation both complete and value-adding.

To address these issues, this chapter gives a synthesis of some commonly used accuracy indicators. There are confusion matrices as the basis for assessing accuracy and classic accuracy indicators, such as OA. Accuracy indicators are categorized to individual-class indicators and all-class indicators.

This is followed by a description of some of the extended methods for accuracy assessment. These include spatial correspondence analysis of multiple map classifications, synthesis of map classifications, accuracy assessment accounting for boundary/interior pixels, and geolocation accuracy assessment.

In a prospective view, with the growth and maturing of GLC products, it is imperative to devise indicators that can accommodate maps of large data volumes and complex data structures. Methods for real-time accuracy assessment should also be explored. Advanced technology will facilitate more comprehensive, reliable, and efficient validation of GLC products.

REFERENCES

Agresti, Alan, Atalanta Ghosh, and Matilde Bini. 1995. "Raking kappa: describing potential impact of marginal distributions on measures of agreement." *Biometrical Journal* 37(7):811–820.

Aickin, Mikel. 1990. "Maximum likelihood estimation of agreement in the constant predictive probability model, and its relation to Cohen's kappa." *Biometrics*, 293–302.

Alganci, Ugur, Baris Besol, and Elif Sertel. 2018. "Accuracy assessment of different digital surface models." *ISPRS International Journal of Geo-Information* 7(3):114.

Ali, Atef Bel Hadj, and François Vauglin. 2000. "Assessing positional and shape accuracy of polygons in vector GIS." *Procedings 4th International Symposium on Spatial Accuracy Assessment in Natural Resources and Environmental Sciences.*

Anderson, James R. 1971. "Land-use classification schemes." *Photogrammetric Engineering*, 37(4): 379–387.

Anderson, James Richard. 1976. *A land use and land cover classification system for use with remote sensor data.* Vol. 964 US Government Printing Office.

Aronoff, Stan. 1982. "Classification accuracy: a user approach." *Photogrammetric Engineering and Remote Sensing* 48(8):1299–1307.

Bishop, Yvonne M, Stephen E Fienberg, and Paul W Holland. 2007. *Discrete multivariate analysis: Theory and practice.* Springer Science & Business Media.

Brennan, Robert L, and Dale J Prediger. 1981. "Coefficient kappa: Some uses, misuses, and alternatives." *Educational and Psychological Measurement* 41(3):687–699.

Chen, Jun, Jin Chen, Anping Liao, Xin Cao, Lijun Chen, Xuehong Chen, Chaoying He, Gang Han, Shu Peng, and Miao Lu. 2015. "Global land cover mapping at 30 m resolution: A POK-based operational approach." *ISPRS Journal of Photogrammetry and Remote Sensing* 103:7–27.

Code, Shannon. 1963. "Information theory and coding."

Cohen, Jacob. 1960. "A coefficient of agreement for nominal scales." *Educational and Psychological Measurement* 20(1):37–46.

Comber, Alexis, Peter Fisher, Chris Brunsdon, and Abdulhakim Khmag. 2012. "Spatial analysis of remote sensing image classification accuracy." *Remote Sensing of Environment* 127:237–246.

Congalton, Russell G. 1991. "A review of assessing the accuracy of classifications of remotely sensed data." *Remote Sensing of Environment* 37(1):35–46.

Congalton, Russell G. 1994. "Accuracy assessment of remotely sensed data: Future needs and directions." Proceedings of Pecora.

Congalton, Russell G, and Kass Green. 2019. *Assessing the accuracy of remotely sensed data: Principles and practices.* CRC Press.

Congalton, Russell G, Richard G Oderwald, and Roy A Mead. 1983. "Assessing Landsat classification accuracy using discrete multivariate analysis statistical techniques." *Photogrammetric Engineering and Remote Sensing* 49(12):1671–1678.

Czaplewski, Raymond L. 1994. *Variance approximations for assessments of classification accuracy.* Vol. 316: US Department of Agriculture, Forest Service, Rocky Mountain Forest and ….

Czaplewski, Raymond L, and Glenn P Catts. 1992. "Calibration of remotely sensed proportion or area estimates for misclassification error." *Remote Sensing of Environment* 39(1):29–43.

Eon, Hong Sung. 2006. "Measuring the Positional Accuracy of GIS Polygon Data." *Journal of Korean Society for Geospatial Information Science* 14(4):3–10.

Everitt, BS. 1968. "Moments of the statistics kappa and weighted kappa." *British Journal of Mathematical and Statistical Psychology* 21(1):97–103.

Fahsi, Ahmed, T Tsegaye, W Tadesse, and T Coleman. 2000. "Incorporation of digital elevation models with Landsat-TM data to improve land cover classification accuracy." *Forest Ecology and Management* 128 (1–2):57–64.

Finn, John T. 1993. "Use of the average mutual information index in evaluating classification error and consistency." *International Journal of Geographical Information Science* 7(4):349–366.

Fleiss, Joseph L, Jacob Cohen, and Brian S Everitt. 1969. "Large sample standard errors of kappa and weighted kappa." *Psychological Bulletin* 72(5):323.

Foody, Giles M. 1992. "On the compensation for chance agreement in image classification accuracy assessment." *Photogrammetric Engineering and Remote Sensing* 58(10):1459–1460.

Foody, Giles M. 2002. "Status of land cover classification accuracy assessment." *Remote Sensing of Environment* 80(1):185–201.

Fox, Garey A, and George J Sabbagh. 2002. "Estimation of soil organic matter from red and near-infrared remotely sensed data using a soil line Euclidean distance technique." *Soil Science Society of America Journal* 66(6):1922–1929.

Grzegozewski, Denise Maria, Jerry Adriani Johann, Miguel Angel Uribe-Opazo, Erivelto Mercante, and Alexandre Camargo Coutinho. 2016. "Mapping soya bean and corn crops in the State of Paraná, Brazil, using EVI images from the MODIS sensor." *International Journal of Remote Sensing* 37(6):1257–1275.

Guptill, Stephen C, and Joel L Morrison. 2013. *Elements of spatial data quality*: Elsevier.

Gwet, Kilem L. 2014. *Handbook of inter-rater reliability: The definitive guide to measuring the extent of agreement among raters*: Advanced Analytics, LLC.

Hay, AM. 1988. "The derivation of global estimates from a confusion matrix." *International Journal of Remote Sensing* 9(8):1395–1398.

Helldén, Ulf. 1980. "A test of landsat-2 imagery and digital data for thematic mapping illustrated by an environmental study in northern Kenya, Lund University." *Natural Geography Institute Report* 47.

Hudson, William D. 1987. "Correct formulation of the kappa coefficient of agreement." *Photogrammetric Engineering and Remote Sensing* 53(4):421–422.

Hughes, Michael L, Patricia F McDowell, and W Andrew Marcus. 2006. "Accuracy assessment of georectified aerial photographs: Implications for measuring lateral channel movement in a GIS." *Geomorphology* 74 (1–4):1–16.

Kempen, Bas, Dick J Brus, Gerard BM Heuvelink, and Jetse J Stoorvogel. 2009. "Updating the 1: 50,000 Dutch soil map using legacy soil data: a multinomial logistic regression approach." *Geoderma* 151 (3–4):311–326.

Klemenjak, Sascha, Björn Waske, Silvia Valero, and Jocelyn Chanussot. 2012. "Automatic detection of rivers in high-resolution SAR data." *IEEE Journal of Selected Topics in Applied Earth Observations Remote Sensing* 5(5):1364–1372.

Koukoulas, Sotlrlos, and George Alan Blackburn. 2001. "Introducing new indices for accuracy evaluation of classified images representing semi-natural woodland environments." *Photogrammetric Engineering and Remote Sensing* 67(4):499–510.

Kounadi, Ourania. 2009. "Assessing the quality of OpenStreetMap data." *Msc geographical information science, University College of London Department of Civil, Environmental Geomatic Engineering*,19.

Landis, J Richard, and Gary G Koch. 1977. "The measurement of observer agreement for categorical data." *Biometrics* 30:159–174.

Liu, Canran, Paul Frazier, and Lalit Kumar. 2007. "Comparative assessment of the measures of thematic classification accuracy." *Remote Sensing of Environment* 107(4):606–616.

Ma, Zhenkui, and Roland L Redmond. 1995. "Tau coefficients for accuracy assessment of classification of remote sensing data." *Photogrammetric Engineering and Remote Sensing* 61(4):435–439.

Mao, Wenjuan, Haitao Zhao, Wenli Han, Hongjing Tu, and Wenchao Gao. 2023. "Quality inspection of remote sensing farmland resource monitoring data achievements." *The International Archives of the Photogrammetry, Remote Sensing Spatial Information Sciences* 48:1453–1458.

Myint, Soe W, Patricia Gober, Anthony Brazel, Susanne Grossman-Clarke, and Qihao Weng. 2011. "Per-pixel vs. object-based classification of urban land cover extraction using high spatial resolution imagery." *Remote Sensing of Environment* 115(5):1145–1161.

Næsset, Erik. 1996. "Use of the weighted Kappa coefficient in classification error assessment of thematic maps." *International Journal of Geographical Information Systems* 10(5):591–603.

Nelson, Ross F. 1983. "Detecting forest canopy change due to insect activity using Landsat MSS." *Photogrammetric Engineering and Remote Sensing* 49(9):1303–1314.

Pebesma, Edzer J, and Cees G Wesseling. 1998. "Gstat: a program for geostatistical modelling, prediction and simulation." *Computers and Geosciences* 24(1):17–31.

Pontius jr, Robert G., and Marco Millones. 2011. "Death to Kappa: birth of quantity disagreement and allocation disagreement for accuracy assessment." *International Journal of Remote Sensing* 32(15):4407–4429.

Pontius jr, Robert G., Emily Shusas, and Menzie McEachern. 2004. "Detecting important categorical land changes while accounting for persistence." *Agriculture, Ecosystems and Environment* 101(2–3):251–268.

Rodriguez-Galiano, Victor Francisco, Bardan Ghimire, John Rogan, Mario Chica-Olmo, and Juan Pedro Rigol-Sanchez. 2012. "An assessment of the effectiveness of a random forest classifier for land-cover classification." *ISPRS Journal of Photogrammetry and Remote Sensing* 67:93–104.

Rosenfield, George H, and Katherine Fitzpatrick-Lins. 1986. "A coefficient of agreement as a measure of thematic classification accuracy." *Photogrammetric Engineering and Remote Sensing* 52(2):223–227.

Short, Nicholas M. 1982. *The Landsat tutorial workbook: Basics of satellite remote sensing.* Vol. 1078: National Aeronautics and Space Administration, Scientific and Technical ….

Smits, PC, SG Dellepiane, and RA Schowengerdt. 1999. "Quality assessment of image classification algorithms for land-cover mapping: a review and a proposal for a cost-based approach." *International Journal of Remote Sensing* 20(8):1461–1486.

Stehman, Stephen V. 1997. "Selecting and interpreting measures of thematic classification accuracy." *Remote Sensing of Environment* 62(1):77–89.

Stehman, Stephen V 2004. "A critical evaluation of the normalized error matrix in map accuracy assessment." *Photogrammetric Engineering and Remote Sensing* 70(6):743–751.

Stehman, Stephen V 2009. "Sampling designs for accuracy assessment of land cover." *International Journal of Remote Sensing* 30(20):5243–5272.

Stehman, Stephen V, and Raymond L Czaplewski. 1998. "Design and analysis for thematic map accuracy assessment: fundamental principles." *Remote Sensing of Environment* 64(3):331–344.

Stehman, Stephen V, and Giles M Foody. 2009. "Accuracy assessment." *The SAGE Handbook of Remote Sensing*:297–309.

Story, Michael, and Russell G Congalton. 1986. "Accuracy assessment: a user's perspective." *Photogrammetric Engineering and Remote Sensing* 52(3):397–399.

Strahler, Alan H, Luigi Boschetti, Giles M Foody, Mark A Friedl, Matthew C Hansen, Martin Herold, Philippe Mayaux, Jeffrey T Morisette, Stephen V Stehman, and Curtis E Woodcock. 2006. "Global land cover validation: Recommendations for evaluation and accuracy assessment of global land cover maps." *European Communities, Luxembourg* 51(4):1–60.

Thomlinson, John R, Paul V Bolstad, and Warren B Cohen. 1999. "Coordinating methodologies for scaling landcover classifications from site-specific to global: steps toward validating global map products." *Remote Sensing of Environment* 70(1):16–28.

Tran, Trung V, Jason P Julian, and Kirsten M De Beurs. 2014. "Land cover heterogeneity effects on sub-pixel and per-pixel classifications." *ISPRS International Journal of Geo-Information* 3(2):540–553.

Tsendbazar, Nandin-Erdene, Sytze De Bruin, Steffen Fritz, and Martin Herold. 2015. "Spatial accuracy assessment and integration of global land cover datasets." *Remote Sensing* 7(12):15804–15821.

Tung, F, and E LeDrew. 1988. "The determination of optimal threshold levels for change detection using various accuracy indexes." *Photogrammetric Engineering and Remote Sensing* 54(10):1449–1454.

Turk, G 2002. "Map evaluation and chance correction." *Photogrammetric Engineering and Remote Sensing* 68(2):123–+.

Türk, Goksel. 1979. "Gt index: a measure of the success of prediction." *Remote Sensing of Environment* 8(1):65–75.

Tveite, Havard. 1999. "An accuracy assessment method for geographical line data sets based on buffering." *International Journal of Geographical Information Science* 13(1):27–47.

Ustin, SL, QJ Hart, L Duan, and G Scheer. 1996. "Vegetation mapping on hardwood rangelands in California." *International Journal of Remote Sensing* 17(15):3015–3036.

Vorovencii, Iosif 2014. "Assessment of some remote sensing techniques used to detect land use/land cover changes in South-East Transilvania, Romania." *Environmental Monitoring and Assessment* 186:2685–2699.

Wang, Ming, Qingquan Li, Qingwu Hu, and Meng Zhou. 2013. "Quality analysis of open street map data." *The International Archives of the Photogrammetry, Remote Sensing Spatial Information Sciences* 40:155–158.

Wickham, James D, Stephen V Stehman, Leila Gass, Jon Dewitz, Joyce A Fry, and Timothy G Wade. 2013. "Accuracy assessment of NLCD 2006 land cover and impervious surface." *Remote Sensing of Environment* 130:294–304.

Ye, Su, Robert G Pontius Jr., and Rahul Rakshit. 2018. "A review of accuracy assessment for object-based image analysis: from per-pixel to per-polygon approaches." *ISPRS Journal of Photogrammetry and Remote Sensing* 141:137–147.

Yu, Bing, and Songhao Shang. 2017. "Multi-year mapping of maize and sunflower in Hetao irrigation district of China with high spatial and temporal resolution vegetation index series." *Remote Sensing* 9(8):855.

Yu, Le, Jie Wang, Nicholas Clinton, Qinchuan Xin, Liheng Zhong, Yanlei Chen, and Peng Gong. 2013. "FROM-GC: 30 m global cropland extent derived through multisource data integration." *International Journal of Digital Earth* 6(6):521–533.

Zhang, Wen-Bin, Yee Leung, and Jiang-Hong Ma. 2019. "Analysis of positional uncertainty of road networks in volunteered geographic information with a statistically defined buffer-zone method." *International Journal of Geographical Information Science* 33(9):1807–1828.

Zhuang, Xin, Bernard A Engel, Xiaoping Xiong, and Chris J Johannsen. 1995. "Analysis of classification results of remotely sensed data and evaluation of classification algorithms." *Photogrammetric Engineering and Remote Sensing* 61(4):427–432.

6 Software Tools for Validation

Shishuo Xu
Beijing University of Civil Engineering and Architecture, Beijing, China

Maria Antonia Brovelli
Politecnico di Milano, Milan, Italy

Gang Han
National Geomatics Center of China, Beijing, China

Yang Zhao
Beijing University of Civil Engineering and Architecture, Beijing, China

Songnian Li
Toronto Metropolitan University, Toronto, Canada

Gorica Bratic
Politecnico Milano, Milan, Italy

Candan Eylul Kilsedar
Politecnico Milano, Milan, Italy

6.1 INTRODUCTION: NEED AND CHALLENGES

Land cover validation is a process of assessing and quantifying the classification accuracy of land cover maps and, in this context, validation can be considered as a suite of techniques for determining the quality of land cover maps (Strahler et al. 2006). Other important aspects, such as sampling design, reference data, and accuracy assessment, have been well discussed in previous chapters in this book. This chapter focuses on the software tools that can be used to support land cover validation, especially the online, web-based tools that have been developed to address some challenges facing land cover mapping and validation at global scale, i.e., the lack of accurate ground-truth data, validation datasets, and reference data, and crowdsourcing citizen's contributions to collaborative validation (Fritz et al. 2009).

In order to effectively organize the validation of large-area land cover data, it is essential to establish and adhere to standardized validation technical specifications and utilize suitable methods and tools to enhance the efficiency and quality of validation efforts (Chen et al. 2018). Over the past years, we have seen the development of validation tools in two streams: (1) as built-in functionality or plugin extension of exiting geographical information system (GIS) and remote sensing software packages and (2) as an online, web-based, or mobile tool.

In the past, land cover data validation primarily relied on stand-alone or offline tools. The tools found in some GIS and remote sensing software are mainly designed for accuracy assessment of image classification results using some reference data, such as the accuracy evaluation module in the commercial remote sensing image processing software ERDAS. This module offers functions such as random generation or addition of sample points and setting reference information. It can calculate producer's accuracy, user's accuracy, overall accuracy, the Kappa coefficient, and generate accuracy assessment reports. However, it does not support cross-region collaborative validation in a networked environment.

As advanced Internet technologies emerge, some online, web-based tools have been developed and adopted for supporting online collaborative land cover validation. These web-based tools mainly focus on collecting and utilizing user-generated content as a supplement to validate the classification of land cover maps or annotate land cover changes, such as the web-based land cover sample collection system VIEW-IT (Clark and Mitchell Aide 2011) and the automatic search of sample information based on geotagged texts over the web (Hou et al. 2015). Recently, the International Institute for Systems Research in Austria (IIASA) developed LACO-Wiki (http://laco-wiki.net/) and Map Accuracy Tools (https://iiasa.ac.at/models-tools-data/map-accuracy-tools), which allow volunteers to upload sample data and conduct single-point accuracy evaluation, etc., but they do not yet support sampling design, resource sharing, and interaction among participants.

Over the last decades, enormous sources of large geospatial datasets became available, such as Google Maps, high-resolution satellite imagery in Google Earth, and existing land cover maps with higher levels of accuracy. At the same time, the volunteer geographic information (VGI)—i.e., geospatial contents generated by nonprofessionals using mapping or similar systems available on the Internet and mobile devices—offers new ways of obtaining input to land cover validation processes. This approach allows nonprofessionals to inspect existing land cover maps to identify land cover changes or to confirm the land cover types of a certain area using reference data and their local knowledge. This is especially important for land cover validation at a global scale, which emerges as collaborative validation.

Despite these promising developments, more efforts are needed to develop the tools that can address some challenges of validating high-resolution global land cover (GLC) data products at a global scale.

Spatial heterogeneity of GLC: High-resolution GLC datasets present strong spatial heterogeneity of land cover over the globe. Traditional sampling methods may not be able to handle spatial heterogeneity efficiently and therefore may not produce credible samples, which can cause problems, such as inappropriate sample sizes for target regions, underrepresented samples for rare classes, and irrational sample distribution (Chen et al. 2016). Thus, taking spatial heterogeneity of land cover into consideration is necessary to allow either higher sample densities or larger sample sizes for regions that are more heterogeneous (Chen et al. 2016; Defourny et al. 2009). With the help of an online tool, the land cover types of the sample points can be verified by experts or volunteers by interpreting high-resolution images or comparing them with large-scale land cover maps and/or in situ measurements.

Integration of data and resources: Land cover products, samples, and reference data and resources are essential for GLC product validation. However, the discrete and dispersible data and resources are too difficult to use to provide valuable information during validation processes. It is therefore important to integrate and visualize all available data and resources from a great variety of sources (Chen et al. 2021). Except for land cover products and samples, other reference data includes very high resolution (VHR) satellite images, geotagged photos and social media contents, high-accuracy thematic maps, in situ measurement data, crowdsourcing and VGI data, and so on (see more details in Chapter 3).

Furthermore, collecting sample data over a large area is costly, time-consuming, and labor-intensive. It may become even more difficult for GLC validation over the entire globe. Continuous advances in Internet technologies have made it possible to develop some web-based tools or systems that enable people from anywhere in the world online access to reference data and other sample information (Stehman et al. 2012; Fritz et al. 2012; Han et al. 2015). However, most of these systems are dedicated to some particular validation steps, not supporting the entire process. For instance,

Geo-Wiki helps interpret imagery reference data but does not support sample size calculation, automatic sample allocation, and other needed functions.

In order to facilitate the use of data and resources, auxiliary tools for integration and visualization are necessary. The major functionality includes the integration and access of reference materials, map zooming and panning, maps overlapping and multisquare box displaying, sample uploading and downloading, and so on. These functions help the experts and volunteers around the world to search and use reference information, upload, judge, and label samples, and annotate errors.

Generation of comprehensive accuracy assessment: There are a number of reasons for the online tools to support comprehensive accuracy assessments of high-resolution GLC data:

- The increasing demand from environmental change studies, monitoring the progress of Sustainable Development Goals (SDGs), and other application areas where the accuracy assessment reports of the land cover datasets that cover large areas or even at a global scale are requested
- The rapidly increasing number of people (experts, practitioners, researchers, and volunteers) working in the area of land cover validation forming a large voluntary validation community with shared interests and their own validation resources

Up to now, a number of software tools have been developed, including desktop tools and online tools, which may be used for supporting some tasks of GLC validation. The desktop tools mainly rely on well-known software—e.g., ArcGIS, QGIS, and ERDAS. The online tools include web-based tools (e.g., Geo-Wiki and LACO-Wiki) and mobile tools (e.g., Land Cover Collector). Although some tools, such as GLCVal, have been designed to support collaborative validation, especially for GLC data products, they are not able to meet all the requirements of collaborative validation of high-resolution GLC products discussed in Section 6.4. Section 6.5 discusses the design, implementation, and example uses of the GLCVal tool in more detail.

As such, developing more advanced tools to enable the complete validation process toward "collaborative validation" is needed. In the following two sections, a number of existing desktop and online tools are briefly reviewed to help understand what is needed for a collaborative validation tool.

6.2 DESKTOP TOOLS

Both free and open-source software (FOSS) and proprietary software packages provide functionalities for supporting land cover data validation. The proprietary software includes ArcGIS and ERDAS IMAGINE, available for Windows OS only, while FOSS software includes QGIS, GRASS GIS, System for Automated Geoscientific Analyses (SAGA), and Orfeo ToolBox, which are available under General Public License (GPL) or compatible licenses and are also available for Linux and MacOSX in addition to Windows OS. These tools are briefly described in the following sections.

6.2.1 QGIS

QGIS[1] is a free and open-source geographic information system. It does not provide any validation tools as part of its built-in functionalities (QGIS Development Team 2018) but does provide a few plugins that can be utilized for land cover validation, as follows:

- Semi-Automatic Classification Plugin
- AcaTaMa
- Accuracy Assessment (AccAssess)

Semi-Automatic Classification Plugin (SCP) is developed for a supervised classification of remote sensing images (Congedo 2016). Nonetheless, the validation is included in the plugin's post-processing group of functionalities. The plugin requires a classified image to be a raster, while a

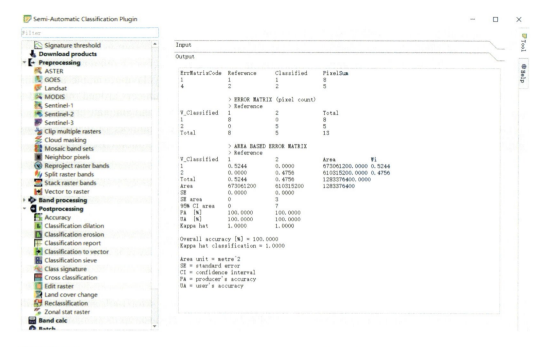

FIGURE 6.1 A capture of accuracy assessment report generated by SCP.

reference image can be either a raster or a vector. However, both classified and reference maps must have the same notation for the classes and must be loaded in the Layer Panel of QGIS. The plugin computes standard error, confidence interval, producer's accuracy, user's accuracy, overall accuracy, and Kappa, and creates a raster representing the spatial distribution of the change. The results can be either visualized using SCP or exported as an additional file (e.g., *.tif and *.vrt).

Before performing the validation operation, defining the band set is required. The classification map is added to the band set and made active. The overall validation process mainly includes random point selection, sample judgment, and accuracy assessment. The details concerning each step are as follows.

- Creation of random points. The number of random points and the band are defined.
- Sample judgment. The sample points created in the previous step are visualized on the base map (e.g., Google Satellite, ESRI imagery, and other high-resolution images), which can be loaded through the Center for Applied GIS of Ho Chi Minh City (HCMGIS) plugin.
- Accuracy assessment. The metrics mentioned earlier (e.g., accuracies and Kappa coefficient) are computed in this step. A capture of the accuracy assessment report is illustrated in Figure 6.1.

Another plugin developed for the accuracy assessment of thematic maps is AcATaMa,[2] which can be used for validating land cover maps (see Figure 6.2 for its main interface). In order to use AcATaMa, the classified image must be a raster, while the reference data must be vector points. The vector points can be generated within the tool by simple or stratified random sampling and should be (re)classified using the classification functionality of the plugin to establish the link between the classes of the reference and the classified map. All input data must be loaded in the Layers Panel. The output includes an error matrix (confusion matrix), an error matrix of estimated area proportion, and a quadratic error matrix of estimated area proportion, as well as overall accuracy, user's accuracy, producer's accuracy indexes, and total class area in hectares. The results are displayed in a dialogue window (see Figure 6.3), but they can also be exported to a CSV (Comma Separated Value) file.

Software Tools for Validation

FIGURE 6.2 The main interface of the AcATaMa tool.

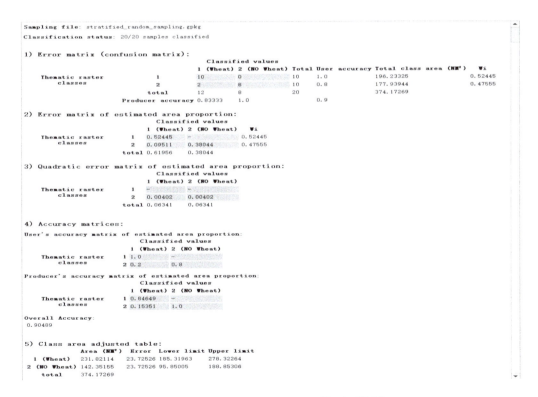

FIGURE 6.3 A capture of accuracy assessment report generated by AcATaMa.

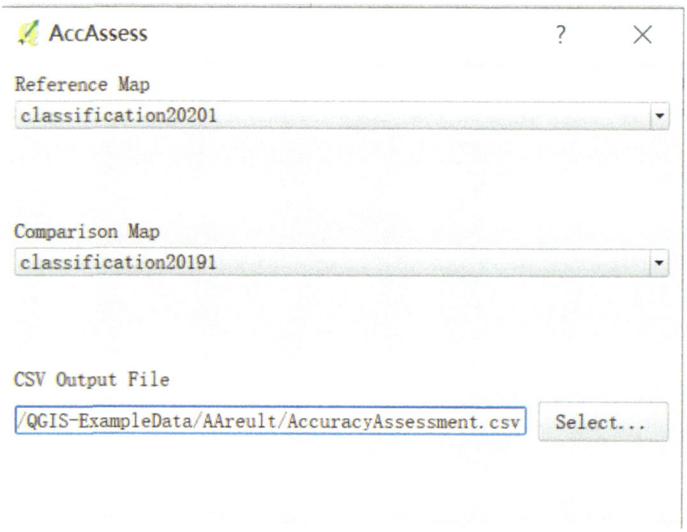

FIGURE 6.4

FIGURE 6.4 The main interface of the AccAssess tool.

classification20201 (ref) vs. classification20191

	1	2	127	Totals	Accuracy
1	1609771	2179660	0	3789431	42
2	2262035	6782298	0	9044333	75
127	0	0	8216572	8216572	100
Totals	3871806	8961958	8216572	21050336	
Accuracy	42	76	100		79
Quantity Disagreement	82375				
Allocation Disagreement	4359320				

FIGURE 6.5 A capture of the accuracy assessment report in the CSV file generated by AccAssess.

Accuracy Assessment (AccAssess) plugin[3] is a very simple validation tool (Kibele 2016). Once the inputs for the tool, two raster datasets (i.e., reference map and comparison map, see Figure 6.4), are inserted, the tool creates an error matrix and a CSV file (see Figure 6.5). Again, the input rasters must be loaded in QGIS to be used within the plugin. The resulting CSV file contains values of the user's accuracy, producer's accuracy, overall accuracy, quantity, and allocation disagreement.

6.2.2 GRASS GIS

GRASS[4] (Geographic Resources Analysis Support System) GIS is another free and open-source geographic information system. It provides an r.kappa module for the accuracy assessment of classification results by calculating the error matrix and Kappa parameter (GRASS Development Team 2021). Based on the two images/rasters (classified and reference) with a consistent classification nomenclature, r.kappa can compute error matrix and related accuracy indexes (overall accuracy, Kappa, Kappa variance, conditional Kappa, omission, commission error, and observed correct; see Figure 6.6). The output is by default shown in the r.kappa dialogue window. If the path and the output file name are specified, it can also be saved as a text file or a csv file.

Software Tools for Validation

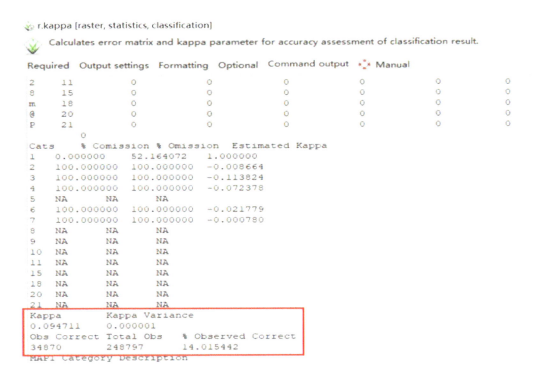

FIGURE 6.6 The result of r.kappa in GRASS dialogue window.

FIGURE 6.7 The main interfaces of the confusion matrix in SAGA (a) The main interface of the confusion matrix using vector polygons and raster grids, (b) The main interface of the confusion matrix using both raster grids.

6.2.3 SAGA

SAGA is a free open-source GIS software. It has two variations of the confusion matrix module: one uses the vector polygon, and the other one uses raster as reference data, and a classified map is raster in both cases (Conrad et al. 2015). The main interfaces regarding these two cases are shown in Figure 6.7(a) and (b), respectively. All validation input data must have a lookup table in which classes and their visualization colors are defined. The lookup table can be built into the software. Both comparison and reference data must have the same categories to be directly comparable for the creation of the confusion matrix. Overall accuracy, user's accuracy, producer's accuracy, and Kappa are the results given by the SAGA Confusion Matrix Module. The output of the module is given as virtual tables (e.g., confusion matrix, class values, and summary) and can be visualized within the software or can be manually saved as csv, txt, or dbf file. SAGA's validation output is also a raster of land cover change.

6.2.4 ORFEO TOOLBOX

Orfeo ToolBox (OTB) is a geospatial library created by the French Space Agency (CNES) (Grizonnet et al. 2017). The Confusion Matrix Computation is an OTB application that allows users to compute a confusion (error) matrix of the classification results by comparing it with the ground-truth data. Figure 6.8 shows the window of the Computing Confusion Matrix Computation app. The ground truth can be provided as raster or vector data. The results represented by a confusion matrix, including precision, recall, F-score, Kappa, and overall accuracy, are displayed either in a log window or exported to a CSV file (see Figure 6.9). For using the Confusion Matrix Computation app, it is required that classified input is a raster, while the reference dataset might also be vector points or a raster image, under the condition that the column of the attribute table related to class values is an integer type. Similar to the aforementioned tools, the classes of reference and classified maps must be consistently denoted.

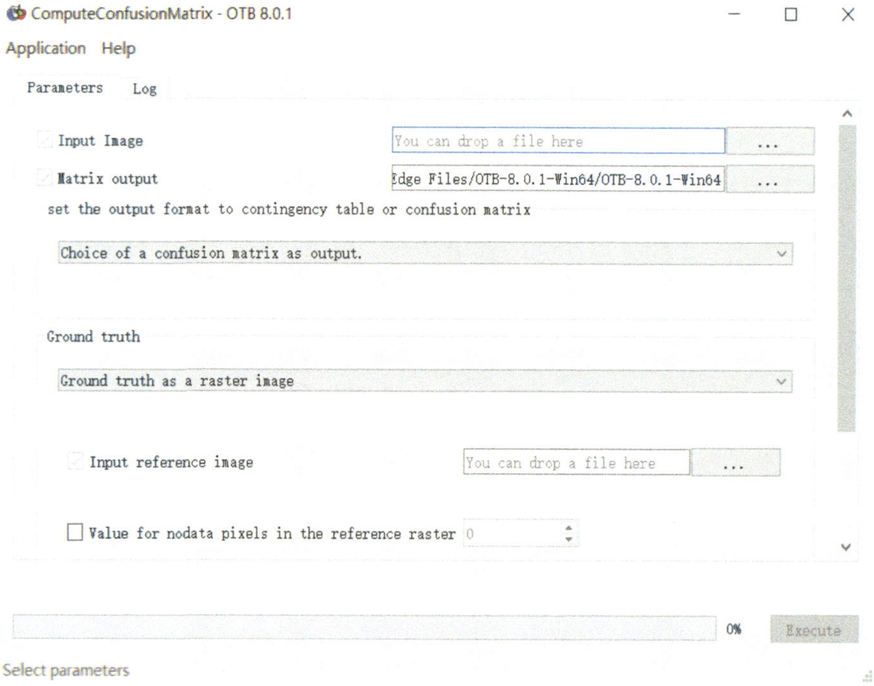

FIGURE 6.8 The window of computing confusion matrix.

Software Tools for Validation

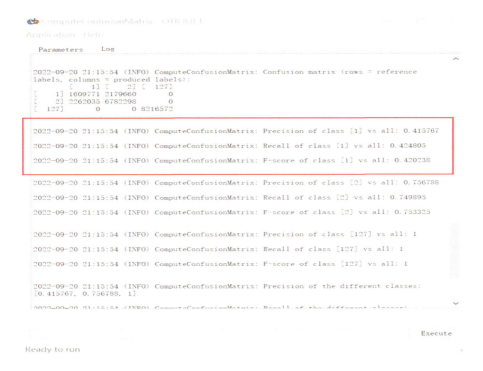

FIGURE 6.9 A capture of Confusion Matrix generated by OTB.

6.2.5 ERDAS IMAGINE

Validation in ERDAS IMAGINE is based on the raster file format (Hexagon Geospatial/Hexagon 2018). The tool used for accuracy assessment is the accuracy assessment tool that belongs to the group of raster supervised classification tools. This tool samples the points of the assigned classified map by random sampling, stratified, or equal random sampling, but the values of the ground truth must be filled manually. The points can be visualized on the top of the ground-truth raster layer to

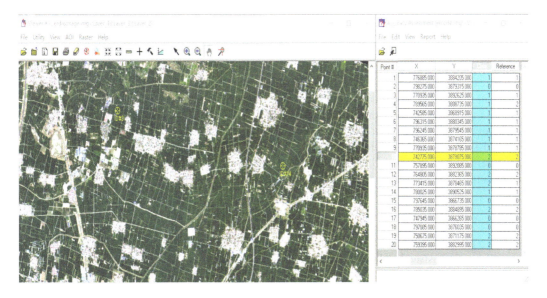

FIGURE 6.10 Accuracy assessment tool operation steps

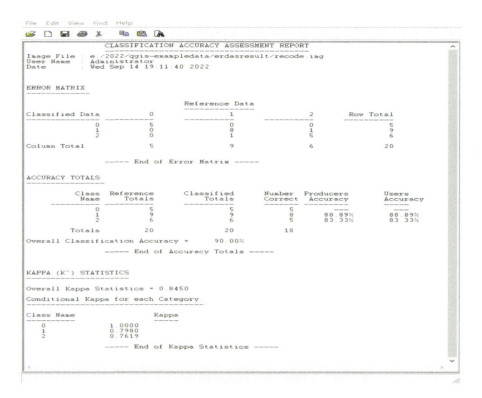

FIGURE 6.11 A capture of the accuracy assessment report generated by ERDAS IMAGINE.

determine corresponding ground-truth values (see Figure 6.10). After these steps, the error matrix and some accuracy indexes can be computed, including overall accuracy, producer's and user's accuracy, Kappa, and conditional Kappa, which are visualized by the software and exported as text or generic .dat file (see Figure 6.11).

6.2.6 ArcGIS

ArcGIS is a complete suite for desktop GIS developed by ESRI.[5] The tool for validation, including functions such as *Compute Confusion Matrix*, *Create Accuracy Assessment Points*, and *Update Accuracy Assessment Points*, can be found in ArcMap under the desktop GIS framework (see Figure 6.12). It is based on the vector that contains two columns—i.e., "Classified" and "Ground Truth," in which both classified and reference map values are stored, respectively. The two columns are cross-tabulated by the Compute Confusion Matrix function to compute the confusion matrix and several accuracy indexes, including overall accuracy, Kappa, user's, and producer's accuracy. Figure 6.13 shows the computation results. The results can be visualized in ArcGIS or exported as Text File, File and Personal Geodatabase Tables, dBase Tables, Info Tables, or SDE Tables. It only supports raster classifications (no polygons) and the sample units are point only.

6.2.7 ArcGIS Pro

ArcGIS Pro, a powerful desktop GIS application, is also developed by ESRI company. ArcGIS Pro follows a consistent validation process and provides similar validation functions to ArcGIS, including Compute Confusion Matrix, Create Accuracy Assessment Points, and Update Accuracy Assessment Points (see Figure 6.14). The classified image can be a raster or a feature class. It requires entering the number of sample points, the location of which are randomly distributed. The attribute table of the randomly generated sample points contains two columns—i.e., "Classified"

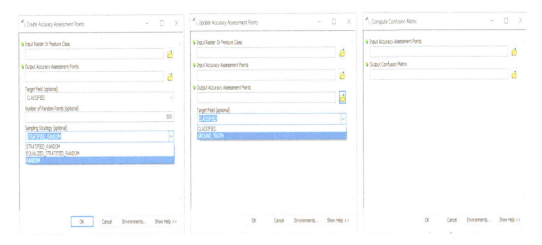

FIGURE 6.12 The interfaces of ArcGIS accuracy assessment functions.

Table

CM123

OBJECTID *	ClassValue	C_0	C_1	C_4	C_5	C_6	C_7	C_8	C_9	C_10	C_13	Total	U_Accuracy	Kappa
1	C_0	4	0	0	0	0	0	0	0	0	0	4	1	0
2	C_1	0	1	0	0	2	0	0	1	0	0	4	0.25	0
3	C_4	0	0	6	0	1	0	0	1	0	0	8	0.75	0
4	C_5	0	0	0	3	4	0	0	0	0	0	7	0.428571	0
5	C_6	0	0	0	0	5	0	0	0	0	0	5	1	0
6	C_7	0	0	0	2	1	1	0	2	0	0	6	0.166667	0
7	C_8	0	0	0	0	2	0	1	5	0	0	8	0.125	0
8	C_9	0	0	0	0	1	0	0	5	0	0	6	0.833333	0
9	C_10	0	0	0	0	0	1	0	0	0	0	1	0	0
10	C_13	0	0	0	0	0	0	0	0	0	2	2	1	0
11	Total	4	1	6	5	16	2	1	14	0	2	51	0	0
12	P_Accuracy	1	1	1	0.6	0.3125	0.5	1	0.357143	0	1	0	0.54902	0
13	Kappa	0	0	0	0	0	0	0	0	0	0	0	0	0.492208

FIGURE 6.13 The computation results of the confusion matrix provided by ArcGIS.

and "GroundTruth." The ground-truth value is manually filled according to the reference map. The output includes overall accuracy, user's accuracy, producer's accuracy indexes, and Kappa. By default, the output is presented in a geodatabase table (see Figure 6.15).

A summary of the software input types and formats for both classified and reference data, as well as the formats of the output file, are shown in Table 6.1 for each of the aforementioned tools. It can be seen that many of the FOSS, as well as the proprietary software tools have a module for accuracy assessment or validation included. The map production at a high rate might lead to the availability of large numbers, not necessarily accurate maps. Easily available validation tools help meet the needs of validation of a large number of maps and determine their potential use for different purposes. However, to be widely used, validation tools should be user-friendly. The tools shown in this section have various levels of complexity in terms of their use and various data types required for input. Since there are multiple tools for validation, users can choose the most suitable one based on their specific needs.

6.3 ONLINE TOOLS

A number of web-based tools and mobile tools have been developed since early 2000. Different from the tools described in Section 6.2, these tools are mostly based on the VGI approach to collect nonprofessionals' contributions to help validate land cover maps.

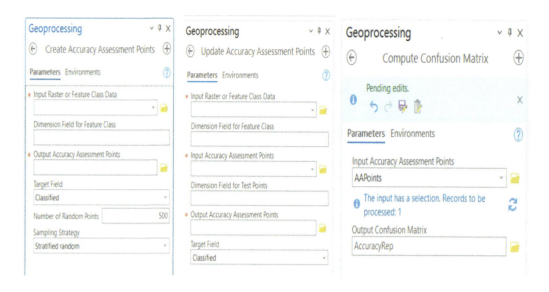

FIGURE 6.14 The interfaces of ArcGIS Pro accuracy assessment functions.

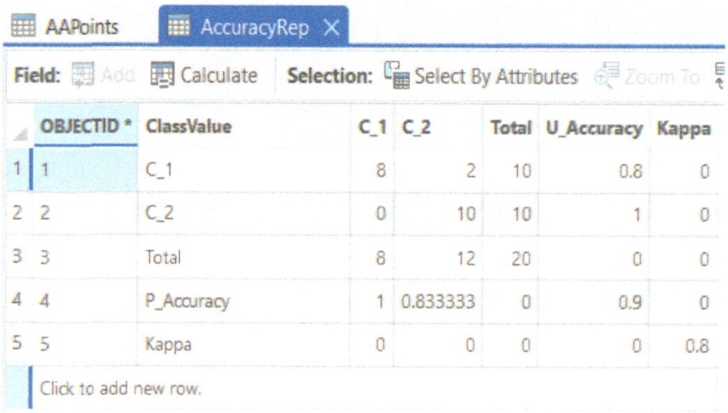

FIGURE 6.15 The computation results of the confusion matrix provided by ArcGIS Pro.

With the development of information technology, online, web-based tools and systems are increasingly drawing attention. They are designed and developed to support collaboration within and among teams to reduce the duplicated effort that may result from validations at disparate sites (Morisette et al. 2002). In the GlobCover project, a tool has been developed for data collection (Defourny et al. 2009). Based on the Google Earth platform, a geospatial Wikipedia named Geo-Wiki was developed, which has been made available to the Internet community interested in spatial validation (Fritz et al. 2009). Geo-Wiki is an online validation platform for visualizing hybrid GLC maps and in situ samples, which integrates high-resolution satellite images from Google Earth with crowdsourcing into a single portal (Fritz et al. 2012). For interpreting reference data from high-resolution images, a browser-based collaborative tool for "crowdsourcing" was developed (Clark and Mitchell Aide 2011). A web-based system was also developed for supporting GlobeLand30 dataset validation (Han et al. 2015).

6.3.1 Geo-Wiki

Geo-Wiki[6] is a web-based platform developed at the IIASA in Laxenburg, Austria, which allows for crowdsourcing GLC and land use and reference datasets (Fritz et al. 2017). The platform provides

Software Tools for Validation

TABLE 6.1
Summary of the Input and Output of Selected Validation tools

Name of the Software and the Tool/Plugin	Classification Input Data Type				Reference Input Data Type				Output					
	Raster	Vector			Raster	Vector			.csv	.txt	.tif	.dbf	other	Visualize in the Software
		Point	Line	Polygon		Point	Line	Polygon						
SCP in QGIS	✓	✗	✗	✗	✓	✓	✗	✗	✗	✗	✓	✗	✓	✓
AcATaMa in QGIS	✓	✗	✗	✗	✗	✓	✗	✗	✓	✗	✗	✗	✗	✓
Accuracy Assessment in QGIS	✓	✗	✗	✗	✓	✗	✗	✗	✓	✗	✗	✗	✗	✗
r.kappa in GRASS GIS	✓	✗	✗	✗	✓	✗	✗	✗	✓	✓	✗	✗	✗	✓
Confusion Matrix in SAGA	✓	✗	✗	✗	✓	✗	✗	✓	✓	✓	✗	✓	✓	✓
Confusion Matrix Computation in OTB	✓	✗	✗	✗	✓	✓	✗	✗	✓	✗	✗	✗	✗	✓
Accuracy Assessment in ERDAS	✓	✗	✗	✗	✗	✓	✗	✗	✗	✓	✗	✗	✓	✓
Compute Confusion Matrix in ArcGIS	✓	✗	✗	✓	✗	✓	✗	✗	✗	✓	✗	✓	✓	✓
Compute Confusion Matrix in ArcGIS Pro	✓	✗	✗	✓	✗	✓	✗	✗	✗	✗	✗	✓	✗	✓

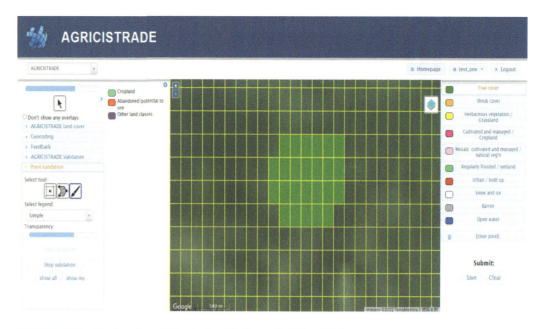

FIGURE 6.16 The interface of the pixel validation tool of Geo-Wiki.

a few tools to support user interaction with the data, including pixel validation, normalized difference vegetation index (NDVI) vegetation profile, identification, time series, photo uploading, and distance measurement on the map. The availability of these tools depends on the type of data. For example, the pixel validation tool is only available when the land cover data is selected. As shown in Figure 6.16, the idea of the pixel validation tool is to validate the land cover type of the selected pixel and provide the correct type of land cover based on the reference image data and contributors' local knowledge.

Users need to register an account in order to contribute. The user interface is intuitive and easy to follow. Users are provided with tools to select pixels by drawing geometries and to select land cover classifications (legend) to be used for validation (as shown in the left panel in Figure 6.16). The grid resolution is provided, including GlobCover 150 m (res:0.001389), forest mask Ukraine 1/4 300m (res:0.000139), forest mask Ukraine 300 m (res:0.000556), and forest change Moscow 1/4000 30 m (res:0.000250). The reference data provided include Sentinel-1, Sentinel-2, Google Maps Satellite, Mapillary, Bing Aerial with Labels, Bing Aerial, ESRI World Imagery, and OpenStreetMap.

In addition to the pixel validation tool, Geo-Wiki also offers other validation tools, such as the AGRICISTRADE Validation tool and global built-up surface validation. Lesiv et al. (2018) used AGRICISTRADE Validation tool to collect reference data on abandoned lands. Users can select a class for validation (as shown in the right panel in Figure 6.17). However, it is not currently available for use in the newest version. The global built-up surface validation is done by comparing Google Maps and Microsoft Bing images to see if the building areas have changed. The classification process is based on the Google Map. It is similar to the pixel validation tool (see Figure 6.18). In 2020, they held the global built-up surface validation competition. Even though the competition is over, it is still possible to validate and give validation scores.

6.3.2 LACO-Wiki

Also developed by IIASA, LACO-Wiki[7] is a web-based tool specially designed for validating land cover and land use maps using a variety of reference layers, including satellite and aerial imagery and OpenStreetMap (See et al. 2015). Different from Geo-Wiki, which is a manual, pixel-by-pixel

Software Tools for Validation

FIGURE 6.17　The interface of AGRICISTRADE Validation tool of Geo-Wiki.

FIGURE 6.18　The interface of global built-up surface validation tool.

validation tool, LACO-Wiki supports the complete validation process by allowing users to create and manage data, generate validation samples, interpret samples, and view validation results—a report with the accuracy assessment. Depending on the type of uploaded data, the platform supports the generation of different types of samples—e.g., Random Point, Random Pixel, Stratified Point, Stratified Pixel, Augmentation Point, and Augmentation Pixel. Figure 6.19 shows the main interface of the platform.

LACO-Wiki allows users to upload their own maps for validation in either vector or raster format (e.g., shapefiles or GeoTIFFs). It supports sample generation using random, stratified, or systematic

![LACO-Wiki interface screenshot]

FIGURE 6.19 The main interface of LACO-Wiki.

sampling strategies and the sample size can be specified by users or the minimum sample size based on the required confidence levels (see Figures 6.20[a] and [b]). The validation can be done by the user or other contributors invited by the user. A customizable accuracy assessment report can then be generated based on the validation results, which can be downloaded as an Excel file.

Extending the online version of LACO-Wiki, a LACO-Wiki mobile application has also been developed as a FOSS (see Figure 6.21) for in situ data collection (See et al. 2019). The mobile version instructs users step by step for validation, including validation session creation, sample point selection, validation, and results uploading.

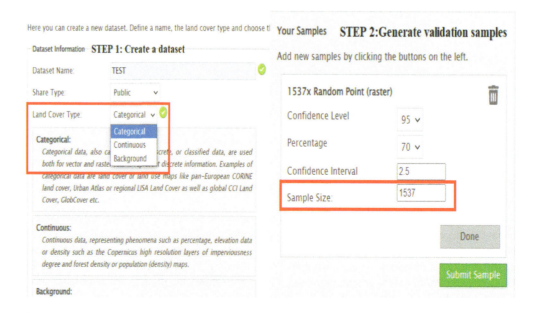

FIGURE 6.20 LACO-Wiki validation. (a) Select the land cover type, (b) Select the sample size.

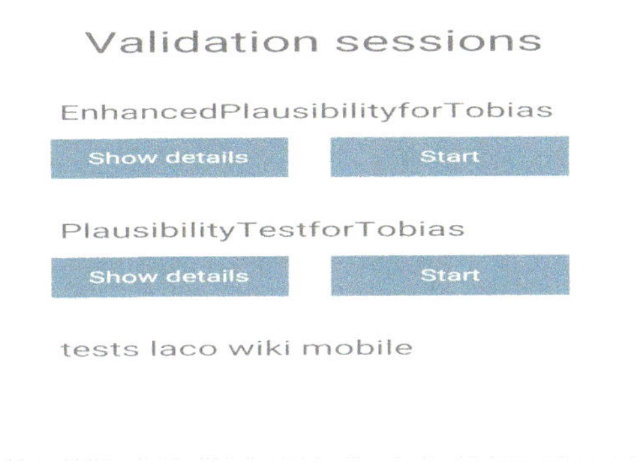

Validation sessions

EnhancedPlausibilityforTobias

[Show details] [Start]

PlausibilityTestforTobias

[Show details] [Start]

tests laco wiki mobile

FIGURE 6.21 The interface of LACO-Wiki mobile application.

6.3.3 MAP ACCURACY TOOLS

The Map Accuracy Tools[8] is an online platform designed for calculating accuracy metrics of land cover maps from remote sensing and analyzing thematic map accuracy (see Figure 6.22), which is the latest tool that was developed by IIASA in 2018. When validating land cover maps, the ground-truth data are compared with the values from the classified map, resulting in the production of a confusion matrix, such as overall accuracy, allocation disagreement, exchange, quantity disagreement, adjusted mutual information, and Kappa coefficient.

The Map Accuracy Tools allow users to upload their confusion matrix to the website, where various accuracy statistics can be calculated. It provides four ways for uploading a confusion matrix, including (1) using the "Load Example" buttons, (2) manually typing values into the cells, (3) copying and pasting a selection of values directly from an Excel document, and (4) importing a matrix as a csv file. Advanced users also have the option to download the R source code and run it locally. Furthermore, the tool can be used in educational settings to help students understand the concepts of map accuracy assessment.

In addition, the Map Accuracy Tools provides two example confusion matrices for users to experiment with, allowing them to add classes, delete classes, and merge classes to observe the effect. For instance, as shown in Figure 6.23, in the "merge classes" window, users can merge the evergreen forest class and the deciduous forest class by moving them from left to right into the same container (empty containers are ignored). In consequence, the window showing the merged matrix and the merged metrics will pop up (see Figure 6.24), where the overall confusion matrix and the overall metrics are compared in an intuitive way. The complete validation results shown in Figure 6.24 can also be downloaded as a csv file.

6.3.4 LAND COVER COLLECTOR

Land Cover Collector[9] was developed to enable ground data collection. The collected data can be used as a land cover reference dataset that is of great importance for validation (See et al. 2015; Fritz et al. 2017). It was designed as a free and open-source system for collecting land cover reference data according to GlobeLand30 nomenclature. The system provided Android and iOS apps, as well as a web-based platform developed based on the Apache Cordova mobile application development framework. As a result, it is a cross-platform. The leaflet library was used for web mapping. The land cover data collected were stored using PouchDB and CouchDB databases, which enabled both offline and online data collection. The source code of the application is protected by the GNU GPLv3 license.

Figure 6.25 shows the main interface of this web-based platform. Upon registration, contributors can log into the system and collect land cover classification data by choosing the "Add" section. The data collection is relatively straightforward, involving five steps: insert the point for data collection,

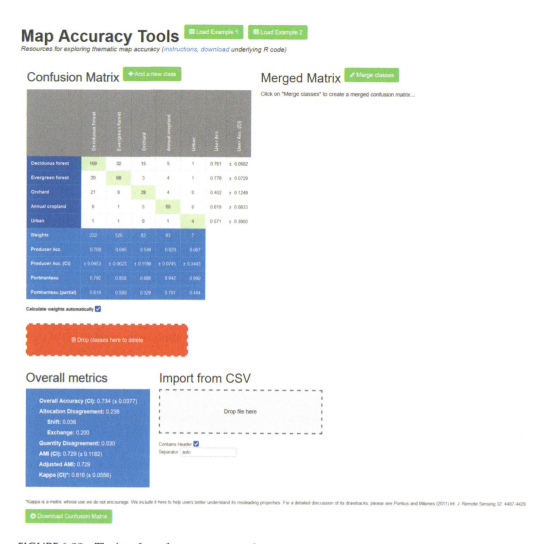

FIGURE 6.22 The interface of map accuracy tools.

choose the land cover class from a predefined list, confirm how much users are certain about the land cover class they select, add optional comments, and add four photos in cardinal directions (in the order of north, east, south, and west). The list of land cover classes is predefined based on GlobeLand30 nomenclature, while the certainty is rated as either 20%, 40%, 60%, 80%, or 100%.

Contributors can view and query the data they have collected using the tool in the "My Map" section and view and query all the points collected by all contributors in the "Everyone" section (see Figure 6.26). The "Information" section provides instructions on how to use the apps of this tool. The data collected using this tool are licensed under the Open Database License (ODbL) v1.0 and can be downloaded in JSON format. The tool supported eight languages at the time it was developed: English, Italian, Arabic, Russian, Chinese, Portuguese, French, and Spanish.

The Land Cover Collector is a simple and intuitive tool for crowdsourcing data points that carry land cover type information. Once collected, the data can be downloaded for validation use. Unfortunately, as of the time this chapter was written, all three types of apps were not available anymore on the web and in the app stores.

In summary, the online tools reviewed in this chapter mostly involve the collection of reference data and local knowledge to support the validation of GLC data products, with the exception of the Map Accuracy Tools and the LACO-Wiki tool. While the Map Accuracy Tools focus on accuracy

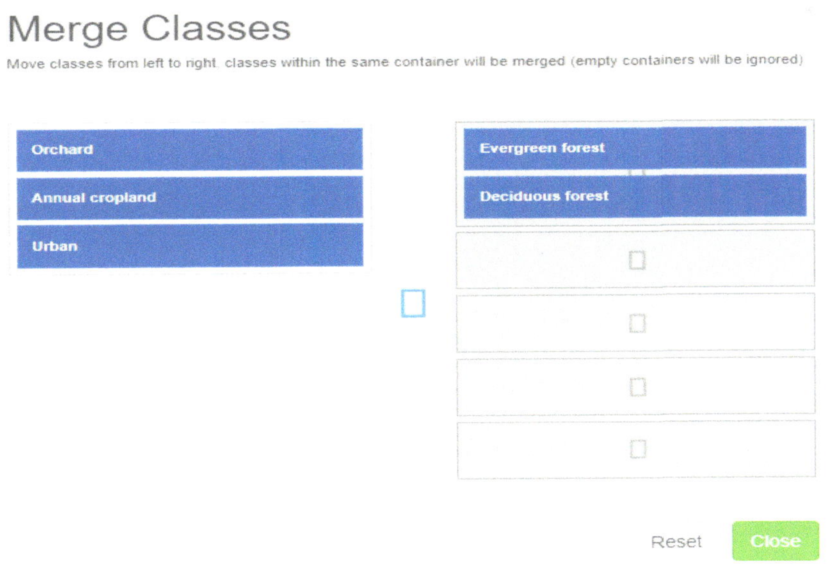

Software Tools for Validation

FIGURE 6.23 The interface of merging different classes.

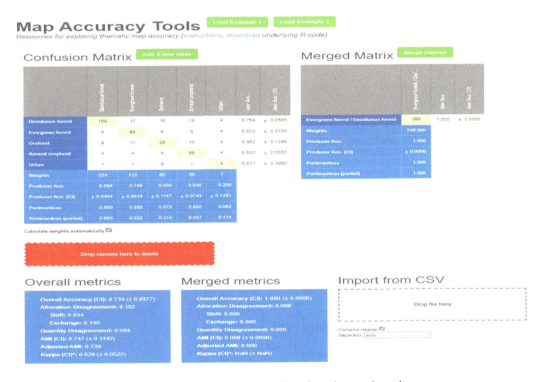

FIGURE 6.24 The validation results, including merged matrix and merged metrics.

assessment, the LACO-Wiki tool provides support for a complete validation process. The development of this kind of online tool can still be considered as in its early stage and requires further investigations in terms of supporting resource sharing and integration through web services, interaction, and collaboration among validation contributors, and various sampling, judgment, and accuracy assessment methods, toward a complete collaborative validation process.

FIGURE 6.25 The main interface of the land cover collector.

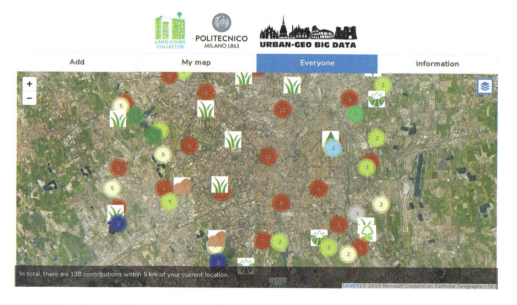

FIGURE 6.26 Points in Milan collected by all users.

6.4 REQUIREMENTS AND CONCEPTUAL DESIGN

Collaborative validation of GLC refers to the joint efforts by contributors, including researchers, experts, and volunteers, from different regions and institutions to assess, validate, and improve the accuracy of maps that depict the land cover. During this collaborative validation process, the Internet and service computing technologies can be integrated and play an important role, which makes it possible to produce more accurate, up-to-date, and globally consistent land cover products that can support environmental management, policymaking, and scientific research at local, national, and international levels (Chen et al. 2018).

Software Tools for Validation 183

As such, a service model is required to support the development of online tools that support the collaborative validation process with functional and, ideally, automation mechanisms through, e.g., web services composition. The land cover validation service model leverages computing technology (Bouguettaya et al. 2017) to encapsulate business data and algorithms related to land cover validation into web services. These services are accessible via the Internet, facilitating the interconnection and sharing of validation information resources. It offers online computing services and supports cross-region, process-oriented collaborative processing online. This approach not only streamlines the validation process but also enhances collaboration and data accessibility across different regions and disciplines. From an online service computing viewpoint, this model hinges on the publication of information data and model algorithms as web services, encompassing four key elements: participation, content service, interactive messaging, and interactive operation (Chen et al. 2017). Figure 6.27 illustrates a four-layer structure of how various services are integrated into the aforementioned four key elements of the collaborative validation process.

The four key elements include participants, content services, interactive messaging, and interactive operations, which are briefly discussed next.

Participants: The model identifies three principal roles—i.e., service publishers, service users, and service managers. Service users further diversify into four specialized participant categories: sampling experts, validation information providers, inspection experts, and validation experts. Each category plays a distinct role, contributing uniquely across various validation stages. This structured organization not only ensures a comprehensive and collaborative approach to the land cover validation process, allowing for seamless integration of expertise at every stage of the process but also guides the software implementation of the overall process of the collaborative validation, shown in Figure 6.28.

Content services: As shown in Figure 6.27, validation tools offer users an array of data and algorithmic services that are essential for land cover validation, supplying both the computational resources (data) and methods necessary for conducting online calculations. Specifically, data services encompass a variety of data and their processing, such as classified data, base map data, classified images, reference surface coverage data, and sample data. Algorithm services, on the other hand, provide functionalities for sample collection, validation, and accuracy assessment, among others (National Technical Committee for Standardization of Geographic Information 2017). Users can access these services via algorithm service interfaces, leveraging relevant data services to perform analyses and obtain results pertinent to different stages of the validation process.

Interactive messaging: It serves as the conduit for exchanging requests and responses among participants and content services within the land cover validation process. These messages are adaptable to various formats, including images, vectors, and texts, and are structured using appropriate data formats, such as JPG, JSON, and XML. They encapsulate diverse types of data, including verification photographs, sample point coordinates, and accuracy validation outcomes. Rooted in the Internet concept, the interplay between participants, content services, and interactive messages lays the foundational infrastructure for operational synergy and the flow of information in Internet-enhanced land cover validation. However, to achieve a fully integrated Internet business model, it is essential to sequentially link these algorithm services via specific interactive operations, thereby enabling effective human-computer interaction functionalities.

Interactive operations: It refers to the process where service users engage with algorithm services by selecting inputs and manipulating the outcomes of both data and algorithmic computations. This includes actions such as choosing specific areas for validation, defining confidence intervals, selecting relevant reference information, conducting consistency comparisons, and inputting validation data. These operations are critical for tailoring the validation process to specific needs, allowing users to precisely control and refine the parameters and data used in land cover validation analyses.

As shown in Figure 6.27, the overall process of collaborative validation of GLC mainly involves four types of experts, and they are engaged in different tasks. They can participate in the

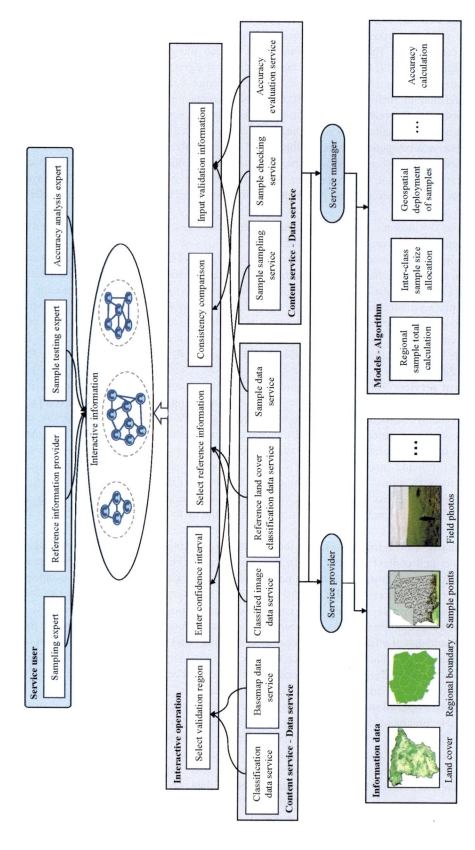

FIGURE 6.27 An Internet-based GLC validation service model. (Adapted from Chen et al. 2018.)

Software Tools for Validation

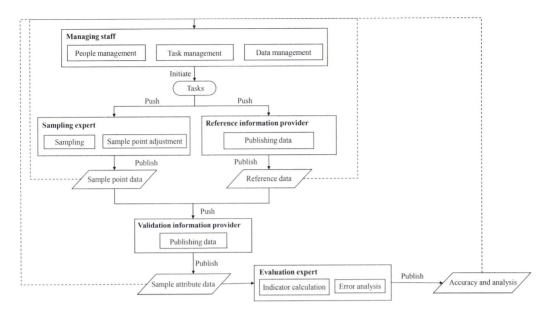

FIGURE 6.28 The overall process of the collaborative validation of GLC data. (Adapted from Chen et al. 2018.)

collaborative validation process according to their roles and responsibilities, which means the validation tools need to have mechanisms to coordinate their collaboration workflows (see Figure 6.28), leveraging message passing and subscribe/publish technologies. To begin with, the service manager initiates the validation task, selects the validation area, and pushes it to the sampling expert and the reference information provider. Subsequently, the sampling expert enters the confidence interval, selects the sampling algorithm service, publishes the sampling results as a sample data service, and pushes it to the validation expert. Accordingly, the validation expert checks the attribute information of sample points based on the reference land cover classification data service, selects reference information, performs consistency comparison and other operations, and pushes it to the validation expert. Finally, the validation expert calculates the accuracy validation index, analyzes the distribution and source of data errors, and pushes the results to the service manager to complete the task.

Furthermore, the collaborative validation tools should meet the actual validation needs by providing online processing functions, such as sample design, interactive validation, and accuracy assessment, through realizing the interconnection of multisource reference resources, adaptive sampling calculation, multirole expert participation, and sharing of sample data and validation results to provide support for collaborative validation of GLC data. Figure 6.29 shows a schematic implementation framework of such tools.

Aligning with Figure 6.29, the collaborative validation tools also need to support the typical validation that follows five steps—namely, selecting validation data, selecting validation area, setting sampling parameters and allocating samples, sample verification, and accuracy assessment. These steps require a number of functions, including sample size estimation, spatial allocation, validation information integration, expert judgment and information management, sample database management, online error information reporting, accuracy calculation, report generation, etc. In addition, the validation tools should ideally provide interfaces and functions to support the tasks in each of the steps shown in Figure 6.29 to complete the entire process of land cover validation.

Reference data collection: The validation process is complicated, especially at the global scale, requiring a huge amount of different types of auxiliary data (Morisette et al. 2002). High-quality reference samples are invaluable for validation. Designed for different applications, validation

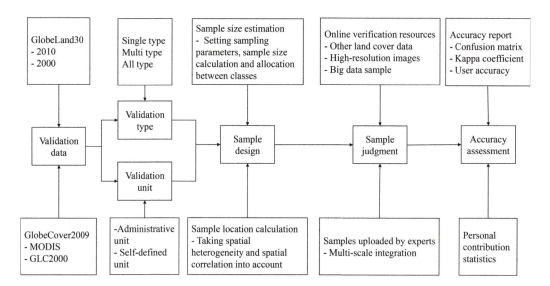

FIGURE 6.29 A schematic implementation framework of online validation tools (Adapted from Chen et al. 2018).

datasets for specific land cover types have also been developed. For example, GOFC-GOLD developed a data access portal for various validation datasets (GOFC-GOLD 2016). For global, continental, or national scales, land cover maps and reference samples are difficult to obtain from field observations, so high-resolution images are often used for interpretation (Wickham et al. 2013). Independent field, airborne, and other satellite data are critical for assessing the quality of land cover products, so they should be collected and used in the validation process (Morisette et al. 2002). In addition, there is also much additional valuable information on the Internet; for example, www.confluence.org provides many in situ pictures of the latitude-longitude confluence points in the world. Online web mapping service providers, such as Google Maps and Microsoft Bing Maps, also provide time-series, high-resolution imagery.

These rich data sources require validation tools to provide efficient access to them by allowing users to upload and share their reference datasets, and access and download existing higher-resolution datasets for validation purposes.

Automatic sampling and sample design: Spatial sampling design is one of the key steps in the land cover validation process (Strahler et al. 2006; Stehman et al. 2012). Over the past many years, several sampling approaches have been developed to determine the sample size and spatial distribution based on some criteria, such as probability, cost-effectiveness, and spatial balance (Zhu et al. 2000; Stehman 2009; Olofsson et al. 2014; Chen et al. 2016). Please refer to the details of these methods in Chapter 2.

However, most of these approaches exist in paper manuscripts or articles. It is not easy for people who take part in the land cover validation process, especially since most of these people are not familiar with computer programming. Even though there are some sampling design algorithms implemented as software, most are dedicated to specified projects and not open for the whole land cover validation user community. Due to high spatial heterogeneity, it is very hard to choose one universal sampling design method at a global scale. The choice of sampling design method should take into account the heterogeneity of the region for validation. As such, adaptive computing sampling design methods should be provided in the online validation platform.

Sample interpretation: For supporting sample judgment, tools that facilitate human-computer interaction should be provided by an online validation platform. A web map and map operation functions are the basis for online validation. Zoom in, zoom out, and pan are the basic supporting

Software Tools for Validation

functions. When users validate a place on Earth's surface, zooming in until it is possible to see sufficient details is necessary. The point on the map indicates the place where the user chooses to undertake the validation. It is convenient for users to identify different statuses of sample points by colors on the map. When a sample point is selected by users on the map, the land cover classification represented by the point in the place of the land cover product should be displayed on the map of the validation platform. Meanwhile, different ancillary datasets corresponding to the place where the point is located can be easily overlaid or shown on another viewport of the map. Different types of validation methods can be implemented for validators to use, such as blind validation, plausibility validation, and enhanced plausibility validation (See et al. 2015).

Accuracy assessment: When a validator finishes the judgment of all sample points in a selected sample dataset, a final accuracy result should be computed to illustrate the quality of the land cover product. A confusion matrix should be computed based on users' sample judgment results, which includes all land cover classes in the product for validation. Also, in the quantitative accuracy assessment result, except for the overall accuracy, the user's accuracy, and the producer's accuracy, the Kappa index also needs to be computed. Online validation tools should have user interfaces and functional tools for supporting the computation of these indicators of accuracy assessment.

Finally, it is important to consider sample integration. The samples are an important output of land cover data validation. Previously related land cover mapping or validation projects are designed and constructed to their own needs, so the standards are often inconsistent, and the sample database is difficult to share with other users. According to the principle of unified standard and sharing, it is important to establish test sample databases of GLC.

In this case, valid samples judged by experts should be standardized and documented according to some standard metadata. Some example metadata of validation sample points are shown in Table 6.2.

6.5 GLCVaL TOOL FOR GLOBELAND30

GLCVal[10] is a tool developed by the National Geomatics Center of China, which was designed and developed to support the international collaborative effort for validating GlobeLand30 data products. The basic idea is to integrate the concept of "Internet plus" and service computing technologies with land cover validation procedures to develop an online collaborative validation tool (Chen et al. 2018). This online tool serves as a platform to assemble land cover validation methods, reference data or access to the data, and workflows into a single portal, including storage and access of available reference data, wizard-guided validation workflow, self-adaptive sampling design, user-friendly sample interpretation based on satellite images, and generation of accuracy assessment report.

The GLCVal tool allows users to carry out their validation tasks following step-by-step guidance. To use the tool, a user account with proper access credentials is required. After login, a new validation task is created with a supporting functional sequence. Users can either select or upload a land cover data product for validation. They can choose an administrative or geographical region of interest or define the area of interest by uploading administrative boundary data for validation. After that, they can choose a sampling approach, such as Landscape Shape Index (LSI)–based sampling, among others, to generate sample points. Users are then able to interpret the land cover type of each sample point by comparing it with higher-resolution images of the same location. After completing the judgment and labeling of all the sample points, an accuracy assessment report is generated automatically. As the labeled samples are one of the important outputs of GLC validation, sample metadata is developed, and all the valid sample points are then integrated and stored for further utilization and sharing.

6.5.1 DESIGN

The design of GLCVal follows the guidelines for the development of a standard-based portal as outlined by the Open Geospatial Consortium (OGC) and Mapbox Vector Tile specification. The system

TABLE 6.2
Metadata of the Validation Sample Points

Name	Code	Type	Mandatory Fields	Description
Sample number	FID	Digital (long integer)	Mandatory	ID of the sample: the coding is adding region code with "+6" and sequence codes. Asia: As, Africa: Af America: Am, Europe: EuAustralia/Antarctica: Au
Sample name of the first level	FNAME	Text	Mandatory	Type name in GLC classification formulated in Subject One
Sample code of the first level	FCODE	Digital (integer)	Mandatory	Type code in GLC classification formulated in Subject One
Sample name of the second level	SNAME	Text		Type name in GLC classification formulated in Subject One
Sample code of the second level	SCODE	Digital (integer)		Type code in GLC classification formulated in Subject One
Other names of the sample	OTHER	Text		Classification system of IGBP and FAO or other type names
Image feature of the samples	SIMG	Text		Include band combination, color, and texture, etc.
Sample acquisition time	SDATE	Date	Mandatory	The time for judging land cover types
Sample acquisition sources	SOURCE	Text	Mandatory	Acquisition methods include reference collection, field investigation, high-resolution images, etc.
Sample collector	SPERSON	Text	Mandatory	
Sample collecting unit	SUNIT	Text	Mandatory	
Sample reliability assessment	CONFIDENCE			A/B/C/D
Vegetation overlay degree	CANOPY	Text		The description of height, density, overlay degree, and cover degree on the forest, shrubland, grassland,
On-site record materials of the samples	DATUMFILE	Text		On-site record materials of the samples
Picture	PICTURE	Text	Mandatory	On-site picture materials of the collecting samples and links to screenshots of other methods
Video	VIDEO	Text		Video material links of on-site collecting samples
Note	MEMO	Text		Other description information about the samples

architecture is based on the principle of Service Oriented Architecture (SOA) and Microservices Architecture (Lewis and Fowler 2014; Microservices, Building 2015; Fowler 2019). A web-based system model is shown in Figure 6.30, which includes four parts—i.e., dataset integration, data and function services, interactive mapping, and validation management—structured in three tiers.

GLC validation requires a huge amount of various data. For GlobeLand30 validation, great effort has been made to collect, integrate, and manage these datasets. Reference data, GLC products, and samples are the three main types of data. The collection and integration of various reference data and resources lay a solid foundation for collaborative validation. VHR images, thematic maps, in situ measurements, crowdsourcing, and VGI data make up the majority of the reference resources. When using these reference data, appropriate evaluation needs to be done to ensure the data quality, such as completeness, logical consistency, positional accuracy, thematic accuracy, temporal quality, and

Software Tools for Validation   189

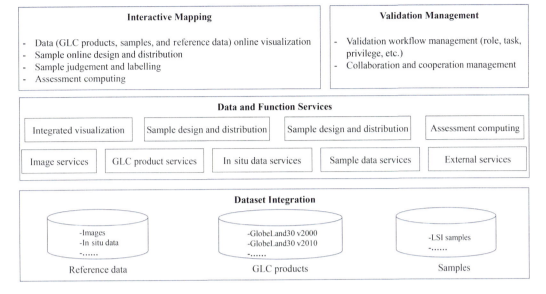

FIGURE 6.30 The system model for GLCVal.

usability. The in situ measurement data acquired from fieldwork is another valuable reference source for GlobeLand30 validation. Since field data collection is time-consuming and labor-intensive, their usage at the global scale is limited due to high cost. As a result, field data collection can act as supplementary means in some cases where qualified images are missing, or the detailed class information is needed. GLC products consist of a series of GlobeLand30 products, such as version 2000, version 2010, and so on. Samples are composed of the points generated by sample design algorithms, such as LSI and data uploaded by users.

Once the data sources discussed earlier are properly integrated and published, functions are required to better utilize them. The basic services of the system can be classified into two types: data services for serving the aforementioned datasets and function services based on the identified functional requirements. Functions for integrated visualization serve a number of purposes for visualizing multisource data and services. Different datasets and various visualization styles are combined to provide useful information to users. Different sample design and distribution algorithms are encapsulated as services that are adopted for Globeand30 validation to guarantee samples with good spatial representation and even spatial distribution. The algorithms are required to follow and satisfy some criteria, such as relatively high sample size or density for heterogeneous landscapes, even spatial distribution, representative sample numbers for all classes, especially rare classes and fragmented classes, and self-adaptive ability to different regions (Chen et al. 2016). Sample judgment and labeling functions provide an online service to help validation experts or volunteers judge and label the land cover types of sample points through the interpretation of high-resolution images or comparison with large-scale land cover maps and in situ measurement data. Some additional supporting information can be provided or uploaded, such as text commentaries and photos of the samples. Accuracy assessment computing service aims to provide an accuracy index of how closely the derived class allocations depicted in the thematic land cover map represent reality. The commonly adopted method is the confusion matrix and its derived basic descriptive measures, including overall accuracy, user and producer accuracy, Kappa, and so on.

Integration of various data and external services is vital for a web-based validation system. Functionality needs to be developed to handle data conversion and re-projection for data publishing and service integration if the external data services are incorporated. Further strategies are required to deal with data and processing service composition if more complex geo-processes need to be

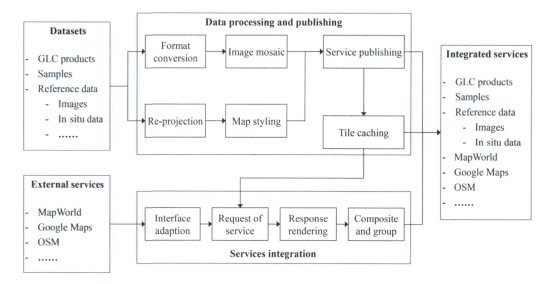

FIGURE 6.31 Integration of data and internal/external services.

formed using the basic processes. An overview of how heterogeneous datasets and disparate external services are integrated is illustrated in Figure 6.31. The procedure includes two main streams of processing—i.e., processing and publishing of heterogeneous data and integration of internal and external services.

Data format conversion and re-projection are important to validation processes. Due to its suitability for global-scale datasets, the geographic projection with WGS84 is used by the most collected reference data; therefore, they are transformed (re-projected) to WGS84. Similar to the projection problem, various data formats also need to be converted to a universal or commonly used format, such as ERDAS IMG and GeoTIFF for imagery.

With regard to the huge volume of images in reference data, it is impossible to publish each scene image as a Web Map Service (WMS), which results in notoriously slow access to services and difficulties in service composition. Based on image mosaic, global images can be combined and published as web services with reasonable granularity. Map styling is based on the pervasive Styled Layer Descriptor (SLD; OGC 2007) or Mapbox Style Specification (Mapbox Mapbox Style Spec | Mapbox Docs 2023), which supports user-defined symbolization and color-coding of geographic features and coverage data.

Service publishing and tile caching follow the OGC Web Services (OWS) standards or specifications, including WMS, Web Map Tile Service (WMTS), and Mapbox Vector Tile (MVT) Service. Images can be published as WMTS services, and geographic vector data can be published as MVT services. To accelerate the browsing speed of geospatial data and enhance the user experience, data partition based on its granularity and map tile caching procedure is considered in data and service publishing.

Besides internal web services, external web services published by other data providers can be integrated. A service integration workflow is designed at the application level of the GLCVal. Based on the workflow, the interface adaption process mainly handles the differences between the application program interfaces (API) of external services. Requests of service are constructed to retrieve WMSs concurrently according to the geographical extent of user requests and service types. The last optional step of this workflow is compositing and grouping map services, where multisource map services can be composited into a map layer so that it can be easily switched on and off in the tool interface. Moreover, a service register tool is developed to facilitate importing multisource map services into GLCVal.

Software Tools for Validation 191

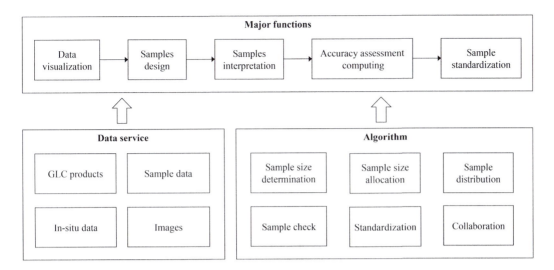

FIGURE 6.32 Functional design of GLCVal.

Major functionalities are illustrated in Figure 6.32 with its supporting data services and algorithms. The web-based online validation system serves as a platform to assemble GLC product validation methods and workflows into a single portal, including the storage and online visualization of all available reference data, "wizard" guided validation workflow, self-adaptive sampling designs, user-friendly interpretation of sample from imagery, and generation of the accuracy assessment report.

6.5.2 Implementation

A schematic diagram of the general system implementation architecture is illustrated in Figure 6.33, which consists of different standard components integrated into a single portal. The data described in the previous section is contained in three separate repositories shown at the bottom of Figure 6.33. The first repository contains images and GLC products. The second repository is a MongoDB database, which stores the tiles generated from the raster data from the first repository. The first two repositories are published as OGC WMTS by Image Server. The third repository is also a database of samples for validation generated by sample design algorithms or uploaded by users. It uses Vector Server as a rendering engine and is stored in a PostgreSQL database with PostGIS extension, which allows for spatial queries on features with spatial extent and attributes. Through these two publishing and rendering engines, a series of OGC WMTSs and MVT services are published and integrated into the validation web portal. As macro-services, these services can even be imported into some stand-alone software in the validation process for GLC products.

The implementation adopts a browser/server balanced configuration, meaning that some processing tasks are allocated on the browser-side computer, and others are deployed on the server side. The web portal operates using Node.js running on an NGINX web server together with Image Server and Vector Server. The client browser loads the system's web interface, which is written in JavaScript based on some server-side and client-side JavaScript Software Development Kits (SDK), such as Node.js and Mapbox GL JS due to their compatibility with OGC web service standards and vector tile service specification. External services, such as MapWorld or OpenStreetMap, are imported and connected on the client side for access to high-resolution imagery.

Performance is always an important issue that needs to be considered when a web-based system is developed. There are many options for improving a web-based system's performance, such as compression, caching, geographic load balancing, adding hardware, and so forth. Optimizing the use of geographic data tiling and caching is often the initial place to start, as configuration changes

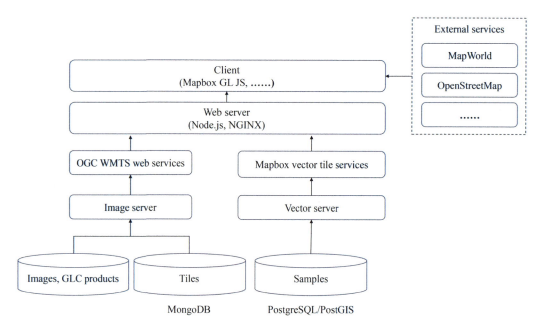

FIGURE 6.33 System implementation architecture.

are generally free and can increase efficiency dramatically. To improve the performance, these two aspects are considered first in the web-based validation system, which is the browser-side cache for map data and tiles generated on the server side.

6.5.3 Wizard (Walk-Through Example)

Step-by-step guidance is provided for users to carry out their entire validation tasks, from choosing a land cover data product from a list of products (currently IGBP DISCover, MODIS, UNID, GLC2000, GlobCover, GLCNMO, FROM-GLC, and GlobeLand30) and validation region, through sampling design and sample judgment, to generating assessment report. Figure 6.34 illustrates the overall workflow of the validation.

To demonstrate the usefulness of GLCVal, an example of how the system can be used to support GLC product validation is described in the following. After logging into the portal, a new validation task can be created with an adaptive supporting functional sequence (see Figure 6.35). To begin with, a name for a new task must be specified, which is suggested to indicate the spatial extent for validation.

In creating a new validation task, it is required to select a GLC product first. When users click the list box on the left side of Figure 6.36, a series of GLC products will appear. After users choose a GLC product, it will be displayed on the interactive map (see the right side of Figure 6.36).

In the next step, users can choose an administrative or geographical region as an interest area or define the area of interest by uploading a specific geographical extent for validation. When a region is selected by clicking the region list box, users can choose a country first, of which the spatial extent will be displayed in the interactive map (see Figure 6.37). Furthermore, users can also specify a state or province by clicking the next list box. Accordingly, the spatial extent displayed on the interactive map will be changed based on the user's selection.

Subsequently, the main function supported by the system is the implementation of sample design and spatial distribution. Users can choose a sampling approach to generate the sample dataset or upload their own sample datasets, which will be visualized on the browser interface. In

Software Tools for Validation

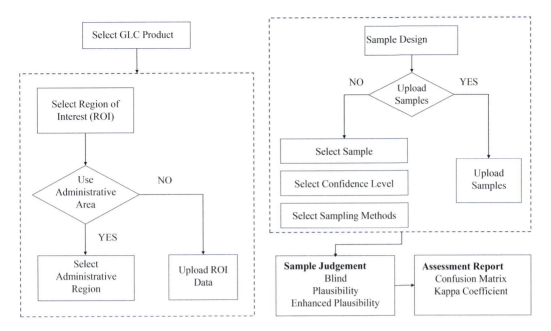

FIGURE 6.34 GLCVal validation workflow.

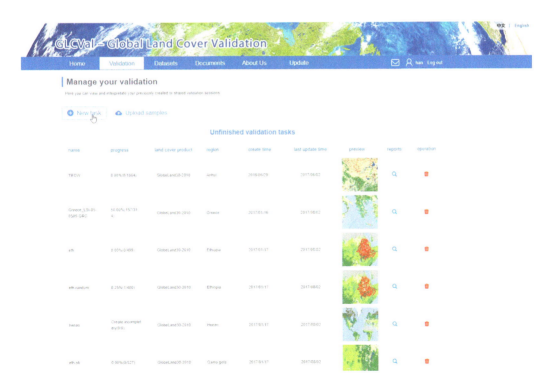

FIGURE 6.35 Creating a new validation task.

FIGURE 6.36 Selection of a GLC product.

FIGURE 6.37 Selecting a region for validation.

this step, users are able to specify the sampling method, the confidence level, and the selection method for the sampling design approach. After the aforementioned three parameters are selected, the samples generated by the sampling design approach will be shown on the interactive map (see Figure 6.38).

Interpreting the land cover type of each sample can be then implemented by users by comparing higher-resolution images of the same location, as shown in Figure 6.39. After the judgment and labeling of all the samples are completed, an accuracy assessment report will be generated automatically (see Figure 6.40).

Software Tools for Validation 195

FIGURE 6.38 Sampling design and distribution.

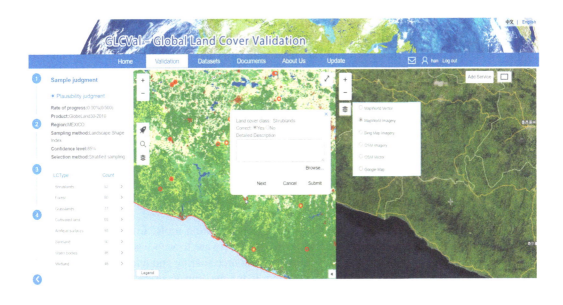

FIGURE 6.39 Sample judgment and labeling.

As the samples are given in this step, users can conduct the validation by interpreting the land cover type of each sample, which can then be implemented by comparing with the higher-resolution image of the same location, as shown in Figure 6.39. By clicking each sample point displayed on the interactive map, users obtain the interface window for validation, where the original land cover type of the sample point from the GLC product is shown. It allows users to identify whether the original land cover type is correct or not by clicking the radio box option. At the same time, the user can also type some text in the commentary input box as a detailed description. Finally, users confirm the

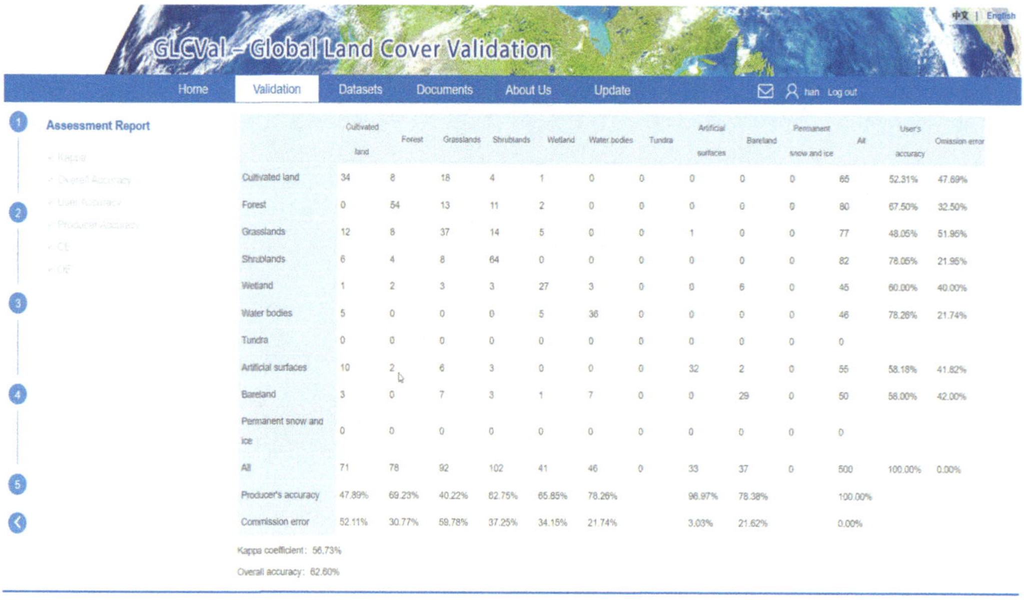

	Cultivated land	Forest	Grasslands	Shrublands	Wetland	Water bodies	Tundra	Artificial surfaces	Bareland	Permanent snow and ice	All	User's accuracy	Omission error
Cultivated land	34	8	18	4	1	0	0	0	0	0	65	52.31%	47.69%
Forest	0	54	13	11	2	0	0	0	0	0	80	67.50%	32.50%
Grasslands	12	8	37	14	5	0	0	1	0	0	77	48.05%	51.95%
Shrublands	6	4	8	64	0	0	0	0	0	0	82	78.05%	21.95%
Wetland	1	2	3	3	27	3	0	0	6	0	45	60.00%	40.00%
Water bodies	5	0	0	0	5	36	0	0	0	0	46	78.26%	21.74%
Tundra	0	0	0	0	0	0	0	0	0	0	0		
Artificial surfaces	10	2	6	3	0	0	0	32	2	0	55	58.18%	41.82%
Bareland	3	0	7	3	1	7	0	0	29	0	50	58.00%	42.00%
Permanent snow and ice	0	0	0	0	0	0	0	0	0	0	0		
All	71	78	92	102	41	46	0	33	37	0	500	100.00%	0.00%
Producer's accuracy	47.89%	69.23%	40.22%	62.75%	65.85%	78.26%		96.97%	78.38%		100.00%		
Commission error	52.11%	30.77%	59.78%	37.25%	34.15%	21.74%		3.03%	21.62%		0.00%		

Kappa coefficient: 56.73%

Overall accuracy: 62.60%

FIGURE 6.40 Accuracy assessment report.

validation result and submit it to the system. Initially, the style of the sample displayed on the map is a hollow circle. After submitting the validation result, the style will be changed to a solid circle. In this validation process, different colors are used to identify the status change from beginning to end. It is easy to capture the validation status of the samples based on the style change. In addition, another reference interactive map window can be opened in parallel with the product window to provide another reference for users.

Finally, after the judgment and labeling of all samples are completed, an accuracy assessment report will be generated automatically (see Figure 6.40). There are many different parameters designated in the report, such as Kappa, overall accuracy, user's accuracy, and producer's accuracy. The report will be changed corresponding to the users' selection of these parameters.

6.6 SUMMARY

This chapter introduces the development of land cover validation tools, from the built-in functionality or plugin extension of GIS and remote sensing software to web-based and mobile tools. However, most of them do not provide full support to resource sharing and integration, information interaction among participants, and various sampling, judgment, and accuracy assessment methods, which are needed for collaborative validation. Aimed at providing a reference for implementing the collaborative validation of land cover products, it further presents the requirements and conceptual design to clarify the overall implementation process, as well as the conceptual framework of designing land cover validation functions. Subsequently, a number of existing tools are surveyed and summarized for validating GLC maps, including the desktop tools (e.g., QGIS, ArcGIS, ArcGIS Pro, and ERDAS IMAGINE) and the online tools (e.g., Geo-Wiki, LACO-Wiki, and Map Accuracy Tools). Especially, those tools developed for Globeland30 are elaborated to indicate the most recent progress made by the National Geomatics Center of China in the GLC validation field.

Some measures that can be taken to promote large-scale land cover validation are considered and discussed, such as allowing users to upload reference datasets through an online validation platform and standardizing all valid sample points according to the metadata property. As advanced Internet technology emerges, the Internet-based GLC validation service that involves participants, content

NOTES

1 https://qgis.org/en/site/
2 https://smbyc.github.io/AcATaMa/
3 https://github.com/jkibele/acc-assess
4 https://grass.osgeo.org/
5 https://www.esri.com/en-us/arcgis/about-arcgis/overview
6 https://www.geo-wiki.org/
7 https://www.laco-wiki.net/
8 https://iiasa.ac.at/models-tools-data/map-accuracy-tools
9 https://github.com/kilsedar/land-cover-collector
10 http://www.globallandcover.com/

REFERENCES

Bouguettaya, A, M Singh, M Huhns, Quan Z Sheng, Hai Dong, Qi Yu, AG Neiat, S Mistry, B Benatallah, and B Medjahed. 2017. "A Service Computing Manifesto: The Next 10 Years." *Communications of the ACM,* 60(4): 64–72.

Chen, Jun, Fei Chen, WU Hao, Lijun Chen, and HAN Gang 2018. "Internet[+] Land Cover Validation: Methodology and Practice." *Geomatics and Information Science of Wuhan University,* 43(12): 2225–32. doi:10.13203/j.whugis20180305

Chen, Jun, WU Hao, and LI Songnian. 2017. "Research Progress of Global Land Domain Service Computing: Take GlobeLand 30 as an Example." *Acta Geodaetica et Cartographica Sinica,* 46(10): 1526.

Chen, Jun, Jun Zhang, WW Zhang, and S Peng. 2016. "Continous Updating and Refinement of Land Cover Data Product." *International Journal of Remote Sensing*, 20: 991–1001.

Chen, Jun, L Chen, F Chen, Y Ban, S Li, G Han, X Tong, C Liu, V Stamenova, and S Stamenov. "Collaborative validation of GlobeLand30: Methodology and practices." *Geo-spatial Information Science,* 24, no. 1 (2021): 134–144.

Clark, ML, and T Mitchell Aide. 2011. "Virtual Interpretation of Earth Web-Interface Tool (VIEW-IT) for Collecting Land-Use/Land-Cover Reference Data." *Remote Sensing,* 3(3): 601–20.

Congedo, L 2016. "Semi-Automatic Classification Plugin Documentation." *Release*, 4(1): 29.

Conrad, C, M Rudloff, I Abdullaev, M Thiel, F Löw, and JPA Lamers. 2015. "Measuring Rural Settlement Expansion in Uzbekistan Using Remote Sensing to Support Spatial Planning." *Applied Geography*, 62: 29–43.

Defourny, P, L Schouten, S Bartalev, S Bontemps, P Cacetta, AJW De Wit, C Di Bella, B Gérard, C Giri, and V Gond. 2009. "Accuracy Assessment of a 300 m Global Land Cover Map: The GlobCover Experience." In Conference Proceedings: 33rd International Symposium on Remote Sensing of Environment, Sustaining the Millennium Development Goals - ISBN 978-0-932913-13-5. Tucson, AZ (United States of America): International Center for Remote Sensing of Environment (ICRSE); 2009. JRC54524

Fowler, M. 2019. "Microservices Guide." *Martinfowler.Com*. August 21. https://martinfowler.com/microservices/

Fritz, Steffen, I McCallum, C Schill, C Perger, R Grillmayer, F Achard, F Kraxner, and M Obersteiner. 2009. "Geo-Wiki.Org: The Use of Crowdsourcing to Improve Global Land Cover." *Remote Sensing,* 1(3): 345–54. doi:10.3390/rs1030345

Fritz, Steffen, Ian McCallum, Christian Schill, Christoph Perger, Linda See, Dmitry Schepaschenko, Marijn Van der Velde, Florian Kraxner, and Michael Obersteiner. 2012. "Geo-Wiki: An Online Platform for Improving Global Land Cover." *Environmental Modelling & Software*, 31: 110–23.

Fritz, Steffen, Linda See, Christoph Perger, Ian McCallum, Christian Schill, Dmitry Schepaschenko, Martina Duerauer, Mathias Karner, Christopher Dresel, Juan-Carlos Laso-Bayas, Myroslava Lesiv, Inian Moorthy, Carl F. Salk, Olha Danylo, Tobias Sturn, Franziska Albrecht, Liangzhi You, Florian Kraxner, and Michael Obersteiner. 2017. "A Global Dataset of Crowdsourced Land Cover and Land Use Reference Data." *Scientific Data,* 4(1): 170075. doi:10.1038/sdata.2017.75

GRASS Development Team. 2021. *Geographic Resources Analysis Support System (GRASS GIS)*. Open Source Geospatial Foundation. https://grass.osgeo.org/

GOFC-GOLD. 2016. GOFC-GOLD Reference Data Portal. http://www.gofcgold.wur.nl/sites/gofcgold_ref dataportal.php

Grizonnet, Manuel, Julien Michel, Victor Poughon, Jordi Inglada, Mickaël Savinaud, and Rémi Cresson. 2017. "Orfeo ToolBox: Open Source Processing of Remote Sensing Images." *Open Geospatial Data, Software and Standards,* 2(1): 1–8. doi:10.1186/s40965-017-0031-6

Han, Gang, Jun Chen, Chaoying He, Songnian Li, Hao Wu, Anping Liao, and Shu Peng. 2015. "A Web-Based System for Supporting Global Land Cover Data Production." *ISPRS Journal of Photogrammetry and Remote Sensing,* 103: 66–80.

Hexagon Geospatial / Hexagon. 2018. "ERDAS IMAGINE 2018." https://supportsi.hexagon.com/s/article/ERDAS-IMAGINE-2018-released?language=en_US

Hou, Dongyang, Jun Chen, Hao Wu, Songnian Li, Fei Chen, and Weiwei Zhang. 2015. "Active Collection of Land Cover Sample Data from Geo-Tagged Web Texts." *Remote Sensing,* 7(5): 5805–27. doi: 10.3390/rs70505805

Kibele, Jared. 2016. "Benthic Photo Survey: Software for Geotagging, Depth-Tagging, and Classifying Photos from Survey Data and Producing Shapefiles for Habitat Mapping in GIS." *Journal of Open Research Software,* 4(1): e10–e10.

Lesiv, Myroslava, Dmitry Schepaschenko, Elena Moltchanova, Rostyslav Bun, Martina Dürauer, Alexander V Prishchepov, Florian Schierhorn, Stephan Estel, Tobias Kuemmerle, Camilo Alcántara, and et al.. 2018. "Spatial Distribution of Arable and Abandoned Land across Former Soviet Union Countries." *Scientific Data,* 5(1): 1–12.

Lewis, James, and Martin Fowler. 2014. "Microservices." *Martinfowler.Com.* https://martinfowler.com/articles/microservices.html

"Mapbox Style Spec | Mapbox Docs." 2023. *Mapbox.* https://docs.mapbox.com//style-spec/guides/

Microservices, Building. 2015. *Designing Fine-Grained Systems.* Sam Newman.–O'Reilly Media, Inc 280.

Morisette, Jeffrey T, Jeffrey L Privette, and Christopher O Justice. 2002. "A Framework for the Validation of MODIS Land Products." *Remote Sensing of Environment,* 83(1–2): 77–96.

National Technical Committee for Standardization of Geographic Information. 2017. "2017 Land Cover Information Services GB/T 35635-2017." https://openstd.samr.gov.cn/bzgk/gb/newGbInfo?hcno=999B 8E24A96BDE1CF035A1F323E4671F&ivk_sa=1024320u

Olofsson, Pontus, GM Foody, M Herold, SV Stehman, CE Woodcock, and MA Wulder. 2014. "Good Practices for Estimating Area and Assessing Accuracy of Land Change." *Remote Sensing of Environment,* 148: 42–57.

Open Geospatial Consortium Inc (OGC). 2007. "Styled Layer Descriptor Profile of the Web Map Service Implementation Specification." Open Geospatial Consortium Inc. https://portal.ogc.org/files/?artifact_id= 22364

QGIS Development Team. 2018. "QGIS Geographic Information System." Open Source Geospatial Foundation Project.

See, L, C Perger, M Hofer, J Weichselbaum, C Dresel, and S Fritz. 2015. "LACO-Wiki: An open access online portal for land cover validation." *ISPRS Annals of the Photogrammetry, Remote Sensing and Spatial Information Sciences* II-3-W5 (August). Copernicus GmbH: 167–71. doi:10.5194/isprsannals-II-3-W5-167-2015

See, Linda, C Perger, C Dresel, M Saad, A Subash, B Mora, M Pascaud, F Ligeard, N Joshi, and S Fritz. "LACO-wiki mobile: An open source application for in situ data collection and land cover validation." In *Geophysical Research Abstracts,* Copernicus Publications vol. 21. 2019.

Stehman, SV. 2009. "Sampling Designs for Accuracy Assessment of Land Cover." *International Journal of Remote Sensing,* 30(20): 5243–72. doi:10.1080/01431160903131000

Stehman, SV, P Olofsson, CE Woodcock, M Herold, and MA Friedl. 2012. "A Global Land-Cover Validation Data Set, II: Augmenting a Stratified Sampling Design to Estimate Accuracy by Region and Land-Cover Class." *International Journal of Remote Sensing,* 33(22): 6975–93. doi:10.1080/01431161.2012.695092

Strahler, AH, L Boschetti, GM Foody, MA Friedl, MC Hansen, M Herold, P Mayaux, JT Morisette, SV Stehman, and CE Woodcock. 2006. "Global Land Cover Validation: Recommendations for Evaluation and Accuracy Assessment of Global Land Cover Maps." *European Communities, Luxembourg,* 51(4): 1–60.

Wickham, JD, SV Stehman, L Gass, J Dewitz, JA Fry, and TG Wade. 2013. "Accuracy Assessment of NLCD 2006 Land Cover and Impervious Surface." *Remote Sensing of Environment,* 130: 294–304.

Zhu, Z, L Yang, SV Stehman, and RL Czaplewski. 2000. "Accuracy Assessment for the US Geological Survey Regional Land-Cover Mapping Program: New York and New Jersey Region." *Photogrammetric Engineering and Remote Sensing,* 66(12): 1425–38.

7 Collaborative GlobeLand30 Validation

Fei Chen
East China University of Technology, Nanchang, China

Huan Xie
Tongji University, Shanghai, China

Jingxiong Zhang
Wuhan University, Wuhan, China

Zhengxing Wang
Shanghai Ocean University, Shanghai, China

Lijun Chen
National Geomatics Center of China, Beijing, China

Vanya Stamenova
Bulgarian Academy of Sciences, Sofia, Bulgaria

Stefan Stamenov
Bulgarian Academy of Sciences, Sofia, Bulgaria

Thomas Katagis
Aristotle University, Thessaloniki, Greece

Ioannis Gitas
Aristotle University, Thessaloniki, Greece

Maria Antonia Brovelli
Politecnico Milano, Milan, Italy

Yifang Ban
KTH Royal Institute of Technology, Stockholm, Sweden

Monia Elisa Molinari
Politecnico Milano, Milan, Italy

7.1 INTRODUCTION

In 2014, China released the GlobeLand30 data products (30 m resolution land cover datasets) for 2000 and 2010. These pioneering datasets provide comprehensive information on ten land coverage types: cultivated land, forest, grassland, shrubland, wetland, water bodies, tundra, artificial surface, permanent snow, and permanent ice (Chen et al. 2015). The datasets were donated to the United Nations to support global sustainable development research and planning.

The validation of GlobeLand30 was conducted as the essential step after its production. Despite significant efforts made in land cover mapping to produce Globeland30 using 30 m resolution imagery, challenges arose when validating land cover data at a global or regional scale rather than locally (Chen et al. 2021). These challenges encompassed determining an appropriate number of sample pixels and their distribution across continents worldwide, as well as collecting reference data and judging whether samples were correct or wrong in different places around the world. Technical difficulties and a scarcity of reliable reference data posed the primary obstacles during Globeland30 validation.

The Group on Earth Observations (GEO) is a renowned inter-governmental organization in the field of earth observation, with some of its members possessing valuable validation techniques and reliable reference data resources. To facilitate a comprehensive and collaborative validation of the Globeland30 products, the GEO Global Land Cover Community Activity organized the project "Collaborative Validation and Service of Global Land Cover Products" from 2015 to 2017, aiming to address issues related to GlobeLand30 validation, such as the absence of internationally agreed-upon validation guidelines and the lack of reliable reference data. The project had three primary objectives: development of technical specifications, establishment of an online platform, and collaborative validation of GlobeLand30 at various scales (as mentioned in Chapter 1).

With the support of this project, the project participants were invited to join collaborative validation practices. They were given the flexibility to choose their validation region of interest and adopt either a uniform or an independent sampling scheme. Additionally, they shared their reference resources, judgment, results, and assessment accuracies using an online validation platform, GLCVal (see Chapter 6 for details of this tool). The validation regions encompassed countries, regions, and even global land masses with the ultimate objective of obtaining accurate assessments for GlobeLand30 across diverse geographical areas.

The global validation of GlobeLand30 was crucial, as it provided users with essential information regarding the overall accuracy and class-level accuracy on a global scale. Conducting a comprehensive accuracy assessment of the global land cover (GLC) dataset not only facilitates data users in comprehending the uncertainty and applicability of the Globeland30 datasets but also empowers data producers to scrutinize and analyze the types, sources, and spatial distribution of errors (Wu et al. 2014; Fritz et al. 2011).

Despite reporting fairly good accuracies in global accuracy assessments, some researchers have observed significantly lower accuracies (ranging from 10% to 50%) in different regions or for specific land cover classes when validating existing land cover maps using diverse reference samples (Gong 2009; Manakos et al. 2014; Sun et al. 2016). From a user's perspective, it remained unclear which land cover dataset should be used for a particular location or problem and why (Herold et al. 2008). Regional-scale and national-scale validation could enhance users' understanding of the spatial distribution of misclassification of GlobeLand30. Consequently, validation practices on smaller regional areas were conducted by the project participants over the past few years, aiming to assess the accuracy of the GlobeLand30 datasets at both regional and national scales.

The workflow of collaborative validation is illustrated in Figure 7.1. Based on the characteristics of these practices, the guiding principles included a co-design scheme, regional division of validation, various forms of data interaction, and resource integration. The co-design scheme involved GEO Working Groups (GEO WGs) proposing the validation implementation plan and jointly developing an online validation tool that has undergone discussion and modification by the project participants. This collaborative optimization ensured the convergence of internationally advanced

Collaborative GlobeLand30 Validation

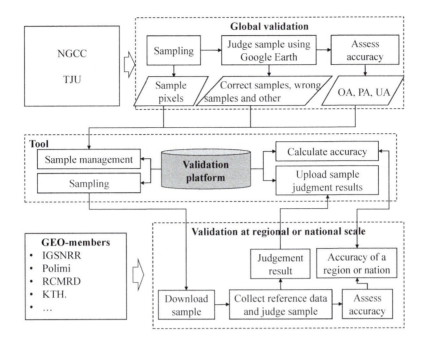

FIGURE 7.1 Workflow of international validation practices.

technologies to guarantee the feasibility and objectivity of validation practices. The regional division of validation was based on the geographical location of the project participants. The technical division of labor constitutes the primary approach for regional and national validation. The National Geomatics Center of China (NGCC), the producer of the data, is responsible for providing the sampling design and sample distribution, while other participants were tasked with collecting reference data and judging samples. Various forms of data interaction and resource integration were facilitated through the development of the online validation tool called GLCVal. This tool provided worldwide sample points supported member states in uploading sample judgment results, calculating accuracy, and managing validation results.

Further, the NGCC and Tongji University (TJU) were responsible for implementing all procedures for global-scale validation practices, while an overall technical specification prepared by them was refined through soliciting advice from other project participants. The validation process entailed the rigorous design of appropriate sample size and distribution, sample judgment supported by multiple reference data, and precise calculation of performance metrics, such as overall accuracy (OA), producer's accuracy (PA), and user's accuracy (UA). Ultimately, all outcomes encompassing sample location maps, judgment results, and indicators of accuracy were effectively managed within the platform GLCVal. At regional and national levels, collaborative validation followed two modes. First, sample pixels were made available on the validation platform GLCVal and generated by the sampling method introduced in Chapter 2. The experts from the project participants then utilized this platform to collect reference data, judge samples, and assess accuracy. Second, the experts had the option to design their sampling scheme for sample selection to complete validation practices. The validation results were then uploaded back to GLCVal. No matter which type of validation was chosen, consistency in technical specifications, validation systems, and processes was essential. For regional and national level validation practices, accuracy reports would be published in various forms by the project participants. The GEO organizations involved in collaborative validation are shown in Table 7.1.

The validation collaborators and partners were recruited through GEO international conferences and workshops. Consequently, numerous organizations participated in the validation practices, including renowned institutions, such as TJU, the NGCC, the Institute of Geographic Sciences and

TABLE 7.1
Partners of International Validation Practices

Participating Part		Partners
Validation practices	Global practices	TJU
		NGCC
	Regional practices	Institute of Geographic Sciences and Natural Resources Research
		Regional Centre for Mapping of Resources for Development
		South African Commission for Rural Development and Land Reform
	National practices	Politecnico Di Milano
		Wuhan University
		TJU
		Bulgarian Academy of Sciences
		Centre for Research and Technology Hellas
		Hydroinformatics Lab, DICA
		Royal Institute Of Technology
Technical specification	The German Space Agency (DLR)	
	Canada's Ryerson University	
	International Institute for Applied Systems Analysis (IIASA)	
	Kenya's Presidential University of Agriculture and Technology	
	Tanzania's National Forestry Research Institute	
	Madagascar's Ministry of Environmental Protection	
	Ukraine's Technical University	
	Brazil's Biodiversity Institute	

Natural Resources Research, the Regional Centre for Mapping of Resources for Development (RCMRD), the South African Commission for Rural Development and Land Reform, Wuhan University, Bulgarian Academy of Sciences, the Centre for Research and Technology Hellas, Politecnico di Milano, and Sweden's Royal Institute of Technology.

Experts from other organizations were involved in the development of the technical specification, including the German Space Agency (DLR), Ryerson University in Canada, the International Institute for Applied Systems Analysis in Austria, the Presidential University of Agriculture and Technology in Kenya, National Forestry Research Institute in Tanzania, Madagascar's Ministry of Environmental Protection, Technical University in Ukraine, Brazil's Biodiversity Institute, among others.

7.2 GLOBAL-SCALE VALIDATION

The assessment of global OA plays a crucial role in enabling map users to evaluate the applicability of land cover maps for their intended purposes. Aiming to ensure the accuracy of GlobeLand30-2010 data on a global scale, an independent evaluation was conducted by TJU in China. A two-rank sampling strategy was employed: the first-rank selected 80 map sheets worldwide, while the second-rank sampling selected over 150,000 samples for all land cover types within the 80 selected map sheets. The OA of this product was estimated at approximately $80.3 \pm 0.2\%$.

The overall process of validating GlobeLand30-2010 is illustrated in Figure 7.2. The methodology for validating GLC products included the sample plan, allocation and judgment of samples, and accuracy assessment. In Chapter 2, the advantages of two-rank sampling were discussed concerning global-scale validation. The first-rank sampling involved global sampling, where the map sheet served as the sample unit. The second-rank sampling involved regional sampling, with pixels serving as the sample unit. Reference data mainly consisted of high-resolution images from Google Earth (GE), Thematic Mapper (TM) images, DCP-verified points, online authentic landscape photos, etc. Three

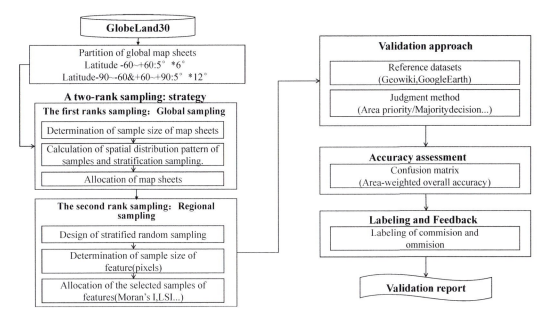

FIGURE 7.2 Flowchart of global validation practices.

groups of technical staff independently judged the sampling points, which were then analyzed and compared with reference data to determine if sampling points were "completely right," "completely wrong," or "uncertain." Finally, a confusion matrix was calculated, and area-weighted OA was used to validate GlobeLand30. Global OA, as well as accuracies for individual types, was obtained.

7.2.1 Sample Distribution

7.2.1.1 First-Rank Sample Distribution

The main task of the first-rank sampling was to determine the sample size and spatial distribution of map sheets. To achieve this, a sample size estimation model based on acceptance sampling (as discussed in Section 2.3.2) was employed. A total of 80 map sheets were selected from a pool of 847 maps from GlobeLand30, with a confidence level of 95% and an acceptance quality level (AQL) of 20% (see Formula 2.4). The methodology for selecting map sheets on a global scale was

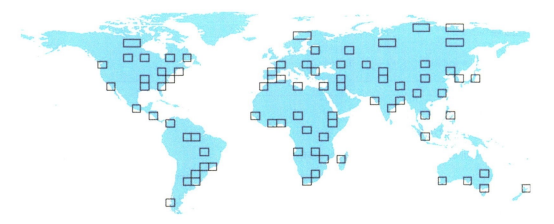

FIGURE 7.3 Distribution of samples map sheet.

204 Global Land Cover Validation

elaborated on in Section 2.4.2. Figure 7.3 illustrates the spatial distribution of these selected map sheets, while Table 7.3 provides continent-wise numbering.

7.2.1.2 Second-Rank Sample Distribution

In the second-rank sampling, the sample size of pixels for each map sheet was determined using a sample size estimation model in acceptance sampling (refer to Section 2.3.2). An illustrative example for calculating the sample size of one map sheet is presented in Table 7.2. The AQL for each class was established based on the product requirements for GlobeLand30. The overall AQL, weighted by area, was used as a replacement parameter in Formula 2.4. With a confidence level of 95% and an AQL of 9%, the computed sample size for this particular map sheet amounted to 3,486 out of a total count of 366,984,105 pixels. To obtain the sample size per single class, this value was multiplied by its corresponding area proportion.

By calculating the sample size for each map sheet and strategically distributing samples based on spatial variability, a substantial number of samples were utilized to assess the accuracy of land cover classification. A total of over 150,000-pixel samples were selected from the 80 map sheets (refer to Table 7.3). The spatial distribution of these samples is illustrated in Figure 7.4.

7.2.2 SAMPLE JUDGMENT

Based on the judgment of multiple levels of samples using high-resolution images (as discussed in Section 4.2), a total of 154,070 samples were accurately classified as "completely correct" or "completely wrong" based on the reference data, accounting for 96.81% of the total samples. Additionally, 5,073 samples could not be definitively categorized due to variations in reference data or subjective judgment, representing 3.19% of the total. Figure 7.4 illustrates these results, with

---

TABLE 7.2
The Calculation of Sample Size for One Map Sheet

Class	AQL (%)	Pixel Number	Area Proportion	OAQL	Sample Size of Map	Sample Size of Single Class
Cultivated land	20	26,775,616	0.073			256
Forest	15	225,695,225	0.615			2,150
Grassland	30	66,255,138	0.181			632
Shrubland	30	24,905,699	0.068			238
Wetland	30	79,662	0.0002	0.099	3486	2
Water bodies	15	2,153,723	0.006			22
		0	0			0
Artificial surfaces	20	1,009,848	0.003			10
Bare land	15	6,446,217	0.018			62
Permanent snow and ice	15	13,049,146	0.036			126

TABLE 7.3
Sample Size of Each Continent

The Number of Map Sheets Assigned to Each Continent

Continent	Asia	Europe	Africa	America	Oceania	Overall
Proportion of the land area	33.01%	7.32%	22.17%	31.19%	6.31%	100%
Number of map sheets	26	6	18	25	5	80
Number of pixels	60,165	12,792	25,656	45,822	9,635	159,143

Collaborative GlobeLand30 Validation

FIGURE 7.4 Samples judgment results for global-scale validation.

green denoting "completely correct" samples, red indicating "completely wrong" ones, and yellow representing indeterminate cases.

7.2.3 Accuracy Assessment

The confusion matrix was established based on the judgment results of 154,070 samples worldwide. However, due to the limited coverage of tundra areas, only a small number of samples were available for analysis and verification of the tundra type. So, the tundra was not being validated. In Table 7.4,

TABLE 7.4
Confusion Matrix of Accuracy Assessment in GlobeLand30-2010

		Classified Type									
		Cultivated land	Forest	Grass land	Shrub land	Wet land	Water bodies	Artificial surfaces	Bare land	Permanent snow and ice	Actual sum
Actual type	Cultivated land	28,016	1,333	1,346	624	125	75	143	415	2	32,079
	Forest	2,373	52,538	3,202	993	769	46	135	223	41	60,320
	Grassland	1,593	2,176	21,241	890	216	54	92	546	57	26,865
	Shrubland	542	821	512	7,404	29	11	15	234	7	9,575
	Wetland	136	1,175	393	68	5,082	229	7	85	1	7,176
	Water bodies	189	503	186	27	138	5,329	13	48	5	6,438
	Artificial surfaces	699	296	265	103	15	29	2,756	94	12	4,269
	Bare land	181	188	472	101	8	14	8	5,657	72	6,701
	Permanent snow and ice	0	4	11	0	0	0	0	13	619	647
	Classified sum	33,729	59,034	27,628	10,210	6,382	5,787	3,169	7,315	816	154,070

Note: The accuracy assessment of the tundra type was not conducted.

TABLE 7.5
The Accuracy of Each Type

	GlobeLand30 in 2010					
Class	User's Accuracy (%)	Producer's Accuracy (%)	Area Percent	Sample Size (Pixel)	Overall Accuracy	Kappa
Cultivated land	83.06	87.33	0.150478			
Forest	89.00	87.10	0.299418			
Grassland	76.88	79.07	0.244509			
Shrubland	72.52	77.33	0.073392			
Wetland	79.63	70.82	0.026222	154,070	83.50% ± 0.19%	0.78
Water bodies	92.09	82.77	0.024507			
Artificial surfaces	86.97	64.56	0.009458			
Bare land	77.33	84.42	0.153051			
Permanent snow and ice	75.86	95.67	0.018964			

GlobeLand30 achieved an OA of 83.50% ± 0.19% in 2010 with a Kappa coefficient value of 0.78. These findings indicated that the GlobeLand30 data product for 2010 exhibits high precision and good quality, as evidenced by an OA exceeding 80% and a Kappa coefficient surpassing 0.75.

The accuracy of each land type is presented in Table 7.5. As depicted in the table, cultivated land, forest, grassland, and bare land collectively accounted for 84% of the global land area. The user's and producer's accuracies of cultivated land and forest exceeded 85%, indicating high-quality classification. The PA of bare land was 84%, higher than its UA. The classification accuracy of grassland was relatively lower, only 79%. Shrubland, wetland, water bodies, artificial surfaces, and permanent snow and ice covered 15% of the continent. Their accuracy varies widely, from 72% to 92%.

7.3 REGIONAL-SCALE VALIDATION

Despite the relatively high accuracies reported at a global scale, the accuracies in different regions of the world remained unknown. To address this gap, the project participants were encouraged to validate GlobeLand30 for their specific areas of interest using diverse sample designs and reference datasets. The experts from institutions such as the Institute of Geosciences and Resources of the Chinese Academy of Sciences, TJU, RCMRD, and the South African Commission for Rural Development and Land Reform accepted invitations to validate GlobeLand30 in both the Roof of the World and Africa.

Wang (2019) conducted an accuracy assessment of GlobeLand30-2010 in the Roof of the World using the GLCVal validation platform. A total of 801 samples were obtained through landscape index (LSI)–based sampling from GLCVal. Visual interpretation on GE revealed a range between 71.72% as the lowest accuracy and 83.86% as the highest accuracy achieved. In order to explore how different sampling schemes affect accuracy, Zhao (2018) from TJU also carried out an accuracy assessment in the Roof of the World using the grid sampling method with a total sample size of 7,210. The results demonstrated that both sampling methods yielded similar accuracies.

In total, about 10,000 samples were distributed in Africa for regional-level validation practices, and the accuracy was assessed through interpretating the samples by the experts using high-quality images from Google. However, the experts lacked familiarity with the surface geography of Africa, posing challenges in sample judgments. To facilitate a comparative analysis, Dr. Luncedo Ngcofe from the South African Commission for Rural Development and Land Reform and Phoebe Oduor from RCMRD were invited to participate in the accuracy assessment for GlobeLand30-2010 in Africa. Ultimately, accuracies were evaluated for eight countries: South Africa, Botswana, Namibia, Rwanda, Tanzania, Uganda, Ethiopia, and Egypt.

7.3.1 Roof of the World

The Roof of the World is the highest geographical region in the world, including the Qinghai-Tibet Plateau, Hengduan Mountains, Himalayas, Hindu Kush Mountains, Pamir Plateau, and so on. It is one of the regions that are most sensitive to global change in the world, with a profound impact on its surrounding environments. Two validation practices were performed based on different samples and different sample judgment methods.

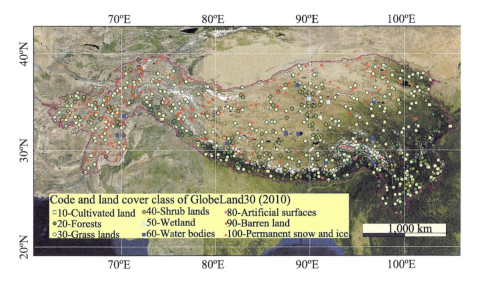

FIGURE 7.5 Spatial distribution of sample points in the roof of the world. (adapted from Wang et al. 2019.)

TABLE 7.6
Sample Sizes of Land Cover Class in the Roof of the World

Land Cover Class	Samples
Cultivated land	24
Forest	127
Grassland	331
Shrubland	116
Wetland	2
Water bodies	10
Artificial surfaces	1
Bare land	171
Permanent snow and ice	19
Total	801

Source: Data from Wang et al. (2019).

TABLE 7.7
The Lowest Accuracy of GlobeLand30-2010 in the Roof of the World

Verification Point Type: Number of Samples of Reference Data

		Cultivated land	Forest	Grass-land	Shrub-land	Wet-land	Water bodies	Artificial surfaces	Bare land	Permanent snow and ice	SUM	UA%
	Cultivated land	20	1	0	1	0	0	1	0	0	23	86.96
	Forest	6	88	14	13	0	0	0	2	0	123	71.54
	Grassland	17	11	226	26	0	1	0	14	5	300	75.33
Verification Point Type: Data Product Samples	Shrubland	6	5	23	70	0	1	0	0	0	105	66.67
	Wetland	0	0	1	0	1	0	0	0	0	2	50
	Water bodies	1	0	0	0	0	8	0	0	0	9	88.89
	Artificial surfaces	0	0	0	0	0	0	1	0	0	1	100
	Bare land	4	1	43	2	0	0	0	91	2	143	63.64
	Permanent snow and ice	0	0	0	2	0	0	0	2	15	19	78.95
	Summary	54	106	307	114	1	10	2	109	22	520	
	PA%	37.04	83.02	73.62	61.40	100	80	50	83.49	68.18		Overall: 71.72%

Source: Data from Wang et al. (2019).

7.3.1.1 LSI Sampling Validation

Through the validation platform GLCVal, 801 sample points were selected using the LSI method. The spatial distribution of the sample point is shown in Figure 7.5 and the number of sample points was distributed as shown in Table 7.6, mainly distributed in forest, grass, shrubs, and bare land areas.

The main reference data was the historical images from GE, which satisfied the validation requirements of GlobeLand30 (2010) in the scarcely populated area of the world's roof. With 30 m × 30 m as the interpretation unit, the nine types of sample points of GlobeLand30 data were interpreted. Three categories of interpretation confidence were used to judge these 801 samples— i.e., high quality, medium quality, and low quality (see Chapter 4 for the definition of these confidence levels). Of the total samples, 479 samples (59.80%), 246 samples (30.71%), and 76 samples (9.49%) were categorized as high-quality, medium-quality, and low-quality, respectively. A total of 725 high and medium-quality samples (91.51%) were eligible to participate in the accuracy assessment of GlobeLand30, while the low-quality samples were excluded from the assessment. The "medium quality" sample was unable to judge their exact land cover class type by visual interpretation of GE and was only indefinitely determined as two class types. This led to two possible validation results, depending on how the "medium quality" samples were judged. One result was the lowest accuracy when all "medium quality" samples were judged into one of the two class types. In this case, the lowest overall classification accuracy of GlobeLand30-2010 land cover data within the geographical area of the Roof of the World was 71.72%, and the Kappa coefficient was 0.62. The confusion matrix, UA, and PA are shown in Table 7.7.

The other result was the highest accuracy, while all "medium quality" samples were judged into two class types. The highest OA of GlobeLand30-2010 land cover data within the Roof of the World was 83.86%, and the Kappa coefficient was 0.78. The confusion matrix, UA, and PA are shown in Table 7.8.

Collaborative GlobeLand30 Validation

TABLE 7.8
The Highest Accuracy of GlobeLand30-2010 in the Roof of the World

| | | Verification Point Type: Number of Samples of Reference Data | | | | | | | | | | |
		10	20	30	40	50	60	80	90	100	SUM	UA%
	Cultivated land	20	1	0	1	0	0	1	0	0	23	86.96
	Forest	6	88	14	13	0	0	0	2	0	123	71.54
	Grassland	17	11	226	26	0	1	0	14	5	300	75.33
	Shrubland	6	5	23	70	0	1	0	0	0	105	66.67
Verification	Wetland	0	0	1	0	1	0	0	0	0	2	50
Point	Water bodies	1	0	0	0	0	8	0	0	0	9	88.89
Type: Data	Artificial	0	0	0	0	0	0	1	0	0	1	100
Product	surfaces											
Samples	Bare land	4	1	43	2	0	0	0	91	2	143	63.64
	Permanent	0	0	0	2	0	0	0	2	15	19	78.95
	snow and ice											
	Summary	54	106	307	114	1	10	2	109	22	520	
	PA%	37.04	83.02	73.62	61.40	100	80	50	83.49	68.18	Overall: 83.86%	

Source: Data from Wang et al. (2019).

TABLE 7.9
Sample Size in Roof of the World

	Pixel Number	AQL (%)	Area Proportion (%)	Sample Size	Total Size
Cultivated land	174,806,079	20	3.78	983	
Forest	626,106,215	30	13.52	899	
Grassland	2,247,192,272	30	48.52	1,182	
Shrubland	107,620,973	30	2.32	612	
Wetland	14,648,078	30	0.32	468	7,210
Water bodies	68,390,096	20	1.48	597	
Artificial surfaces	4,263,953	20	0.09	615	
Bare land	1,223,832,872	20	26.43	1,046	
Permanent snow and ice	164,328,018	20	3.55	808	

Source: Data from Zhao (2018).

7.3.1.2 Grid Sampling Validation

The sampling design included sample size calculation and spatial distribution. A total of 7,210 samples were calculated using the sample size estimation model in acceptance sampling (see Section 2.2). The AQL of each class type was determined from prior knowledge, as shown in Table 7.9. The sample size of each class type was calculated: 983 cultivated land samples, 899 forest samples, 1,182 grassland samples, 612 shrubland samples, 468 wetland samples, 597 water bodies samples, 615 artificial surface samples, 1,046 bare land samples, and 808 permanent snow and ice samples.

Grid sampling was used to validate GlobeLand30 in the Roof of the World to ensure balanced sampling, and the grid was meshed by 1 × 1 degrees. The number in each grid for a class was calculated by multiplying the sample size of the class type and the ratio of the grid in the total area. The spatial distribution of the samples is shown in Figure 7.6.

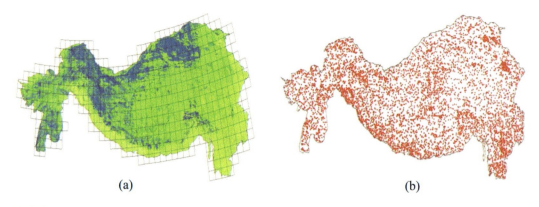

FIGURE 7.6 The samples distribution based on 1×1 degree grid. (a) 1×1 degree grid. (b) Sample distribution. (Adapted from Zhao 2018).

TABLE 7.10
Sample Credibility in the Roof of the World

Credibility Rating	The Degree of Trust	Sample Size	Proportion (%)	Effectiveness	Sample Size	Proportion (%)
T	Completely true	5,709	79.18	Effective	6,867	95.24
F	Completely false	1,158	16.06			
NG	Not clear	148	2.05	Ineffective	343	4.76
NY	No time	195	2.75			

Source: Data from Zhao (2018).

TABLE 7.11
Confusion Matrix of the Roof of the World for GlobaLand30-2010 Data

Classification Type

	Cultivated land	Forest	Grassland	Shrubland	Wetland	Water bodies	Artificial surfaces	Bare land	Permanent snow and ice	Total of Land Type
Cultivated land	766	74	43	24	2	1	26	6	0	942
Forest	23	783	24	24	1	0	0	5	12	872
Grassland	16	76	901	50	10	1	1	35	40	1,130
Shrubland	18	41	55	417	3	1	1	20	10	566
Wetland	0	2	18	0	406	13	0	9	0	448
Water bodies	1	2	0	2	45	517	0	8	5	580
Artificial surfaces	79	6	6	5	0	0	492	14	1	603
Bare land	45	5	43	18	2	3	80	719	48	963
Permanent snow and ice	0	6	9	9	5	0	0	26	708	763
Total of classification	948	995	1,099	549	474	536	600	842	824	6,867

Source: Data from Zhao (2018).

Collaborative GlobeLand30 Validation

Visual interpretation of GE images was used to judge samples according to the principles of area priority, majority decision, and confidence level of sample judgment (see more details in Section 4.5); 6,867 out of 7,210 samples were judged completely correct or completely wrong, accounting for 95.24% of the total samples, as shown in Tables 7.10 and 7.11.

The OA of GlobeLand30-2010 land cover data within the geographical area of the Roof of the World was 83.14%, and the Kappa coefficient was 0.81. The confusion matrix, UA, and PA are shown in Table 7.12. The consistency between the land cover data and the real data in the Roof of the World area was high.

The PA of cultivated land, forest, grassland, wetland, water bodies, permanent snow or ice was higher than 90%, but the shrub, artificial land cover, and bareland was relatively low, with an accuracy of about 85%, as presented in Table 7.12. The shrubland was spread sparsely on the Roof of the World, with high fragmentation and low accuracy. The UA of wetland, water bodies, bare land, and permanent snow or ice was higher than 85%, which were naturally dispersed and easy to distinguish. However, the UA of cultivated land, grassland, forest, and shrubland was relatively low.

7.3.1.3 Comparison Analysis

Table 7.13 presents a comparative analysis of UA, PA, and OA of the land cover data covering the Roof of the World area, obtained by both LSI and grid sampling methods.

A significant disparity in the classification accuracy of different land cover types can be seen in GlobeLand30-2010 land cover data within the geographical area of the Roof of the World. Moreover, the accuracy varied considerably depending on the different methods employed. Generally, according to sampling theory, a higher number of samples can lead to more accurate assessment results; however, this also increases validation costs. Conversely, too few samples could reduce assessment accuracy. Therefore, determining an appropriate number of samples was crucial for effective sampling.

Table 7.13 reveals that wetland, artificial surface, and bare land were among the land cover types with substantial differences in UA, which exceeded 10%. With the LSI sampling method, fewer sample points were selected, particularly for wetlands and artificial surfaces (only two and one sample point, respectively), which resulted in nonrepresentative ground coverage. For example, only two sample points chosen for wetlands yielded a very low UA at just 50%, while its PA remained high at 100%. Similarly, water bodies and permanent snow/ice with limited sample points exhibited extreme mapping accuracies, while artificial surfaces demonstrated a similar extreme effect with only one selected sample point. Consequently, due to the inadequate selection of sample points, the OA within the Roof of the World region was severely impacted.

TABLE 7.12
Results of Grid Sampling Accuracy of Roofs in the World

Class	User's Accuracy (%)	Producer's Accuracy (%)	Commission Error (%)	Omission Error (%)	Sample Size	Overall Accuracy (%)	Kappa
Cultivated land	80.80	91.30	8.70	19.20			
Forest	78.69	92.43	7.57	21.31			
Grassland	81.98	92.04	7.96	18.02			
Shrubland	75.96	85.34	14.66	24.04			
Wetland	85.65	92.41	7.59	14.35	6,867	83.14	0.81
Water bodies	96.46	94.66	5.34	3.54			
Artificial surfaces	82.00	83.75	16.25	18.00			
Bare land	85.39	89.10	10.90	14.61			
Permanent snow and ice	85.92	95.81	4.19	14.08			

Source: Data from Zhao (2018).

TABLE 7.13
Comparison of Grid Sampling With LSI Sampling in the Roof of the World

| | Grid Sampling | | | LSI Sampling | | | | |
| | | | | Lowest | | Highest | | |
Class	User's Accuracy	Producer's Accuracy	Number of Samples	User's Accuracy	Producer's Accuracy	User's Accuracy	Producer's Accuracy	Number of Samples
Cultivated land	80.80%	91.30%	983	86.96	37.04	86.96	37.04	24
Forest	78.69%	92.43%	899	71.54	83.02	71.54	83.02	127
Grassland	81.98%	92.04%	1182	75.33	73.62	75.33	73.62	331
Shrubland	75.96%	85.34%	612	66.67	61.4	66.67	61.4	116
Wetland	85.65%	92.41%	468	50	100	50	100	2
Water bodies	96.46%	94.66%	597	88.89	80	88.89	80	10
Artificial surfaces	82.00%	83.75%	615	100	50	100	50	1
Bare land	85.39%	89.10%	1046	63.64	83.49	63.64	83.49	171
Permanent snow and ice	85.92%	95.81%	808	78.95	68.18	78.95	68.18	19
Overall		83.14%		71.72%		83.86%		

Through the comparative analysis, the overall accuracies of the two sampling methods revealed similar numerical values. The main factor contributing to error classification was the higher altitude and sparse artificial surface coverage in the Roof of the World region. Additionally, due to their dispersed distribution and smaller area coverage, shrubland was allocated relatively fewer sample sizes, resulting in lower accuracy; however, this did not significantly impact the OA. Grasslands and bare lands exhibited relatively low accuracy as they were predominantly found in arid areas where confusion during determination was common. Moreover, forests, grasslands, and shrubs share similar spectral characteristics, leading to mapping errors, while distinguishing between forests and shrubs proved challenging for interpreters. Nevertheless, the comparative analysis indicated that the validation based on the grid sampling method outperformed the validation based on the LSI sampling method in terms of the number of sample points and their spatial location distribution for better OA evaluation.

7.3.2 AFRICA

Africa, being one of the largest continents, posed significant challenges and high expenses for the continental validation of GlobeLand30. Consequently, a country-by-country approach was adopted to carry out the validation practice in Africa. Experts were invited to select one or more countries they were familiar with and interested in, enabling an accurate assessment of each country's data. The NGCC delivered African sample distribution using an LSI-based sampling method. Finally, accuracy assessments were successfully conducted in eight countries: South Africa, Botswana, Namibia, Ethiopia, Rwanda, Tanzania, Uganda, and Egypt.

Figure 7.7 illustrates the collaborative validation process that primarily involved sampling and sample judgment stages. Spatial heterogeneity was taken into account during sample size calculation and distribution through the LSI-based method that facilitated downloading samples from the validation platform GLCVal. Dr. Luncedo Ngcofe, Phoebe Oduor, and their teams judged these samples using local reference resources and then uploaded their judgment results back onto the validation platform for accuracy assessment on a country-by-country basis.

Collaborative GlobeLand30 Validation

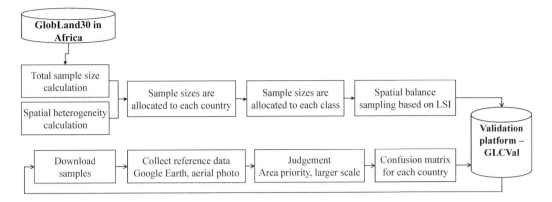

FIGURE 7.7 Workflow of Africa validation.

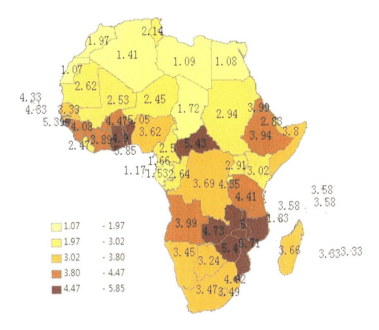

FIGURE 7.8 rLSIs of each country in Africa. (Adapted from Chen 2017.)

1. *LSI-based sampling*

The sample size of 9,295 was determined for Africa validation based on a simple sample size estimation model and validation cost. The sample size and sample distribution were determined using the three-level LSIs-based sampling method. First, the rLSIs were calculated in each country to determine the ratio of sample densities in different countries. The sample size per country was then calculated using its rLSI and area. Figure 7.8 illustrates the rLSI in each country, while Table 7.14 presents the sample sizes and densities (per 10,000 km^2) in each country. Samples were more densely distributed in countries with significant spatial heterogeneity. It is worth noting that some countries with small areas had underrepresented sample sizes, such as Rwanda which only had two samples due to its very small area (only 4,033 km^2, just 1 over 10,000 of the African total area). Therefore, a minimum sample size of 43 samples was defined for these countries based on a simple sample size estimation model. Zambia and Tanzania exhibited high sample density (more than five sample sites per 10,000 km^2) due to their high landscape heterogeneity, whereas Egypt, Western

TABLE 7.14

Sample Sizes and Densities (per 10,000 km²) per Country

Country	Size	Density	Country	Size	Density
Algeria	337	1.4	Libya	185	1.1
Angola	509	4.1	Madagascar	222	3.8
Benin	60	5.2	Malawi	58	4.9
Botswana	**192**	**3.3**	Mali	297	2.4
Burkina Faso	126	4.6	Mauritania	154	1.5
Burundi	43	15.9	Mauritius	43	236.6
Cameroon	120	2.6	Morocco	132	3.3
Cape Verde	43	129	Mozambique	468	6
Central African Republic	346	5.6	**Namibia**	**295**	**3.6**
Chad	225	1.8	Niger	294	2.5
Comoros	43	281.8	Nigeria	457	5
Congo	63	1.8	**Rwanda**	**43**	**17**
Congo the Democratic Republic of the	699	3	Sao Tome And Principe	43	456.5
Cote Divoire	134	4.1	Senegal	95	4.8
Djibouti	43	19.8	Seychelles	43	1,771.2
Egypt	**110**	**1.1**	Sierra Leone	43	5.9
Equatorial Guinea	43	16	Somalia	217	3.4
Eritrea	43	3.5	**South Africa**	**500**	**3.7**
Ethiopia	**493**	**4.4**	Sudan	707	2.8
Gabon	43	1.6	Swaziland	43	25
Gambia	43	39.9	**Tanzania**	**407**	**4.3**
Ghana	119	5	Togo	43	7.5
Guinea	132	5.4	Tunisia	43	2.8
Guinea-Bissau	43	12.9	**Uganda**	**73**	**3**
Kenya	186	3.2	Western Sahara	43	1.6
Lesotho	43	14.2	Zambia	403	5.4
Liberia	43	4.5	Zimbabwe	221	5.7

Source: Data from Chen (2017).

Sahara, and Libya had the lowest sample density (less than two sample sites per 10,000 km²) owing to their low spatial heterogeneity.

Subsequently, the landscape shape index of land cover class (call for cLSI) was computed for each land cover type within every country to quantify the heterogeneity of land cover types and assess the spatial dispersion of their distribution. The allocation of sample size to individual land cover types in a given country commonly employed the Neyman optimal allocation formula, wherein cLSI replaced the standard error. Consequently, an effective execution of sample size allocation from national samples to respective land cover types was achieved. The landscape shape index of the geographic grid unit (uLSI) was calculated for each land cover type, and the appropriate grid units were selected to establish sampling sites based on the uLSI values.

Second, the cLSIs were introduced into the classical Neyman optimal allocation formula as the standard error of the land cover classes. Because rare classes had higher cLSIs, this approach led to a reasonable increase in their sample sizes. Third, a specific curve was created from the uLSIs (called the uLSI curve) to select optimal geographical units at which sample sites would be located. Figure 7.9 shows a spatial distribution map of African samples.

Collaborative GlobeLand30 Validation

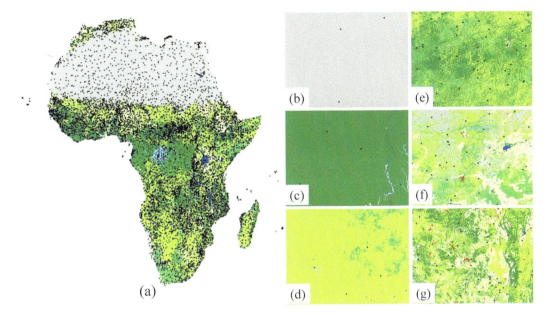

FIGURE 7.9 Spatial distribution map of African samples (a) sample distribution throughout Africa, (b) sample distribution for homogeneous desert, (c) Sample distribution for homogeneous forest, (d) sample distribution for homogeneous grassland, (e) sample distribution in the transition zone of rainforest steppe, (f) distribution of samples at the desert edge, (g) sample distribution in mixed vegetation areas. (Adapted from Chen 2017.)

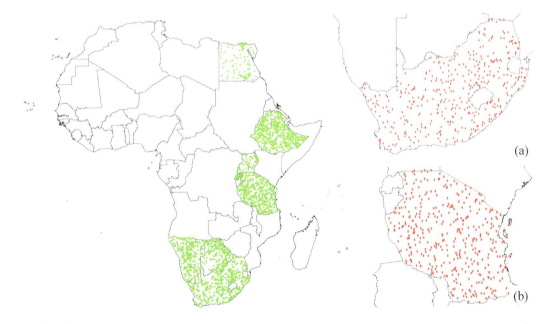

FIGURE 7.10 Location and sample distribution of eight African countries location of eight African countries (a) South African (b) Tanzania.

(Continued)

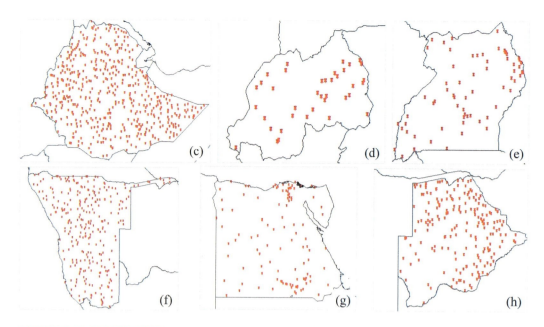

FIGURE 7.10 (CONTINUED) (c) Ethiopia (d) Rwanda (e) Uganda (f) Namibia (g) Egypt (h) Botswana.

TABLE 7.15
The Kappa Coefficient and Accuracy

	Kappa	Accuracy (%)
South African		83.4
Botswana	0.83	86.9
Namibia	0.73	79
Rwanda	0.72	76.7
Tanzania	0.7	76.4
Uganda	0.81	83.56
Ethiopia	0.81	85.19
Egypt	0.81	86.3

2. Accuracy assessment of eight countries

Dr. Luncedo Ngcofe and his team selected South Africa as the region of interest. Phoebe Oduor and her team expressed a greater inclination toward Botswana, Namibia, Rwanda, Tanzania, Uganda, Ethiopia, and Egypt. Figure 7.10 illustrates the distribution of samples in these countries. The chosen eight countries predominantly lie in eastern Africa with a complex landscape pattern of land cover. Only the landscape pattern in Egypt exhibits a high degree of homogeneity.

The OA for the eight countries is provided in Table 7.15. The highest accuracy of GlobeLand30-2010 was observed in Egypt and Botswana, surpassing 86%, which exceeds the OA of GlobeLand30-2010 on a global scale. Ethiopia ranked second with an accuracy of 85.19%. South Africa and Uganda maintained accuracies above 80%, specifically at 83.4% and 83.56%, respectively. However, the OA of Namibia (79%), Rwanda (76.7%), and Tanzania (76.4%) are all below the global OA of 83%.

7.4 NATIONAL-SCALE VALIDATION

In addition to the Roof of the World and Africa, some experts were more interested in the countries where they had lived. To date, more than a dozen countries have conducted accuracy assessments from diverse perspectives with different methods, such as Siberia (Zhang et al. 2015); Kyiv Oblast, Ukraine (Kussul et al. 2015); Iran (Jokar Arsanjani et al. 2016a, 2016b), Kenya (See et al. 2017); Nepal (Cao et al. 2016); Germany (Jokar Arsanjani et al. 2016a, 2016b); Portugal (Mozak 2016); and East Africa (Jacobson et al. 2015), with an OA ranging from 74% to 93%.

Under the GEO-led collaborative validation project, the project participants from Wuhan University, TJU, Bulgarian Academy of Sciences, the Centre for Research and Technology Hellas, Politecnico di Milano, and Sweden's Royal Institute of Technology validated GlobeLand30 using diverse sample designs and reference datasets. The main validation areas that they were interested in included China, Bulgaria, Greece, Italy, and Sweden.

Xie (2022) and Wang (2018a and 2018b) carried out three accuracy assessments in China using the LSI-based sampling, grid sampling, and spatial-balance sampling methods. The results demonstrated that the three methods yielded a similar level of accuracy. Experts from Bulgaria, Greece, and Sweden utilized the online validation platform GLCVal, along with the LSI-based sampling method, to generate samples of land cover data for Bulgaria, Greece, and Sweden for accuracy assessments of GlobeLand30. The validation conducted in Italy by Brovelli (2015) employed measuring agreement and disagreement with existing land cover (LC) maps instead of sampling validation. Due to the unavailability of high-resolution Italian regional data in certain provinces, only eight provinces were included in this validation practice.

7.4.1 China

Regional and country-wide accuracy assessments for GlobeLand30 have also been carried out by researchers from many countries. However, there have been few accuracy assessment efforts for GlobeLand30 in China. Wang (2018a and 2018b) conducted two nationwide validation practices in China employing the LSI-based sampling and grid sampling methods to facilitate the use and production of GlobeLand30 datasets. After that, in order to analyze the impact of the sampling design and sample judgment on the accuracy assessment, Xie (2022) implemented the third validation practice using a spatial-balance sampling design. Through these validation practices, the influence of sampling and sample judgment on validation results was proved, albeit with minimal impact.

7.4.1.1 Validation Using LSI-Based Sampling

In terms of LC composition, the combined area of cultivated land, forests, grasslands, and bare lands constituted approximately 94% of China's total land area. The remaining class types occupied a relatively small portion, accounting for only about 6% of the country's land area. However, employing traditional stratified sampling may lead to an inadequate representation of this low sample size, which can lead to unreliable UA (Stehman et al. 2012). Landscapes in southeastern China are complex and fragmented, where heterogeneous landscapes, spectral mixture pixels, and other factors often make it easier to misclassify. The LSI-based sample method can allocate more sample numbers to areas with greater spatial heterogeneity and more land class types while using the Neyman allocation method to determine the optimal sample size for individual regions and land class types. Therefore, the LSI-based sampling method was adopted to conduct a national-scale accuracy assessment of China's GlobeLand30 dataset for the year 2010 (Wang et al. 2018a). A total of 1,000 sample pixels were selected in China, and the sample allocation and location are shown in Figure 7.11 and Table 7.16 (Wang et al. 2019). It should be noted that the area of tundra in China was very small, so it was not included in the validation.

TABLE 7.16
Validation Sample Allocation for GlobeLand30 2010 LC Over China

Class	Area Proportion (%)	Sample Size
Cultivated land	21.34	135
Forest	22.23	142
Grassland	29.6	146
Shrubland	1.05	116
Wetland	0.43	62
Water bodies	1.56	96
Artificial surfaces	1.81	100
Bare land	20.9	124
Permanent snow and ice	1.08	79
Total	100	1,000

Source: Data from Wang et al. (2018a).

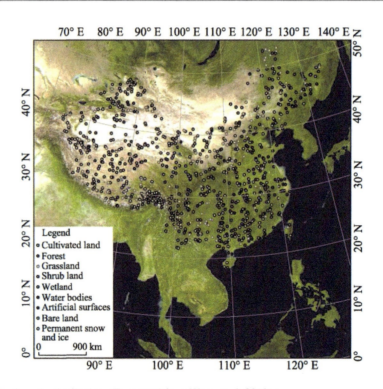

FIGURE 7.11 Sample distribution. (Excerpted from Wang et al. 2018a.)

As shown in Figure 7.11, a certain number of sample pixels were allocated for rare types of LC, such as shrubs, wetlands, water bodies, artificial surfaces, and permanent snow/ice. The spatial distribution of these sample units was also good. Compared with traditional hierarchical sampling design, the LSI-based sampling method handled rare classes and class types with different spatial heterogeneities, thus constituting a more effective validation and data refinement strategy because the easily misclassified locations were well positioned (Chen et al. 2016).

The next step was the sample judgment. The experts determined reference class labels for the 1,000 sample pixels primarily by visual interpretation based on high-resolution satellite images

Collaborative GlobeLand30 Validation 219

from GE. Images taken during the growing season in 2010 were preferred. The reference data interpretation and judgment were based on some basic elements (i.e., shape, size, pattern, shading, tone, texture, site, temporal-spatial context, and elevation). The sample evaluation unit was a 30 × 30 m square. To consider the influence of the neighborhood on accuracy and enhance the reliability of interpretation, a 300 m × 300 m square was also used to assist visual image interpretation.

Among the experts (i.e., interpreters), the decision made by the project manager holds paramount importance due to their extensive interpretation experience. Following an initial round of sample judgment, any samples categorized as "somewhat confident" or "not confident" were reevaluated by the project manager to minimize potential errors in visual interpretation. In cases where there was inconsistency among team members' judgments on a particular sample, precedence was given to the judgment of the project manager. Auxiliary data such as photos, maps, and Landsat images might be utilized for informed judgment. Ultimately, a confusion matrix was compiled summarizing the results of sample judgments, encompassing OA along with the user's and producer's accuracies. Table 7.17 presents the error matrix comprising 1,000 sample pixels (Wang et al. 2018a). Considering that sample pixels from distinct class types bear different weights, a revised equation for estimating accuracy based on stratified sample data was employed to calculate China's OA. The outcomes for OA, UA, and PA are listed in Table 7.17. The PAs of bare land, cultivated land, permanent snow/ice, and grassland all exceed 80%, making them dominant LC types in China except for permanent snow/ice. With lower PAs, shrublands, artificial surfaces, wetlands, and water bodies exhibited high heterogeneity and fragmentation, which made their accurate classification challenging. The UAs of LC classes other than cultivated land and artificial surfaces generally surpassed 75%, indicating a high level of precision in most categories.

7.4.1.2 Grid Sampling Validation

Wang et al. (2018b) proposed a geographical and category-stratified random sampling method, which involved dividing the study area (China) into ten regions and employing LC stratified random sampling within each region. Detailed explanations of these ten regions can be found in Figure 7.12. The validation range did not include permafrost areas, as they were almost nonexistent in China.

TABLE 7.17
The Error Matrix for GlobeLand30 2010 LC in China

Class Type	Cultivated land	Forest	Grass land	Shrub land	Wetland	Water-bodies	Artificial Surfaces	Bare Land	Permanent Snow/Ice	Total	UA
Cultivated land	106	15	2	4	0	2	6	0	0	135	78.5
Forest	8	108	14	8	0	0	2	1	1	142	76.1
Grassland	6	13	102	12	2	2	4	5	0	146	69.9
Shrubland	2	0	19	88	1	0	0	6	0	116	75.9
Wetland	1	1	1	0	55	3	1	0	0	62	88.7
Water bodies	6	1	6	4	4	72	1	1	1	96	75.0
Artificial surfaces	14	5	3	6	0	3	67	2	0	100	67.0
Bare land	1	1	7	3	0	1	0	111	0	124	89.5
Permanent snow/ice	0	1	0	0	0	0	0	8	70	79	88.6
Total	144	145	154	125	62	83	81	134	72	1,000	
PA	84.7	76.1	83.7	13.7	44.5	54.8	36.7	93.1	84.7	OA: 77.6%	

Source: Data from Wang et al. (2018a).
Notes: The OA is 77.6%.

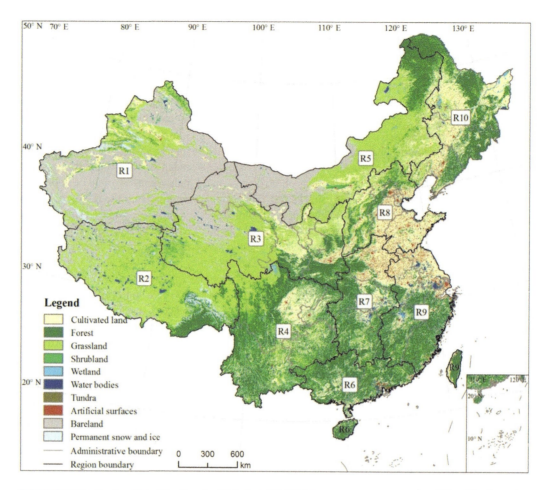

FIGURE 7.12 Regional stratification for GlobeLand30 2010 accuracy assessment over China. The boundaries of ten geographic strata are shown in black. The labels "R1–R10" identify the regions used to geographically stratify sample data (e.g., R1 = Region 1). (Excerpted from Wang et al. 2018b).

Regions R1-R4 were found to have permanent snow distribution. It was determined that each type within each region required 100 samples based on the calculation formula for random sample size, described in Section 2.2, assuming a true UA rate of 50%, a confidence level of 95%, and an expected standard error of 0.05. In total, 8,400 sample pixels were selected in China.

The reference data for sample judgment was sourced from platforms such as GE, Bing Maps, Yahoo Maps, and the Chinese Environmental and Disaster satellite. Three experienced interpreters meticulously judged the reference class labels for all 8,400 sample pixels. Each sample was assigned both the primary and alternative LC class type labels. Additionally, a credibility grade was assigned to each sample with three levels: "confident," "somewhat confident," and "not confident."

Regional overall accuracies for GlobeLand30 2010 over China are shown in Figure 7.13. Accuracy assessment was conducted using a confusion matrix for each geographical area. The OA ranged from 76.0% to 90.3% for the ten regions. The region with the highest accuracy was R1, achieving a remarkable rate of 90.3%. Following closely were R10 (88%), R5 (87.7%), and R3 (85.9%), which ranked second to fourth, all exhibiting an accuracy above 85%. Notably, these four regions were situated in the northern area. Positioned in the middle tier were R2 (82%), R8 (82.4%), and R9 (84%), with OA surpassing 80%. Further, the southern regions comprising R4, R6, and R7 ranked at the lowest, with an average accuracy approximately 8.6% lower than that of the other

Collaborative GlobeLand30 Validation

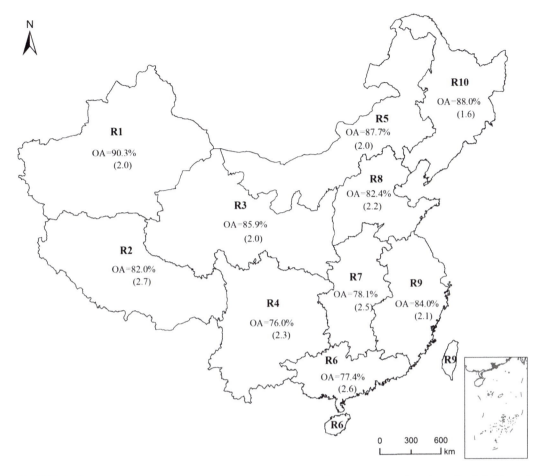

FIGURE 7.13 Regional overall accuracies for GlobeLand30 2010 over China based on the definition of agreement as a match between the map class and either the primary or alternate reference class. Standard errors for the overall accuracies are in parentheses. (Excerpted from Wang et al. 2018b).

seven regions combined. These variations could be attributed to factors such as species composition, classification system intricacies, and landscape heterogeneity.

Moreover, the OA at a nationwide scale based on an 8,400-pixel sample set was calculated according to the geographical confusion matrix. First, the ten geographical confusion matrices were aggregated into a national error matrix. Within this national error matrix, the value $P(i,j)$ in the i row and j column represented the weighted sum of $P(i,j|h)$ (values from individual geographical error matrix) and their respective weights (proportional area coverage within China).

$$\hat{P}_{i,j} = \sum_{h=1}^{H} W_h \hat{P}_{i,j|h} \qquad 7.1$$

In Formula 7.1, h represents the total number of geographical regions in the study area (h = 10 in this case), while w_h denotes the proportion of the region's area for the entire study area. $P(i,j|h)$ signifies cell (i,j) within the error matrix specific to region h. These $P(i,j)$ collectively constitute the national error matrix. Subsequently, upon the national error matrix, it became feasible to calculate national OA, UA, and PA.

TABLE 7.18
The Country-Wide Error Matrix, Cell Entries Are Expressed as a Percent of the Area

	Cultivated Land	Forest	Grass land	Shrub land	Wet land	Water Bodies	Artificial Surfaces	Bare Land	Permanent Snow and Ice	Total	UA
Cultivated land	18	1	0.5	0.6	0.2	0.2	0.7	0.1	0	21.3	84(1)
Forest	0.3	19.3	0.6	1.9	0	0.1	0	0.1	0	22.2	87(1)
Grassland	0.9	1	23.2	2.8	0.5	0.1	0.1	1.1	0	29.6	78(2)
Shrubland	0	0.1	0.2	0.7	0	0	0	0	0	1	70(2)
Wetland	0	0	0	0	0.4	0	0	0	0	0.4	82(2)
Water bodies	0	0	0	0	0.1	1.5	0	0	0	1.6	94(1)
Artificial surfaces	0.2	0.1	0	0	0	0	1.4	0	0	1.8	80(2)
Bare land	0	0	1.6	0.4	0.1	0	0	18.8	0	20.9	90(2)
Permanent snow and ice	0	0	0.1	0	0	0	0	0	0.9	1.1	88(2)
Total	19.5	21.5	26	6.5	1.2	1.9	2.3	20.2	1	100	
PA	93(1)	90(1)	89(1)	11(1)	30(5)	79(5)	62(4)	93(1)	93(4)		

Source: Data from Wang et al. (2018b).
Notes: UA and PA are reported with standard errors (SE) in parentheses. OA is 84.2% (0.7%).

The national error matrix is presented in Table 7.18. When the agreement at a sample pixel was defined as the match between the map label and the primary or alternate reference label, the OA of Globeland30-2010 in China was 84.2%, indicating that the accuracy was relatively high. When the consistency at a sample pixel was more strictly defined as the match between the map label and the primary reference label, the overall national accuracy was estimated at 81.0%.

7.4.1.3 Spatial-Balance Sampling Validation

A two-stage stratified spatially balanced sampling method (as described in Chapter 2.4) was employed for the third validation practice in China. According to the sample size estimation method (as described in Section 2.3), a total of 160 primary sample units and 7490 sample points were selected. The spatial distribution of sample sites across different regions is shown in Figure 7.14.

The sample was visually judged using methods for interpreting high-resolution GE images. The findings revealed that 6.84% of the 7,490 samples were deemed unreadable due to insufficient image resolution or characteristics such as dark tones, blurred textures, and absence of shadows. By conducting statistical analysis on the GE images utilized in the sample judgment, it was determined that these images spanned a period of 17 years, from 2000 to 2017. Specifically, 1.50% of the images were captured during the early 2000s (2000–2005), while a majority of them (88.10%) were recorded between 2006 and 2014. A detailed distribution by year can be found in Table 7.19. Further statistical analysis of the month when these images were taken indicated a relatively even distribution across all months except for October and December, as presented in Table 7.20. It is noteworthy that only a minority (24.73%) of the images corresponded to vegetation growth periods that occurred between June and September.

The results of the sample judgment indicated that 248 samples exhibited low quality and were excluded from the construction of the confusion matrix (including 124 from the Qinghai-Tibet Plateau, 20 from Northwest China, and 104 from East China). There was a total of 906 samples with moderate credibility and 6,336 samples with high credibility, accounting for approximately 96.7% of the overall dataset. The samples with moderate credibility were predominantly distributed in the

Collaborative GlobeLand30 Validation 223

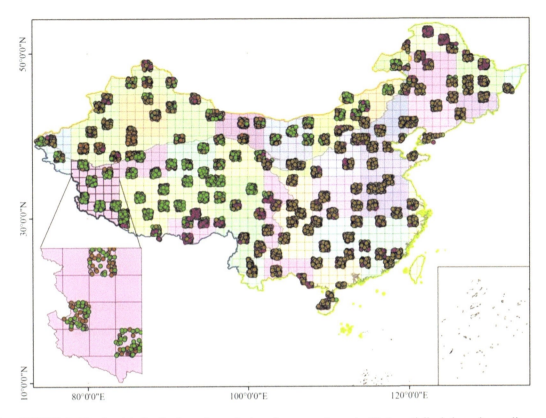

FIGURE 7.14 Spatial distribution of sample based on two-stage stratified spatially balanced sampling. (Excerpted from Xie et al. 2022.)

TABLE 7.19
Statistics of the Yearly Distribution of the Images

Region	2000–2002	2003–2005	2006–2008	2009–2011	2012–2014	2015–2017	No Image	Total
Tibetan Plateau	2	29	59	1,369	494	27	218	2,198
Northwest China	3	33	61	1,039	620	76	135	1,967
East China	7	38	164	1,350	1,442	165	159	3,325
Total	12	100	284	3,758	2,556	268	512	7,490

Source: Data from Xie (2022).

Qinghai-Tibet Plateau region (with a count of 416 in this area, followed by 274 in Northwest China and 216 in East China) (Table 7.21).

The sample judgment was conducted within a 90 m × 90 m sampling unit, adhering to standardized procedures. In the confusion matrix, classifications that match either the primary or secondary labels were considered accurate. Table 7.22 presents the confusion matrix derived from 7,242 samples, yielding an OA of 80.46% and a Kappa coefficient of 0.76 for GlobeLand30 in China. Notably, croplands exhibited the highest PA at 89.80%, while wetlands had the lowest at 56.12%. Performant snow demonstrated exceptional UA with a value of 88.50%, whereas shrubs displayed relatively lower UA at only 53%.

TABLE 7.20

Statistics of the Yearly Distribution of the Images

	Month													
Region	1	2	3	4	5	6	7	8	9	10	11	12	No Image	Total
Tibetan Plateau	99	198	234	117	63	25	49	104	111	244	169	567	218	2,198
Northwest China	48	71	177	156	149	115	195	176	215	276	97	157	135	1,967
East China	205	167	305	347	238	225	164	186	287	353	261	428	159	3,325
Total	352	436	716	620	450	365	408	466	613	873	527	1,152	512	7,490

Source: Data from Xie et al. (2022).

TABLE 7.21

Sample Distribution Based on Two-Stage Stratified Spatially Balanced Sampling

Region	Area Ratio	Primary Sampling Units	Sampling Points
Tibetan Plateau	28.68%	47	2,198
Northwest China	26.30%	42	1,967
East China	45.01%	71	3,325
Total	1	160	7,490

Source: Data from Xie et al. (2022).

TABLE 7.22

Confusion Matrix of GlobeLand30 in China

	Cultivated Land	Forest	Grass land	Shrub land	Wet land	Water Bodies	Artificial Surfaces	Bare Land	Permanent Snow/Ice	Total
Cultivated land	1,286	53	25	8	1	7	30	22	0	1,432
Forest	71	1,132	55	18	5	9	3	12	1	1,306
Grassland	143	197	1,445	50	10	14	16	165	5	2,045
Shrubland	4	1	4	106	1	1	0	45	0	162
Wetland	7	3	18	3	55	6	0	6	0	98
Water bodies	12	5	22	3	4	256	0	15	0	317
Artificial surfaces	34	6	5	3	0	2	271	5	0	326
Bare land	25	18	158	9	2	9	2	1,176	7	1,406
Permanent snow and ice	0	9	14	0	0	2	0	25	100	150
Total	1,582	1,424	1,746	200	78	306	322	1,471	113	7,242
UA (%)	81.29	79.49	82.76	53.00	70.51	83.66	84.16	79.95	88.50	–
PA (%)	89.80	86.68	70.66	65.43	56.12	80.76	83.13	83.64	66.67	–

Source: Data from Xie et al. (2022).

Collaborative GlobeLand30 Validation

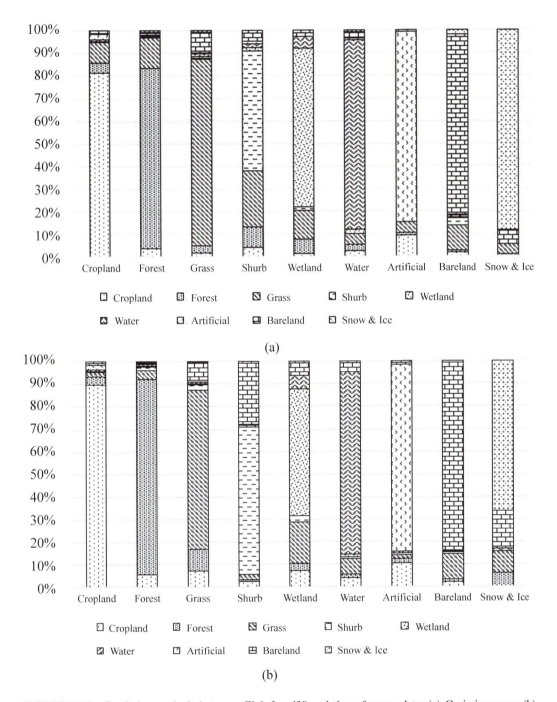

FIGURE 7.15 Confusion analysis between GlobeLand30 and the reference data, (a) Omission error. (b) Commission error. (Excerpted from Xie et al. 2022).

The confusion situation between GlobeLand30 and the reference data was analyzed in Figure 7.15. Omission error represented the proportion of LC in the reference data misclassified as other land cover types, as depicted in Figure 7.15a. Here, the horizontal axis represents the LC type in the reference data, and the vertical axis represents the missed error. Commission error signified the percentage of

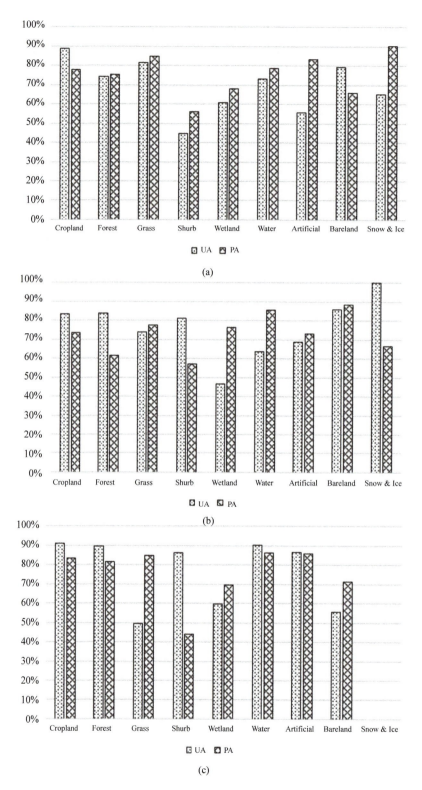

FIGURE 7.16 Classification accuracy in different regions of China. (a) Tibetan Plateau. (b) Northwest China. (c) East China. (Excerpted from Xie et al. 2022.)

Collaborative GlobeLand30 Validation

LC in GlobeLand30 classified as other land cover types, as shown in Figure 7.15b, where the horizontal axis denotes the land cover type in GlobeLand30.

The results indicated that confusion between forest, grassland and shrubs, wetlands and water, as well as grassland and bare land, significantly impacted the classification accuracy of GlobeLand30.

The accuracies of user's and producer's classifications for the Tibetan Plateau, Northwest China, and East China are presented in Figure 7.16. The classification accuracies for cropland in these regions consistently hovered around 80%, indicating a stable performance. In the Tibetan Plateau and Northwest China, the limited proportion of cropland and susceptibility to image quality made classification challenging. Seasonal rivers were mistakenly classified as wetlands, leading to reduced accuracy. In East China, irrigated croplands might be misclassified as wetlands or water bodies, while tea and fruit trees were often confused with forests. Additionally, in Northwest China, low-crowned arid zone vegetation with sparse distribution along with forests and shrubs posed difficulties in achieving high classification accuracies.

7.4.2 BULGARIA

The validation of the GlobeLand30 in the territory of Bulgaria was carried out within the framework of the international task for the validation of GLC data at 30 m resolution, which was part of the Earth Observations Group (GEO) component SB-02-C2 Global Land Cover Validation and User Engagement. The GLC datasets validated in the GEO-led international task included the 30 m resolution GlobeLand30 dataset of the NGCC, the 300 m resolution GlobCover dataset of the European Space Agency, the 25 m resolution forest/nonforest map dataset of Japan, and the 30 m resolution Global Forest Observation dataset of the University of Maryland, USA. The study aimed to propose a comprehensive method for validating GLC products and apply it to the validation of the aforementioned four GLC datasets for the territory of Bulgaria.

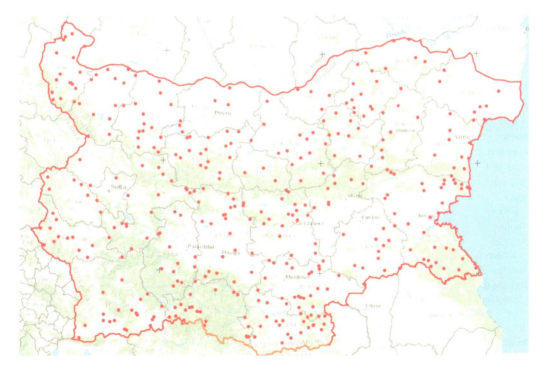

FIGURE 7.17 Distribution of sample points for the territory of Bulgaria. (Excerpted from Stamenova and Stamenov 2022.)

7.4.2.1 Validation Method

In Bulgaria, the LC exhibits significant fragmentation, with some predominant LC types, including urban areas, agricultural areas, forests, bare lands, water bodies, and wetlands. The online verification platform GLCVal offered a sampling function that effectively addressed landscape heterogeneity in regions. By utilizing the GLCVal platform at a 95% confidence level, a total of 382 sample points were generated. Figure 7.17 presents the distribution of sample points for the territory of Bulgaria.

Various reference data sources, such as high-resolution satellite images, aerial photos, and additional information about the LC of the territory of Bulgaria, were employed to interpret these samples within the context of Bulgarian LC. The primary source of reference information was color orthophoto maps of Bulgaria with spatial resolutions 0.5 m and 0.4 m, acquired in 2006 and 2010–2011, respectively. This dataset was provided by the Ministry of Agriculture, Food and Forestry of the Republic of Bulgaria exclusively for scientific purposes. Additionally, an ArcGIS online platform offering high-resolution satellite imagery served as another valuable reference information source. Each sample point was judged for its accurate LC type through visual interpretation based on these reference sources. Both direct elements (such as shape, size tone) and indirect elements (including relationships and location) were used in judgment. Validation accuracy was presented through an error matrix format whereby OA, UA, and PA were calculated for each class within every GLC dataset.

7.4.2.2 Results Analysis

The results of the validation were presented as an error matrix in Table 7.23. The estimated OA of the Globeland30 after performing the validation was 79.84%. The producer's and user's accuracy were given for each of the LC classes.

TABLE 7.23
Error Matrix for GlobeLand30 Dataset

Land Cover	Cultivated Land	Forest	Grass land	Shrub land	Wet land	Water Bodies	Tundra	Artificial Surfaces	Bare Land	Permanent Snow and Ice	All	UA
Cultivated land	75	10	7	10	0	0	0	4	0	0	106	70.8
Forest	5	154	4	4	0	0	0	0	0	0	167	92.2
Grassland	2	9	26	7	0	1	0	1	0	0	46	56.5
Shrubland	1	3	2	23	0	0	0	0	0	0	29	79.3
Wetland	0	0	0	1	1	0	0	0	0	0	2	50.0
Water bodies	0	0	1	0	1	5	0	0	0	0	7	71.4
Tundra	0	0	0	0	0	0	0	0	0	0	0	0.0
Artificial surfaces	1	0	1	1	0	0	0	21	0	0	24	87.5
Bare land	0	1	0	0	0	0	0	0	0	0	1	0.0
Permanent snow and ice	0	0	0	0	0	0	0	0	0	0	0	0.0
All	84	177	41	46	2	6	0	26	0	0	382	0.0
Producer's accuracy	89.3	87.0	63.4	50.0	50.0	83.3	0.0	80.8	0.0	0.0	0.0	79.8

Source: Data from Stamenova and Stamenov (2022).

The LC classes with the highest PA (over 80%) were the cultivated lands, forests and water bodies, and artificial surfaces, while the lowest results were observed in the shrubland and wetland classes. The UA showed that the class with the highest values was forest (92.2%), followed by the artificial surfaces. The cultivated lands, shrublands, and water bodies were assessed with UA over 70%, while the lowest values were again observed for the wetland LC class.

A comparison between GlobeLand30 data and the other three LC datasets (GlobCover2009, the global forest/nonforest product of the JAXA, and the tree canopy cover product) is presented in Table 7.24. The GlobCover2009 LC dataset has 22 LC classes; 17 are presented on the territory of Bulgaria. The result from validation is also presented in Table 7.24, and the estimated OA was 51.04%. The global forest/nonforest product of the JAXA has the smallest pixel size of 25 m and contains three LC classes—forests, nonforests, and water areas. The estimated OA for this product was 80.1%. The calculated UA was higher than the PA for all the classes. The lowest value of per-class accuracy was observed for the PA of the water class. The tree canopy cover product for 2010 shows the tree cover in percentages from 0 to 100 for trees with a height higher than 5 m. This dataset has a 30 m resolution and only one class – forest cover. The validation for this dataset was made through visual interpretation and assigning true or false values, i.e., 1 and 2, respectively, depending on the availability of forest for each sample unit. The estimated OA for the tree canopy cover product was very high: 96.07%.

For GlobeLand30, the misclassified pixels were predominantly concentrated in the mountainous and lowland regions of Southern and Western Bulgaria, particularly in the central areas of Rhodope, Kraishte, and Stara Planina, where there was a notable heterogeneity in landscape and LC. GlobCover2009 exhibited a higher occurrence of mixed pixels with a relatively uniform distribution of misclassified pixels throughout Bulgaria. Additionally, errors in tree canopy cover 2010 were observed specifically within the Rhodope Mountains.

7.4.3 GREECE

The GLC validation activities for Greece followed the general guidelines reported in the corresponding chapters of the technical specification document, initially composed by the NGCC. A thematic accuracy approach was adopted for validating the various LC types that were selected in the sampling process. This approach referred to the comparison between the GlobeLand30 map and the reference (or ground truth) data and was based on the generation of a confusion matrix, which was one of the most common methods for assessing the accuracy of thematic maps. Four main descriptive measures were derived from the matrix—namely, the PA, the UA, the OA, and the Kappa coefficient.

These included a total of 157 sample points covering eight LC types, according to the GlobeLand30 nomenclature. These types were forests (F), grasslands (GL), shrublands (SL), wetlands (WL), cultivated lands (CL), water bodies (WB), artificial surfaces (AS), and barren land (BL). The spatial distribution of the 157 samples is shown in Figure 7.18, and the number assigned to each continent is shown in Table 7.25.

TABLE 7.24

Comparison of OA

	GlobeLand30	GlobCover2009	Forest/Nonforest	Tree Canopy Cover Product
Class	10 classes	22 classes	3 classes	Only forest
OA	79.8	51.04	80.1	96.07

Source: Data from Stamenova and Stamenov (2022).

FIGURE 7.18 Distribution of the 157 Selected Samples for Greece.

TABLE 7.25
The LC Types and Number of Sample Points Assigned for the Validation Exercise in Greece

Code	LC Type	No of Samples
10	Cultivated land	25
20	Forests	25
30	Grassland	24
40	Shrubland	25
50	Wetland	8
60	Water bodies	14
80	Artificial surfaces	20
90	Barren land	16

The reference data used for the sample validation comprised both GE imagery and VHR orthophotos. The VHR data was freely provided by the National Cadastre and Mapping Agency S.A. of Greece through a Web Mapping Service (WMS). These orthophotos were acquired during a campaign from 2007 to 2009 for the whole of Greece and have a spatial resolution of 20 cm for urban areas and 50 cm for the rest of the country.

The procedure initially comprised photointerpretation of the area surrounding the sampling points over the VHR or GE images by using a 30 m × 30 m rectangular buffer zone around the point. In order to assign an LC class label to the reference zone, a threshold criterion was applied, meaning that the LC covering the largest area within the zone was selected. In addition, a majority decision rule was applied where two experts in photo interpretation had to agree on the same LC type. More

Collaborative GlobeLand30 Validation

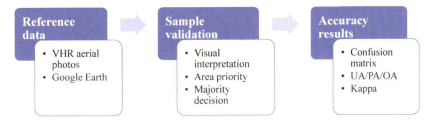

FIGURE 7.19 Workflow of the main steps followed for the validation of the LC samples.

TABLE 7.26
The Reference Orthophotos (VHR) and GE Images Used in the Validation Process

SID	Globeland-30	LC (Reference)	Validation	Reference Data Thumbnail
N34352010-539781	SL (40)	Shrub (VHR)	True	
N34352010-540620	CL (10)	Forest (GE)	False	

specifically, the experts conducted an independent interpretation of the reference data and assigned LC labels, without prior knowledge of the GobeLand30 classification. In case of label disagreement between the two decisions, the majority rule was applied. The final step included the generation of the confusion matrices and the derivation of the statistical measures. The specific workflow is presented in Figure 7.19.

An example of the reference data used for the photointerpretation of the area surrounding the sample points is given in Table 7.26. The common coordinate system (CS) for both the sample vector points and the background reference images was the WGS84 (GCS).

The descriptive statistics derived from the confusion matrix are presented in Table 7.27. More specifically, OA and the Kappa coefficient, as well as PA and UA, for each class were estimated. In addition, confidence intervals (CI) were computed for each metric. The lowest UA (48%) was estimated for cultivated land, where only 12 samples were correctly mapped out of the 25. The large commission error observed here was mainly attributed to confusion with shrub areas, which could be expected since agricultural land was often abandoned and not treated. The highest UA was estimated for water bodies (78.57%), followed by forests class (76%), with 19 out of 25 samples being correctly mapped. For the other vegetated classes, grasslands and shrubs, errors occur due to confusion between these two classes and often shrublands were erroneously mapped as forests (6 out of

TABLE 7.27
Accuracy Assessment Results for the GlobeLand30 Validation in Greece

LC type	PA (%)	User's Accuracy (%)	95% CI	
			Lower Limit	Upper Limit
Cultivated land	63.16	48.00	0.217	0.563
Forests	55.88	76.00	0.318	0.637
Grasslands	73.68	58.33	0.299	0.671
Shrublands	45.71	64.00	0.228	0.523
Wetland	85.71	75.00	0.309	0.910
Water bodies	84.62	78.57	0.415	0.879
Artificial surfaces	100.00	65.00	0.409	0.837
Barren land	58.82	62.50	0.239	0.651
Kappa	0.59		0.50	0.673
OA	64.33 %		0.563	0.717

25 samples). These errors could be expected since in the Mediterranean region, transitional vegetation areas were often occupied by leathery, broad-leaved evergreen shrubs, the maquis, or small trees. For the rest of the classes, most errors were attributed either to the classifier or to uncertainties due to the geolocation of the samples, often positioned in borders between classes.

The OA result, about 64%, of the GlobeLand30 validation for Greece did not appear quite satisfactory, however, there were a few factors that need to be considered before final conclusions. First of all, the Mediterranean landscape was quite fragmented, with larger patches of land transforming into smaller ones and eventually becoming isolated or surrounded by larger patches of different LC types. This fragmentation could cause confusion and uncertainties in land interpretation, especially when area threshold criteria have to be applied by the interpreter to assign a label to the reference data. Such a case could be the one displayed in Figure 7.21. Then, several samples selected for this study were located, as mentioned previously, at the edge of adjacent LC types where it would be possible that any geolocation error could cause confusion. Such an example can be seen in Figure 7.20, where the point was classified as a water body; however, in the reference image, it appears within a vegetated stripe across the river. Similar errors were observed for the AS, resulting in low accuracy. The reference data selected for the validation was acquired nearly the same period that the GlobeLand30 was created, and we assume that no major LC changes occurred.

A solution could be to assign levels of certainty for each sample and then compute a weighted confusion matrix instead of the crisp validation within the buffer zone. A better suggestion would be to additionally perform spatial filtering during the sampling process to select more appropriate sampling points. This filtering could reduce confusion due to fragmentation or geolocation issues, and fuzzy weight assignment could be avoided since it does not always ensure a more robust validation.

7.4.4 ITALY

This section contributes to the classification quality assessment of the GlobeLand30 dataset by summarizing the methodology and results obtained from the first comprehensive thematic accuracy assessment of Globeland30 for Italy.

7.4.4.1 Validation Method

The objective of this study was to conduct a comprehensive thematic accuracy assessment of the GlobeLand30 dataset covering Italy. Two commonly utilized methods were employed for LC dataset

Collaborative GlobeLand30 Validation 233

FIGURE 7.20 Location of a GlobeLand30 sample, which is initially classified as a water body.

FIGURE 7.21 Location of a GlobeLand30 sample, which is shrub land but initially classified as cultivated land.

accuracy assessment as proposed by Yang (2017): (1) cross-validation of sample pixels within a confusion matrix and (2) measuring agreement and disagreement with existing LC maps or statistical LC information. The latter approach was employed for validation in Italy. The accuracy assessment involved comparing the Italian map with GlobeLand30 at the pixel level, resulting in a confusion matrix and derived statistical data such as OA, UA, and PA.

High-resolution LC maps had been generated for the majority of regions in Italy and were accessible to users as open data. At the time of validation, 8 out of 20 regions had available data, and four of them provided data for comparison with both year's GlobeLand30 maps. The comparable regions included Lombardy, Liguria, Sardinia, Emilia-Romagna, Veneto, Friuli Venezia Giulia, Bolzano and Trento, and Abruzzo. However, it was important to note that this geographical distribution did not encompass the entirety of Italy. All regional datasets could be obtained from ESRI (Environmental Systems Research Institute) and were created in different years using diverse reference systems and sources. Most datasets aligned closely with the Chinese Classification of GlobeLand30, a three-tier hierarchical classification system based on LC types. The initial level of the LC class comprises five categories: artificial surfaces, agricultural areas, forests and semi-natural areas, wetlands, and water bodies.

TABLE 7.28

First and Second Reclassification Methods: Correspondence Between GlobeLand30 and Italian Regional Data

First Legend	Second Legend	Italian Regional Data	Globeland30
1. Artificial surfaces		Artificial surfaces	Artificial cover
2. Agricultural areas		Agricultural areas	Cultivated land
3. Forests and semi-natural areas	3.1 Forests	Forests	Forests
	3.2 Grass and/or herbaceous vegetation associations	Scrub and/or herbaceous vegetation associations	Shrublands, grasslands
	3.3 Open spaces with little or no vegetation	Open spaces with little or no vegetation	Barren lands
	3.4 Glaciers and perpetual snow	Glaciers and perpetual snow	Permanent snow and ice
4. Wetlands		Wetlands	Wetland
5. Water bodies		Water bodies	Water bodies

Source: Data from Brovelli et al. (2015).

GlobeLand30 and the Italian regional map differ in format, legend, scale, and reference system. Before calculating the confusion matrix, data preprocessing was necessary. The following steps were performed for each dataset: (1) to rasterize the Italian vector dataset and (2) to unify the spatial resolutions of the two datasets at 30 m and 5 m for Globeland30 and the regional map, respectively, and to reclassify the two datasets. Due to the utilization of different classification systems in GlobeLand30 and Italian datasets, a direct comparison between them was not feasible. Therefore, it was necessary to convert the classification of both datasets into a unified system. Two specific methods were considered for this purpose.

The first reclassification method employed the first legend to classify both GlobeLand30 and Italian regional data into five LC categories—namely, artificial surfaces, agricultural areas, forests and semi-natural areas, wetlands, and water bodies. The second reclassification method utilized the subsequent legend to further categorize the forest and semi-natural areas into four subclasses: forests, grasses, and/or herbaceous vegetation associations; open spaces with little or no vegetation; glaciers; and perpetual snow. Table 7.28 illustrates the correspondence between the LC types of GlobeLand30 dataset and those of the Italian regional data.

GlobeLand30 and the Italian regional map were compared twice. The first comparison was a pixel-by-pixel comparison. The second comparison was then performed to evaluate the influence of the GlobeLand30 co-location tolerance on the classification quality, as it was evaluated by the data producers (70 m). This means that a pixel could be located in a buffer of 70 m to its true position. All cells belonging to a buffer of 70 m around GlobeLand30 classes border were eliminated, and the confusion matrix and statistics were calculated on the other pixels.

7.4.4.2 Accuracy Assessment

The OA, allocation, and quantity disagreement for each region's dataset using the first legend reclassification without buffer are presented in Figure 7.22a. The OA of GlobeLand30 ranged from 81% (Liguria) to 92% (Autonomous Province of Bolzano). In most regions, the proportion of disagreement between allocation disagreement (AD) and quantity disagreement (QD) was nearly equal. However, Liguria and Abruzzo stand out as exceptions with significantly higher AD than QD. Considering the implementation of a buffer, there was potential for improving OA by approximately 3%–4%. Figure 7.22b illustrates OA, AD, and QD using the second legend reclassification. Compared to the first legend reclassification, accuracy values generally decreased, ranging from 67% (Sardinia) to 81% (Emilia-Romagna) (Table 7.29). Forest classification achieved the highest

Collaborative GlobeLand30 Validation

TABLE 7.29
Comparison Illustration

	Italian Regional Data	GlobeLand30	
Comparison 1-1	Reclassification into first legend	Reclassification into first legend	No buffer
Comparison 1-2	Reclassification into first legend	Reclassification into first legend	Buffer
Comparison 2-1	Reclassification into second legend	Reclassification into second levels	No buffer
Comparison 2-2	Reclassification into second legend	Reclassification into second levels	Buffer

Source: Data from Brovelli et al. (2015).

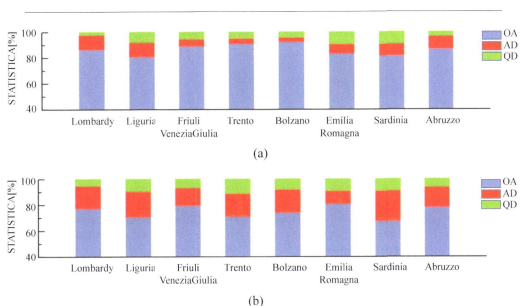

FIGURE 7.22 Comparison between GlobeLand30 2000 and Italian regional datasets (no buffer case): OA, AD, QD for each dataset. (a) First legend reclassification. (b) Second legend reclassification. (Excerpted from Brovelli et al. 2015.)

accuracy among subclasses, with most parts of Italy exhibiting forest classification accuracy above 70%. Finally, under buffer conditions, an improvement in OA by about 4%–5% could be observed.

Figures 7.23a and 7.23b depict the analysis of the UA and PA values, respectively. Forests and semi-natural areas exhibited high detection rates across all regions, with an accuracy consistently above 80%. Other class types demonstrated varying levels of accuracy across different regions, and water bodies and wetlands were particularly noteworthy in this regard.

According to the first legend reclassification method and no buffer, Figure 7.24 (a) illustrates the comparison results of the first legend reclassification between GlobeLand30 and the five available datasets for Italy in 2010. The OA ranged from 81% (Sardinia) to 86% (Lombardy). By implementing the buffer method, an improvement of approximately 3%–4% could be achieved in terms of OA. Furthermore, Figure 7.24 (b) presents comprehensive statistical data obtained through the application of the second legend reclassification method. It highlighted a significant increase of disagreement values that entail a decrease in OA percentages: new values range from 62% (Sardinia) to 80% (Lombardy and Emilia-Romagna). Incorporating buffer zones led to an overall enhancement of OA by about 3%–5%.

Among subclasses, forests demonstrated superior detection effectiveness, with UA and PA values exceeding 75% in most cases (in Figure 7.25). Comparing with the validation results of Globe

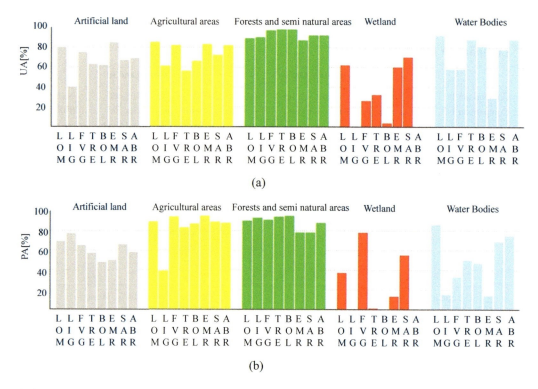

FIGURE 7.23 Comparison Between GlobeLand30 2000 and Italian Regional Datasets (First Legend Reclassification, No Buffer Case): User's Accuracies and Producer's Accuracies for Each Class and Dataset. (a) User's accuracies. (b) Producer's accuracies. (Excerpted from Brovelli et al. 2015).

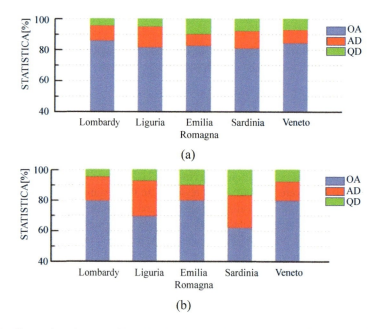

FIGURE 7.24 Comparison between GlobeLand30 2010 and Italian regional datasets (no buffer case): OA, AD, QD for each dataset. (a) First legend reclassification. (b) Second legend reclassification. (Excerpted from Brovelli et al. 2015).

Collaborative GlobeLand30 Validation

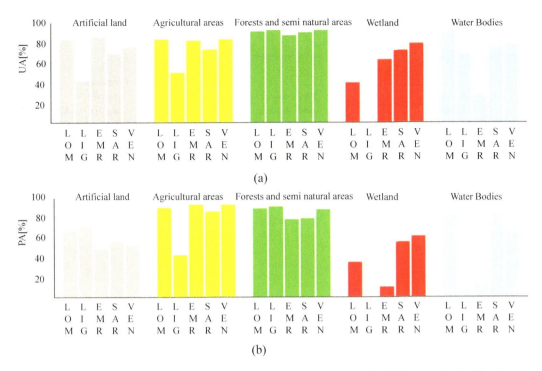

FIGURE 7.25 Comparison between GlobeLand30 2010 and Italian regional datasets (reclassification, no buffer case): user's accuracies for each class and dataset. (a) User's accuracies. (b) Producer's accuracies. (Excerpted from Brovelli et al. 2015).

FIGURE 7.26 Distribution of the LSI-based sample points in Sweden.

TABLE 7.30

Error Matrix for GlobeLand30 Dataset in Sweden

LC	Cultivated Land	Forest	Grass Land	Shrub Land	Wet Land	Water Bodies	Tundra	Artificial Surfaces	Bare Land	Permanent Snow and Ice	All	UA (%)
Cultivated land	40	6		2							48	83.3
Forest		51		7							58	87.9
Grassland	1		10	1							12	83.3
Shrubland		3		49							52	94.2
Wetland		1		2	44			1			48	91.7
Water bodies		2			1	45					48	93.8
Tundra		1					37		2		40	92.5
Artificial surfaces	1	1			2			31	1		36	86.1
Bare land								1	22		23	95.7
Permanent snow and ice							5		4	20	29	69.0
All	42	65	10	61	47	45	42	33	29	20	394	0.0
PA (%)	95.2	78.5	100	80.3	93.6	100	88.1	93.9	75.9	100	0.0	OA: 88.58%

Land30-2000 (Figure 7.22), it was evident that forests and semi-natural areas exhibited consistently high accuracy across all datasets, while other class types displayed variable accuracies.

7.4.5 SWEDEN

Sweden, located in the northernmost region of the world, spans 1,574 kilometers from north to south and 499 kilometers from east to west, encompassing a total area of 450,295 square kilometers. As the third largest country within the European Union, approximately 70% of its land is covered by forests primarily situated in the northern mountains. Conversely, only 8% of its territory consists of farmland distributed across agricultural plains in the southern regions.

The accuracy assessment for GlobeLand30-2010 in Sweden was conducted by a research assistant affiliated with KTH Royal Institute of Technology. The sample points were generated using the LSI-based sampling method provided in the GLCVal system (Figure 7.26). With a confidence level set at 95%, a total of 394 sample points were produced for evaluation purposes. Visual interpretation served as the primary criterion for judging these samples, based on high-resolution satellite images obtained from GE during the period between 2009 and 2011 (with spatial resolution ranging from 0.6m to 2.5 m). Additional auxiliary data sources included local orthophotos, high-resolution satellite imagery, and national LC data.

Table 7.30 presents the accuracy assessment results for GlobeLand30 in Sweden. The OA achieved by GlobeLand30 stands at an impressive rate of 88.58% while exhibiting a Kappa coefficient value of approximately 0.87—indicating substantial agreement between observed and predicted classifications. Notably, it should be emphasized that Sweden boasts around 100,000 lakes characterized by diverse conditions. Remarkably enough, GlobeLand30 demonstrated excellent PA for water bodies when all 45 samples were correctly classified with an exceptional producer's accuracy reaching up to100%. This remarkable achievement underscored GlobeLand30's efficacy within this specific context. Furthermore, apart from forests and bare lands, all other LC classes attained a PA exceeding 80%. Regarding UA, all LC categories except permanent snow exhibited an accuracy surpassing 80%. This discrepancy may be attributed to the spectral resemblance between tundra, bare lands, and artificial surfaces.

7.5 CONCLUSION

Under the support of the GEO project, the validation of GlobeLand30 has yielded remarkable outcomes thus far, with over two dozen validations conducted across various scales and at different levels: global, regional, and national. The accuracy assessment plays a pivotal role in enabling map users to evaluate the suitability of LC maps for their intended purposes. GlobeLand30-2010 achieved an estimated OA of approximately 80.3 ± 0.2%, based on over 150,000 samples encompassing all LC types, as conducted by TJU.

Regional-scale and national-scale validations were carried out in several regions and countries, including the Roof of the World, Africa, China, Bulgaria, Greece, Italy, and Sweden, with accuracies ranging from 75% to 90%. These validation practices substantiated the nonuniform accuracies (ranging from 5% to 10%) in different regions or for specific LC classes.

To date, more than 210,000 samples have been acquired through collaborative validation practices of GlobeLand30 at global, regional, and national levels. The validation sample resource represents the largest dataset in terms of quantity, coverage, typicality, and representativeness for GLC data at a spatial resolution of 30 m. This resource fills a critical gap both domestically and internationally, playing an essential role in the validation of GlobeLand30.

REFERENCES

Brovelli, M.A., Molinari, M.E., Hussein, E.S., Chen, J., Li, R. 2015. The First Comprehensive Accuracy Assessment of GlobeLand30 at a National Level: Methodology and Results. *Remote Sensing*, 7, 4191–4212.

Cao, X., Li, A., Lei, G., Tan, J., Zhang, Z., Yan, D., Xie, H., Zhang, S., Yang, Y., Sun, M. 2016. Land cover mapping and spatial pattern analysis with remote sensing in Nepal. *Journal of Geo-information Science*, 18(10), 1384–1398.

Chen, F., Chen, J., Wu, H., Hou, D., Zhang, W., Zhang, J., Zhou, X., Chen, L. 2016. A landscape shape index-based sampling approach for land cover accuracy assessment. *Science China Earth Sciences*, 59, 2263–2274.

Chen, F. 2017. *A landscape index-based sampling approach for land cover accuracy assessment.* Central South University.

Chen, J., Chen, J., Liao, A., Cao, X., Chen, L., Chen, X., He, C., Han, G., Peng, S., Lu, M., Zhang, W., Tong, X., Mills, J.P. 2015. Global land cover mapping at 30 m resolution: A POK-based operational approach. *ISPRS Journal of Photogrammetry and Remote Sensing*, 103, 7–27.

Chen, J., Chen, L., Chen, F., Ban, Y., Li, S., Han, G., Tong, X., Liu, C., Stamenova, V., Stamenov, S. 2021. Collaborative validation of GlobeLand30: Methodology and practices. *Geo-spatial Information Science*, 24, 134–144.

Fritz, S., See, L.M., Mccallum, I., Schill, C., Obersteiner, M., van der Velde, M., Boettcher, H., Havlík, P., Achard, F. 2011. Highlighting continued uncertainty in global land cover maps for the user community. *Environmental Research Letters*, 6, 044005.

Gong, P. 2009. Assessment of GLC map accuracies using Fluxnet location data. *Progress of Natural Sciences*, 19, 754–759.

Herold, M., Mayaux, P., Woodcock, C.E., Baccini, A., Schmullius, C. 2008. Some challenges in global land cover mapping : An assessment of agreement and accuracy in existing 1 km datasets. *Remote Sensing of Environment*, 112, 2538–2556.

Jacobson, A.P., Dhanota, J., Godfrey, J., Jacobson, H., Rossman, Z.T., Stanish, A., Walker, H., Riggio, J. 2015. A novel approach to mapping land conversion using Google Earth with an application to East Africa. *Environmental Modelling & Software*, 72, 1–9.

Jokar Arsanjani. J., See, L.M., Tayyebi, A. 2016a. Assessing the suitability of GlobeLand30 for mapping land cover in Germany. *International Journal of Digital Earth*, 9, 873–891.

Jokar Arsanjani. J., Tayyebi, A., Vaz, E. 2016b. GlobeLand30 as an alternative fine-scale global land cover map: Challenges, possibilities, and implications for developing countries. *Habitat International*, 55, 25–31.

Kussul, N., Shelestov, A., Basarab, R., Skakun, S., Kussul, O., Lavrenyuk, M. 2015. Geospatial intelligence and data fusion techniques for sustainable development problems. *International Conference on Information and Communication Technologies in Education, Research, and Industrial Applications*.

Manakos, I., Chatzopoulos-Vouzoglanis, K., Petrou, Z.I., Filchev, L.H., Apostolakis, A. 2014. Globalland30 Mapping Capacity of Land Surface Water in Thessaly, Greece. Land, 4, 1–18.

Mozak, S. 2016. Comparing global land cover datasets through the eagle matrix land cover components for continental Portugal. Master's Thesis, Nova Information Management School, Universitat Jaume, Lisbon, Portugal.

See, L.M., Bayas, J.C., Schepaschenko, D., Perger, C., Dresel, C., Maus, V., Salk, C.F., Weichselbaum, J., Lesiv, M., Mccallum, I., Moorthy, I., Fritz, S. 2017. LACO-Wiki: A new online land cover validation tool demonstrated using GlobeLand30 for Kenya. *Remote Sensing*, 9, 754.

Stamenova, V., Stamenov, S. 2022. Integrated method for global land cover products' validation on the example of Bulgaria. *Acta Geographica Slovenica*, 62(3): 64–83.

Stehman, S.V., Olofsson, P., Woodcock, C.E., Herold, M., Friedl, M.A. 2012. A global land cover validation data set, II: Augmenting a stratified sampling design to estimate accuracy by region and land cover class. *International Journal of Remote Sensing*, 33, 6975–6993.

Sun, B., Chen, X., Zhou, Q. 2016. *Uncertainty Assessment of GlobeLand30 Land Cover Data Set Over Central Asia*. ISPRS - International Archives of the Photogrammetry, Remote Sensing and Spatial Information Sciences, 1313–1317.

Wang, Y., Zhang, J., Liu, D., Yang, W., Zhang, W. 2018a. Accuracy assessment of GlobeLand30 (2010) over China with a landscape shape index-based sampling approach. *Journal of Global Change Data & Discovery*, 2(1), 29–34.

Wang, Y., Zhang, J., Liu, D., Yang, W., Zhang, W. 2018b. Accuracy assessment of GlobeLand30 2010 land cover over China based on geographically and categorically stratified validation sample data. *Remote Sensing*, 10, 1213.

Wang, Z. X., Liu, T., Tun W.N. 2019. Google Earth images based land cover data validation dataset for GlobeLand30 (2010) in the region of roof of the world. *Journal of Global Change Data & Discovery*, 3(3): 259–267.

Wu, X., Xiao, Q., Wen, J., Liu, Q., Li, X. 2014. Advances in uncertainty analysis for the validation of remote sensing products: Take leaf area index for example. *Journal of Remote Sensing*, 18(5), 1011–1023.

Xie, H., Wang, F., Gong, Y., Tong, X., Jin, Y., Zhao, A., Wei, C., Zhang, X., Liao, S. 2022. Spatially balanced sampling for validation of GlobeLand30 using landscape pattern-based inclusion probability *Sustainability*, 14: 2479.

Yang, Y., Xiao, P., Feng, X., Li, H. 2017. Accuracy assessment of seven global land cover datasets over China. *ISPRS Journal of Photogrammetry and Remote Sensing*, 125, 156–173.

Zhang, YS. 2015. Characteristics of land cover change in Siberia based on Globe Land30, 2000–2010. *Progress in Geography*, 34(10): 1324–1333.

Zhao, H. 2018. Research on spatial sampling and accuracy validation of high resolution land cover data – Taking the roof of the world as an example. Master's thesis. Tongji University.

8 Conclusions

Xiaohua Tong
Tongji University, Shanghai, China

Jun Chen
National Geomatics Center of China, Beijing, China

8.1 SUMMARIES

Global land cover (GLC) data validation, which estimates the accuracy of GLC products based on selected samples and reference data, is an important prerequisite for the effective application of GLC data products. With the availability of finer-resolution remote sensing data and improved computing and storage capacity, GLC data products have evolved from coarse resolution to finer resolution. However, the spatial heterogeneity of large-area land cover poses challenges to the methods and processes for validating finer-resolution GLC products compared with coarse-resolution products. The objective of this book was to present the theories, key technologies, methods, and software tools for validating finer-resolution GLC data products. A collaborative validation initiative launched by the GEO for validating Globeland30 was also presented, including its general framework and the major tasks. Online validation tools have been developed to support not only these validation tasks but also, hopefully, global validation of finer-resolution GLC data products.

This book can be used as a reference book for graduate students, researchers, engineers, and technicians in surveying and mapping science and technology, geospatial information science, resources and environment, global change remote sensing, land and resources survey, and other emerging interdisciplinary fields.

8.1.1 GLC Products Validation Approaches

In order to address the complexity and difficulties of the validation of 30 m resolution GLC data products, a number of methodological issues were examined, and appropriate approaches were developed or improved, from sampling design through using reference data selection to sample judgment and accuracy assessment.

Sampling design, as presented in Chapter 2, is the initial stage in the sampling-based GLC product validation. It includes three parts: design of sampling plan, sampling method, and allocation of samples. The main purpose of sampling design is to determine the sample size from the population, including classical methods, methods based on acceptance probability, and methods based on landscape shape index. Design of the sampling method means to define strata and sample size in stratified sampling. Different sampling methods can be used according to the characteristics of GLC products, such as the classical stratified methods, two-rank sampling methods, landscape shape index (LSI)-based sampling methods, and modified spatially balanced sampling distribution methods. The purpose of sample allocation is to determine the spatial locations of samples. Various allocation methods are available, such as classical allocation methods, two-rank-based allocation, LSI-based allocation, and grid allocation methods. The sampling methods have been gradually developed from simple random sampling to systematic sampling, stratified design, cluster sampling, spatiotemporal stratified sampling, and so on. The key to sampling design is to choose an adequate and appropriate design.

DOI: 10.1201/9781003557791-8

Selection and exploitation of reference data is the foundation for achieving reliable validation results of GLC products. The exploration of various sources of reference data, which include very high resolution (VHR) remote sensing images, high-resolution land cover products, large-scale accurate maps, crowdsourcing data, and in situ data, can be found in Chapter 3. VHR imageries provide highly detailed and discernible land cover features for validating GLC products. Finer-resolution land cover products offer an accurate depiction of land cover characteristics in a granular view of Earth's surface. Large-scale maps provide land cover and thematic maps in a wide range. Crowdsourced data provide reference information with high temporal accuracy in a wide range. In situ data provide accurate ground-truth information about land cover characteristics in some specific locations. Different kinds of reference data have their own characteristics and limitations. Therefore, it is necessary to consider the integration of different reference data to provide more comprehensive and reliable references for the validation of GLC products.

After the establishment of the sampling plan and methods and determination of the reference data, the sample judgment is followed. Sample judgment, as discussed in Chapter 4, is an important step in collaborative validation of GLC products, which is usually performed by validation experts or volunteers to make subjective judgments on the sample attribute information through validation tools and reference data. Sample judgment should be conducted according to the interpretation principles and interpretation methods. The reliability of the validation samples can be graded based on the availability of reference data and the subjectivity of validation experts. In addition, the crowd-source information, such as points of interest, social media information, and geo-tagged photos, provides additional usable samples of land cover data.

Accuracy assessment approaches, which ultimately estimate the accuracy of the GLC products based on appropriate accuracy indicators, were presented and discussed in Chapter 5. The selection of appropriate accuracy indicators has a great influence on the estimated accuracy of the GLC products. Various common accuracy indicators and several other accuracy assessment methods, which take into account multiclass, multitemporal, multiscale, change detection, and geolocation accuracy of the features, were introduced and analyzed in Chapter 5. Considering the large data amount and complex data structure of the GLC products, further research on accuracy indicators capable of handling complex data structures on a large scale and dynamic indicators providing real-time accuracy assessment is needed.

8.1.2 Online Tools

In order to overcome the difficulties in the collection and judgment of sample data over a large area or the entire globe, online tools are required to enable experts and volunteers from different parts of the world to conduct validations collaboratively. In the past, some land cover data validation tools were developed from the built-in functionality or plug-in extension of the geographical information system and remote sensing software to the web-based and mobile tools, as analyzed in Chapter 6. However, most of them do not well support resource sharing and interaction among participants from different parts of the world. An online GLC validation tool, GLCVal, developed for supporting the collaborative validation of Globeland30, was presented with details as an example. It provides the functions of data collection, sampling design, sample judgment, and accuracy assessment. This tool can be used to verify the accuracy of GLC data products and provide highly reliable data accuracy information for global change research.

8.1.3 GlobeLand30 Accuracy Assessment at Multiscales

In order to organize the GEO-led collaborative validation activity, a technical specification was formulated to describe the approaches and procedures used for sampling design, response, and analysis protocols at 30 m resolution and global scale. A series of validation practices for the GlobeLand30

Conclusions **243**

were carried out at global, regional, and national scales with the guidance of the technical specification and the support of the developed online tool.

At the global scale, the two-rank sampling strategy was employed. GlobeLand30-2010 achieved an estimated overall accuracy of better than 80%, based on over 150,000 samples encompassing all land cover types. Despite the relatively high overall accuracy, it is still needed to verify the accuracy in different regions of the world.

Therefore, the GlobeLand30 accuracies in different regions and nations of the world were further validated by international participants. Regional-scale and national-scale validation was carried out for areas including the Roof of the World, Africa, China, Bulgaria, Greece, Italy, and Sweden, with accuracies ranging from 75% to 90%. In total, 217,494 samples have been acquired through collaboration with experts and scholars from different countries. These practices not only confirm the accuracy of GlobeLand30 from different perspectives but also testify to the existing validation approaches.

8.2 FUTURE DIRECTIONS

8.2.1 NEW VALIDATION APPROACHES FOR HIGHER-RESOLUTION GLC PRODUCTS

This book mainly addressed the challenges for the validation of 30 m resolution GLC data products. With the availability of higher-resolution satellite remote sensing data and the development of new GLC data products, more challenges arise from the validation of these higher-resolution GLC products with stronger spatial heterogeneity and correlation regarding the strata separation, appropriate sample size determination, reasonable allocation of samples, and so on. Therefore, the continued development or improvement of validation approaches for supporting new/higher-resolution GLC products will be in demand.

8.2.2 DEVELOPMENT OF GLOBALLY LABELED SAMPLE DATABASES

In general, large amounts of samples are needed to achieve a reliable validation result of the finer-resolution GLC products. This makes the sample collection and judgment the most labor-intensive and time-consuming process in the validation. It is therefore important to document the labeled samples generated in the GLC validation and share them with other validation activities. At the same time, some new technologies, such as artificial intelligence (AI)-based approaches should be introduced to generate the labeled sample databases at global and regional scales. Therefore, it is necessary to establish labeled sample databases for GLC validation. Appropriate standard approaches need to be developed to classify or label samples, and a technical specification should be further formulated.

8.2.3 DEVELOPMENT OF INTERNATIONAL STANDARDS FOR LAND COVER VALIDATION

Due to the spatial correlation and heterogeneity, as well as the uncertainty in sample judgment, the assessed accuracy derived in different scales or using different validation methods may vary greatly. Although there are some international standards that focus on the general sampling procedures (ISO2859 2020; ISO3951 2013), the standards for land cover validation are still lacking in specifying the principles of validation, approaches of sampling design, reference data selection, sample judgment, and accuracy assessment. Therefore, it is necessary to develop and formulate an international standard for finer-resolution GLC product validation.

8.2.4 NEW TOOLS FOR VALIDATING GLC DATASETS

Validation tools are essential components for facilitating the validation of the GLC products, which have a large range and amount of data. Appropriate validation tools can enable collaborative validation and greatly improve validation efficiency. There is a high demand for efficient and intelligent verification tools for the validation of GLC products. For instance, the sampling design and judgment are often conducted manually to take into account the heterogeneity of the region for validation. The future validation tools should be more intelligent and smart, providing adaptive sampling design capacities for automatic sampling design, supporting more intelligent sample judgment, and facilitating the integration and access of multisource sampled data.

Index

Pages in *italics* refer to figures and pages in **bold** refer to tables.

A

AccAssess, *167–168*, 168
acceptance quality level (AQL), 31–32, 203
accessibility, 44
    data, 183
accuracy, 136–137, 148–149, 151, 155
    assessment, 2, 11–15, *16*, 23, *24*, 26–27, 31, 36, 47, 56, 59, 83, 105, 130–131, 136–139, 141, 148, 151–152, 156–159, 163–168, 171–174, **175**, 177–179, 183–187, 189, 191, 194–196, **205**, 206–208, 212, 217, 220, **232**, 233–234, 238–239, 241–243
    evaluation, 25–26, 48, 131, 164
    forest classification, 235
    indicators, *138*, 159
    Kappa coefficient and, 216
    normalized, 150
    overall, 136, 149, 151
    positional, 15, 65, 71, 147–148, 157, 188
    producer's, 137, 149, 171
    thematic, 95, 189
    user, 51, 131, 137, 141–142, 149, 164, 166, 168, 170–173, 187, 196, 201, **206**, **211–212**, 228, **232**
aerial photo interpretation, 108
allocation options, 51
artificial surfaces, 8, 111, 209
attribute accuracy, 148–149

B

balance
    data redundancy, 50
    errors, 124
    sampling, 42, 44–45, 209
    spatial, 9, 26, 42–47, 59, 186, 222, 241
bare land, 8, 209
binary map, 122–123, 125
Bing Maps, 70, **71**, 72, 186, 220
biodiversity, 1, 6, **7**, 14, 87, **202**

C

carbon dioxide, 1
cause and consequence analysis, 1
change detection, 6, 136
change in entropy (EC), 145–146
class-level LSI (cLSI), 40, 45
classification, 95, 136, 148
classification success index, 142, 149
cluster sampling, 9, 11, 27, 29, 38, 51–53, 56, 59, 241
collaboration, 13–14, 16, 91, 93, 108, 134, 163–165, 174, 187, 243
collaborative validation, 13–17, 134, 163–165, 181–183, *185*, 189, 200–201, 209, 217, 239, 241–244

    need for, 196
    process, 212
    tools, 187
completeness, 148–149
confidence level, 209
confusion matrix, 137, **139**, 148, 171
Copernicus Global Land Services (CGLS), 52
costs, **10**, 11, 29, 32, 59, 189
    control, 25
    data collection, 30–31, 52, 164
    effectiveness, 9, 26, 59, 73, 134, 157, 186
    reductions, 9, 73
    validation, 12, 211, 213
crowdsourcing, 13
    data from, 89, 91, 99, 180, 242
cultivated land, 3, **6**, 8, 56–58, 82, 111, **118**, **121**, 132–133, 200, **204–210**, 211, **212**, 217, **218**, 219, **222**, **224**, **228**, 229–234, **238**

D

data
    accessibility, 183
    collection, 13, 99
    credibility, 105
    crowdsourced, 91, 99, 180
    geospatial, 7, 33–34, 73, 80, 92, 127, 164, 190
    in situ, 99, 209
    processing, 91
    quality, 95, 189
    spatial, 6–8, 66, 209
digital elevation models (DEMs), 107–108
digital imagery, 3
DISCover, 2, 9, **10**, 11–12, 24–25, 31, 36, 51, 105, 192
drones, 73–74

E

Earth system modeling, 1
economic growth, 1
ecosystem, **7**, 87, 94, 112
    services, 1
    sustainability, 6
effective land area ratio, 53–54
environment, 97, 115, 120–121, 207
    analysis, 85
    impacts on, 1, 87
    management, 87, 99, 182
    monitoring, 88, 94
    networked, 164
    protection, 66, 202
    studies, 1, 14, 165
ESA-S2-LC20, 3, **4**, 80, *81*, *83*
ESRI, **10**, 80
European Space Agency (ESA), **2**, 74, 227

F

field-collected data, *see* in situ data
field sampling, 186, 209
forest, 6, 111, 209
fragmentation, 44, 211, 219, 228, 232
  landscape, 47

G

generic statistical analysis, 1
GEO (Group on Earth Observations), 3, 6, 200
Geo-Wiki, 14, 89, **90**, 156, 165, 174, *176*
GeoEye, 66–67, **69**
Geographic Resources Analysis Support System (GRASS), 165, 168, *169*
geospatial data, 7, 33–34, 73, 80, 92, 127, 164, 190
GLC2000, 2, **10**, 11–14, **24**, 25, 31, 38, 156, 192
GLC-SHARE, **10**, 12–13, **24**, 25, 36
global kernel density estimation, 53–54
global land cover (GLC), 6, 136
  data, 65
  datasets, **2**
  mapping, 2–3
  products, 3, 6–7, **24**, 136, 148
Global Land Cover Community Activity, 14, 200
global sampling, 35, 38–39, 202
Globcover, 2, **10**, 11–13, **24**, 25, 31, 122, 174, 176, 192, 227, 229
GlobeLand30, 3, **4**, *5*, 6, **7**, *28*, 39, *108, 110–111*, 118, 121–122, 125, **127**, 174, 179–180, 187–189, 192, 196, 200, 204, **205**, 206–209, 211, **218**, *220*, 222–229, 232–239, 243
  accuracy, 13–14, 54, 202, 220–221
  application, **7**
  data, 4, 27
  validation of, 36, 200, 203, 212, 216–217, 241–242
Google, 13, 73, 206
  Dynamic World, 74
  Earth, **10**, 11–12, 25, 69–72, 74, 82, 93, **96–97**, 108, **109**, *111*, 113, 117, 133, 164, 174, 202
  Earth Engine, 74
  Map Maker, 13
  Maps, 70, 93, 164, 176, 186
  Satellite, 166, 176
grassland, 8, 111, 209
greenhouse gases (GHGs), 1
Gray-Level Co-occurrence Matrix (GLCM), 48, 50
grid sampling, 209
ground observation, 209
ground truth, 95, 136
ground-truth
  data, *68*, 73–74, 83, 94, 163, 170–173, 179, 242
  information, 99
Group on Earth Observations (GEO), 3, 6, 200
gross domestic product (GDP), 7

H

heterogeneity, 13, 26–27, 39–41, 47–48, 56, 58, 219
  landscape, 57, 221, 228–229
  spatial, 8, 14–15, 56, 59, 157, 164, 186, 209, 212–214, 217, 241, 243–244
humans, 1, 31, 73, **77**, 85

activities, 3, 6, 87–89, **109**, 111, 120, 127, 130
  impacts of, 3, **109**
  resources, 2
  welfare, 1

I

ice, 3, **6**, 9, 12, 25, *51*, **77**, 80, 82, **88**, **109**, 112, 114, 116, **118**, 121–122, 132, 200, **204**, 206, 209; *see also* permanent snow and ice
IKONOS, 66–67, **69**, 93
in situ
  data, 15, 66, 94, 98–99, 104, 178, 186, 189, 209, 242
  measurements, 164, 188–189
  samples, 174
inclusion probability, 42, 44–47
interactive
  comparison, 108
  tools, 105, 107–108
International Geosphere and Biosphere Project (IGBP), 9
International Organization for Standardization (ISO), 8
International Society for Photogrammetry and Remote Sensing (ISPRS), 3
interpretation
  context-based, 113
  modes, 104
  process, 105–106
  reference class, 104
  rules, 209

J

Java OpenStreetMap Editor, 91

K

Kappa coefficient, 82, 87, 136, 139–141, 143, 151–154, 164, 166, 179, 206, 208, 211, 223, 229, 231, 238
  accuracy and, **216**
kernel density estimation, 53–54

L

land cover, 1, 3, 42, 95, 136, 174
  codes, 82
  modifications to, 1
  mosaics, 107
  types, 6, 82
  validation, 8–9
land cover and change (LCC), 1–2
land use, 3, 42, 136, 209
land use and land cover (LULC), 74–76, **77**, 80, 93, 97–98
Land Use/Cover Area Frame Survey (LUCAS), 97–98
land use efficiency, 7
Landsat, 3–4, 11, 38, 70, 74, 219
  WRS-2, 25
landscape fragmentation, 44, 47, 211, 219, 228, 232
landscape shape index (LSI), 15, 39–41, 45–47, 56–59, 187, 189, 206, 211, **212**, 214, 217–218, *237*, 238, 241
  sampling, 209
  validation, 208
LiDAR, 99
linear address, 42–44

Index

247

M

man-made structures, 1, **6**
Mapillary, 66, 89, **90**, 92, 176
mapping, 1–3, **4**, 7, 11, 24, 26–27, 36, 67, 70, 74, 83,
　　　85–86, 94, 105–106, 127, 143, 148, 157,
　　　163–164, 179, 187, 200, 241
　accuracy, 143, 211
　communities, 14, 91–92
　errors, 212
　GLC, 2
　interactive, 188
　military, 67
　open-source, 91
　remapping, 16, 120–121
　techniques, 36
　thematic, 6, 86, 209
　topographic, 6, 85, 209
　tools, 13
　web, 69, 72–73, 186, 230
Margfit, 150, 152
MCD12Q1, **2**, **10**, 11–13
MERIS, 2
MOD12, **2**, 11, **24**
MODIS, 2, 8, **10**, 11–13, 25, 70, 93, 139, 156, 192
Morton, 55
　code, 42, *43*
　order, 42
　reversed, 42

N

normalization, 47, 150, 152

O

online tool, 163, 178, 209
online validation, 13–15, *16*, 174, 186–187, 191, 196,
　　　200–201, 217, 241
open-source, 6
　mapping, 91
　system, 179
overall accuracy, 136, 149, 151

P

Pareto boundary, 122–125
permanent snow and ice (PSI), 8, **121**, **126**, **204–206**,
　　　208–210, 211, **212**, 218–219, **222**, **224**, **228**,
　　　234, **238**
pixel analysis, 95, 158
PlanetScope, 68
Points of Interest (POIs), 13, 91–92, 127, 242
Poisson distribution, 30, 32, 34
population, 11, 23, 27, 29, 32, **33**, 36–37, 47, 137, 147,
　　　151, 241
　density, 38
　increases, 1, 7
positional accuracy, 15, 65, 71, 147–148, 157, 188, 209
primary sampling units (PSUs), 12, 25, 36, 52, **224**
probability sampling, 27, 47, 51
producer's accuracy, 137, 149, 171
products, GLC, 9, **10**, 11–13, 23, **24**, 25–27, 31–32, 38, 58, 82,
　　　122, 136, 157–159, 165, 188–189, 191–192, 243

accuracy, 36, 241–242
demand for, 3
development of, 3
validation, 7, 11, 16, 37, 51, 131, 202, 227, 241–242,
　　　244

Q

quality control, 209
QuickBird, 66–67, **69**, 93

R

random sampling, 9, **10**, 12, **24**, 25, 27, *28*, 130–131, 142,
　　　157, 171
　methodology, 57
　semi-, 29
　simple, 27, 36, 38, 51, 59, 241
　stratified, 9, 11, 25, 27, 38, 48, 56, 58, 166, 219
reference classification, 118, 120, 146
reference data, 3, 9, **10**, 12–16, 23, 25, 31, 38, 48, 52, 65,
　　　70, 85, 89, 92–95, 98–99, 105–106, 108, 115,
　　　119–120, 131–132, 136, 140, 153, 155–157,
　　　163–166, 170, 173–174, 176, 179–180,
　　　187–191, 196, 200
　accuracy, 105, 137
　availability, 134
　collection, 185–186, 201–204, 206, **208**, 209, 217,
　　　219–220, 225, *226*, 228, 230–232, 241–243
　selection, *66*
　sources, 104
reforestation, 114–116
regional landscape shape indexes (rLSI), 40–41, 213
regions of interest (ROI), 11
remapping, 16, 120–121; *see also* mapping
remote sensing, 6, 136, 209
resolution, 42, 209
resources, 1, 14–15, 164–165, 185, 188
　computational, 183, 200, **202**, 241
　financial, 2, 26
　information, 183
　international, 16
　management, **7**
　natural, 99
　reference, 200
　sharing, 13
　storage, 3
　water, 6, 46, 87

S

sample
　allocation, 209
　assessment unit, 209
　judgment, 104, *105*, *193*, 209
　size, 209
sample assessment units (SAUs), 106–108, 119
sampling, 23–24, 209
　area-guided, 29–30
　balanced, 42, 44–45, 209
　cluster, 29, 53, 59
　design, 12, *24*, 25, *29*, 42, 59, 209
　global, 39
　grid, 209

methods, 27, *28*
nonspatial, 29
probability, 27
random, 25, 27, *28*
semi-random, 29
size estimation, 30–31, 37, 40, 59
spatial, 29
stratified, 27, *28*, 30, 36–37
systematic, 29–30, 59
satellites, 3–4, 66–68, **69**, 74, 76, 228, 238, 243
data, 74
imagery, 70–72, **90**, 93, 95, 98, 136, 164, 174, 176, 187, 218, 238
secondary sampling units (SSUs), 12, 36, 52
shrubland, 8, 111, 209
simple random sampling without replacement (SRSWOR), 52
SinoLC-1, 82–83, *84*
snow, 3, **6**, 12, 25, *51*, **77**, **80**, **82**, **88**, 94, **109**, 112, 114, 116–118, 121–122, **126**, 132, 200, 204–212, **218**, 219–220, 222–**224**, **228**, 234, 238–239; *see also* permanent snow and ice
Social Benefits Areas (SBAs), 6
spatial
balance, 9, 26, 44, 59, 186, 209, 217, 222
heterogeneity, 8, 13–15, 27, 40–41, 48, 56, 59, 157, 164, 186, 209, 212–214, 217, 241, 243
stratification, 36, 48
spatial data, 6, 66, 209
errors in, 8
uncertainty of, 7–8
usability of, 8
spatially balanced sampling, 42–46, 59, 209, 222–223, **224**, 241
stratified sampling, **10**, 11–13, 24, 27, *28*, 30, 36–42, 47–48, *49*, 51, 53–55, 59, 98, 157, 209, 217, 241
structures, man-made, 1, **6**
System for Automated Geoscientific Analyses (SAGA), 165, *169*–170, **175**
systematic sampling, 25, 27, 29–30, 51–52, 59, 241

T

task force, 14–16
Tau, 153–154
temporal
information, 114
quality, 8, 15, 65, 188
thematic
accuracy, 95, 189
maps, 13, 15, 38, 70, 83, 86, 99, 164, 166, 188, 229, 242

thermodynamics, 145
Tianditu, 72–73
topographic maps, 6, 83, 85–86, 99, 209
total disagreement, 155
tundra, 8, 209
Two-Rank Acceptance Sampling Plan (TRASP), 33–34
two-rank sampling, 35–36, 38–39, 59, 202, 241, 243

U

uncertainty, 6, 209
University of Maryland (UMD), **2**, 3, 9, **10**, 11–12, **24**, 25, 227
unmanned aerial vehicles (UAVs), *see* drones
urban
construction, 89
development, 6, **7**
evolution, 6
heat islands, 1
growth, 1
land cover, 30, 47–48
planning, 85
US Geological Survey (USGS), 9, 31, 85
user accuracy (UA), 51, 131, 137, 141–142, 149, 164, 166, 168, 170–173, 187, 196, 201, **206**, **211–212**, 228, **232**

V

validation, 136–137, 185, *193*
collaborative, 16
GLC, 24, 51
online, 13–14, *16*
methods, 185, 209
process, 98–99
schemes, 13
tools, 15
validation principle, **10**, 11–12, 24
Very High Resolution (VHR) imagery, 15, 66–67, 98, 164, 242
Volunteered Geographic Information (VGI), 13, 209

W

water bodies, 8, 209
webcams, **90**, 93–94
wetland, 3, **6**, 8, 11, 38, **77**, 87–88, 97, 112, **118**, **126**, 209
seasonal, 114, **127**, 200, **204**, 205–212, 218–219, 222, **224**, 228–230, **232**, **234**, **238**
types, **126**
WGS84, 4, 70–72, 80, 190, 231

9781032903989